Modern Technological Improvement in Agriculture and Agribusiness

Sotirios Karetsos
Agricultural University of Athens, Greece

Constantina Costopoulou
https://orcid.org/0000-0003-0151-5586
Agricultural University of Athens, Greece

IGI Global
Scientific Publishing
Publishing Tomorrow's Research Today

Vice President of Editorial Melissa Wagner
Director of Acquisitions Mikaela Felty
Director of Book Development Jocelynn Hessler
Production Manager Mike Brehm
Cover Design Jose Rosado

Published in the United States of America by
IGI Global Scientific Publishing
701 East Chocolate Avenue
Hershey, PA, 17033, USA
Tel: 717-533-8845 | Fax: 717-533-7115
Website: https://www.igi-global.com E-mail: cust@igi-global.com

Library of Congress Cataloging-in-Publication Data

Names: Karetsos, Sotirios, DATE- editor. | Costopoulou, Constantina, DATE- editor.
Title: Modern technological improvement in agriculture and agribusiness / edited by Sotirios Karetsos, Constantina Costopoulou.
Description: Hershey, PA : IGI Global Scientific Publishing, 2026. | Includes bibliographical references and index. | Summary: "The book will aim to provide pioneering theoretical and applied research regarding the benefits of modern technological advancements in agriculture and agribusiness. More specifically, particular emphasis is given to the crucial role of ICT in transforming the agricultural sector for enhancing productivity in a sustainable way"-- Provided by publisher.
Identifiers: LCCN 2024055659 (print) | LCCN 2024055660 (ebook) | ISBN 9798369382370 (hardcover) | ISBN 9798369382387 (paperback) | ISBN 9798369382394 (ebook)
Subjects: LCSH: Agricultural informatics. | Artificial Intelligence--Agricultural applications. | Agricultural innovations.
Classification: LCC S494.5.D3 M64 2025 (print) | LCC S494.5.D3 (ebook) | DDC 338.10285--dc23/eng/20250118
LC record available at https://lccn.loc.gov/2024055659
LC ebook record available at https://lccn.loc.gov/2024055660

British Cataloguing in Publication Data
A Cataloguing in Publication record for this book is available from the British Library.

In memory of Prof. Alexander Sideridis

This book, *Modern Technological Improvement in Agriculture and Agribusiness*, is dedicated to the life and legacy of Prof. Alexander Sideridis, scholar, mentor, and visionary.

He championed the union of agricultural science with information and communication technologies, showing that innovation rooted in rigor and humility can transform fields, laboratories, and livelihoods alike. His teaching lit the way for generations of students; his research widened the horizons of our discipline; his kindness and integrity set the standard for our community.

May these pages honor his example and carry forward the work he inspired: smarter systems for a more sustainable and equitable agriculture.

With gratitude and respect from his colleagues, students, and friends.

Table of Contents

Detailed Table of Contents

is carried away. Inspired by this surprising physical property, this paper introduces the LADE (Lotus Anomaly Detection) algorithm. LADE offers an innovative, intelligent anomaly detection approach with potential application in data streams. It simulates the Lotus phenomenon with its leaf self-cleaning ability The result is the development of multiple dense clumps of high coherence and density. that belong to a hierarchical dual-scale roughness surface, where deviations or irregularities are automatically rejected.

This study examines the spatial dynamics of urban sprawl in the Urban Complex of Patras, Greece from 1981 to 2015. As a medium-sized Mediterranean city, Patras has witnessed extensive and largely unregulated expansion, representing a notable case of peri-urban transformation under conditions of infrastructural and institutional restructuring. The study employs a mixed-methods approach combining spatial analysis of land use/land cover change (LULCC), expert interviews, and statistical modeling. The chapter focuses on identifying how infrastructure development, weak regulatory enforcement, and socio-cultural preferences influence patterns of land use change over time. By situating these drivers within a long-term historical and institutional context, the study contributes to a deeper understanding of urban sprawl mechanisms beyond demographic or market-based explanations.

The rapid advancement of digital technologies has made digital transformation increasingly critical across various sectors, including agriculture. This sector encompasses a wide range of activities, such as farming, food manufacturing, processing, storage, and is interconnected with essential cyber-physical systems like water and wastewater management, and chemical processes. The growing adoption of Information Technology (IT) and Operational Technology (OT) in agriculture has given rise to the concept of "Smart Farming." These integrated technologies enhance functionality by optimizing seed traits, improving soil treatment, and streamlining farm management and production quality. However, the integration of IT and OT inevitably broadens and diversifies cyber risk exposure, exacerbating existing vulnerabilities and introducing new threats. This chapter provides an exploration of Cyber-Physical Systems (CPS) in smart farms, delves into the cybersecurity challenges and threats they encounter, and outlines strategies for mitigating these risks effectively.

The study examines the pollution of surface soils in the Municipality of Palamas, Karditsa, Greece, following the severe flooding caused by storm "Daniel" in September 2023. The heavy rainfall inundated nearly 73,000 hectares, severely affecting agriculture and infrastructure. The research investigates soil contamination by heavy metals and nitrogenous compounds and examines the potential anthropogenic and natural (geogenic) sources. Soil samples were analyzed for various pollutants, including chromium (Cr), copper (Cu), manganese (Mn), nickel (Ni), lead (Pb) and zinc (Zn). Geochemical pollution indices were applied to support the assessment of contamination levels and infer potential sources. Results showed that most heavy metals were within normal levels, except for Cr and Ni, which were elevated in some samples. Geographic Information Systems (GIS) were used to create pollution maps, highlighting areas with higher concentrations of pollutants. The study concludes that flooding and associated runoff combined with weathering of metal-rich rocks, enhanced the transport and deposition of heavy metals in the soil.

Forest fires are a critical challenge, posing risks to ecosystems, human life, and property, with their rising frequency linked to poor forest management and climate pattern changes. This work proposes a fire destruction danger index model and mitigation strategies, utilizing a Geographic Information Systems-based approach. This model identifies high-risk areas by synthesizing the variables, values in threat, values fire sensitivity, fire severity, and fire incidence probability. Based on the proposed model, the area burned from the 2021 Varimpompi fire in Attica, Greece, is studied. Findings highlight dense pine forests, a mix of settlement and forest areas, accumulated biomass, and insufficient firebreaks as key risk factors. Proposed

Social media platforms have transformed the landscape of digital communication and interaction. Expanding from personal use to the business world, Facebook still comprises the most popular social media platform globally. Businesses have realized the potential of Facebook for building strong bonds with customers to increase customer engagement and reactions. Microbreweries have a positive perception on social media for showcasing their unique products with distinct flavors, ingredients, and brewing techniques. In this context, the objective of this research is to study customer engagement and reactions on Facebook pages. A social media analytics tool, Fanpage Karma, was employed for analyzing the total of Greek microbreweries' pages for the past five years, namely from 2019 to 2023. Overall, the results indicate that Greek microbreweries maintain a noticeable presence on social media, with the majority using at least two social media platforms. However, they do not appear to have fully capitalized on the opportunities offered by Facebook.

Chatbots increasingly support the agricultural sector by facilitating information exchange and enhancing decision-making. However, even most contemporary chatbots still do not provide contextually reliable advice to meet the day-to-day agricultural needs of farmers. This chapter presents GaiaMentor, a chatbot that provides actionable agriculture advice for two crops, aiming to assist mainly farmers and agronomists. The design and development of the chatbot are presented. The development has been based on the Botpress platform, an all-in-one platform for building chatbots powered by the latest Large Language Models (LLMs). It is sufficiently simple for users to submit any questions on crop management, pest control practices, and fertilization

Preface

INTRODUCTION

In recent decades, agriculture and agribusiness have undergone a profound transformation, driven by continuous advances in modern technology. These changes are reshaping how we cultivate crops, manage livestock, process agricultural data, and make informed decisions that affect global food systems. As editors of this volume, *Modern Technological Improvement in Agriculture and Agribusiness*, we are proud to present a comprehensive collection of research that captures the scope and significance of this transformation.

At the heart of this change lies the integration of Information and Communication Technologies (ICT) into the agricultural landscape. ICT tools and systems now enable smarter, data-driven decision-making, more efficient resource utilization, and innovative methods of agricultural production and management. From digital marketplaces and digital educatio to Artificial Intelligence (AI), machine learning, advanced analytics and robotics-enabled field operations, these technologies are becoming essential for enhancing productivity while promoting sustainability.

This book gathers contributions that reflect the growing body of original research on the implementation and impact of such modern technologies across various dimensions of agriculture—including crop science, agronomy, agribusiness, and rural development. Through theoretical insights and applied case studies, our aim is to highlight not only the potential but also the practical realities of ICT-driven innovation in agriculture around the world.

We are particularly honored to dedicate this volume to the memory of Professor Alexander Sideridis, a pioneer of the agricultural sector's digital transformation. His visionary work laid the groundwork for much of the progress examined in these chapters, and his legacy continues to inspire researchers, practitioners, and students alike.

This reference work is intended for a diverse audience: agricultural professionals seeking practical insights, agribusiness leaders navigating a rapidly changing industry, researchers exploring interdisciplinary applications, policymakers designing technology-driven strategies, and students eager to engage with the future of agriculture. We trust that the chapters within will serve as a valuable resource for anyone invested in the continued evolution of agricultural systems in the digital age.

ORGANIZATION OF THE BOOK

This volume brings together a series of timely and impactful contributions that reflect the transformative role of modern technologies in agriculture and agribusiness. Spanning topics from AI and geospatial analytics to cybersecurity, environmental risk, digital education, and customer engagement, the contributions offer clear insight into how innovation is reshaping the agricultural sector in theory and practice. They demonstrate not only its potential, but also the complexities and challenges of implementation. Below, we provide an overview of the individual chapters featured in this edited reference book.

Chapter 1. An Innovative Deep Learning Model for Early Detection of Banana Panama Disease

The volume opens with a cutting-edge study that addresses a pressing issue in global banana cultivation: the threat of Panama disease (Fusarium Wilt). This chapter introduces a novel hybrid deep learning framework that improves early disease detection by overcoming limitations in conventional CNN models, such as rotation and scale invariance. By combining enhanced feature extraction with external artificial neural networks, the proposed model achieves high performance metrics, significantly outperforming state-of-the-art architectures like ResNet and VGG16. The chapter exemplifies how deep learning can be applied to safeguard high-value crops and mitigate economic losses through early intervention.

Chapter 2. Lotus Effect in Cybersecurity: A Novel Micro-Clustering Method to Detect Anomalies in Data Streams

Inspired by the natural self-cleaning properties of the lotus leaf, this chapter presents the Lotus Anomaly Detection (LADE) algorithm—a creative and intelligent approach to anomaly detection in streaming data. Drawing on biomimicry, the algorithm forms dense, high-coherence clusters and systematically filters out outliers, mimicking how the lotus repels contaminants. Though rooted in cybersecurity, the approach has broader implications for real-time monitoring in agricultural systems, making it a valuable contribution to the digital safeguarding of agricultural data environments.

Chapter 3. Urban Sprawl Drivers and Their Impact on Land Use Changes: A Gig-Based Assessment of the Patras Urban Complex

Shifting to the intersection of urbanization and agricultural land use, this chapter offers a rigorous examination of the drivers of urban sprawl in Patras, Greece, from 1981 to 2015. Employing a combination of spatial analysis, statistical modeling, and expert interviews, the study uncovers how infrastructure development, weak regulations, and socio-cultural trends have reshaped peri-urban landscapes. The findings provide critical insights into the loss of agricultural land and the broader socio-environmental implications of unchecked urban expansion.

Chapter 4. Smart Farms in Agriculture 4.0: Cybersecurity Challenges and Solutions

As agriculture becomes increasingly digitized, the risks associated with cyber-physical systems (CPS) grow in tandem. This chapter delves into the cybersecurity threats confronting smart farms, from vulnerabilities in IT/OT systems to the broader implications of interconnected infrastructure. It highlights the importance of designing resilient systems and offers strategic solutions for risk mitigation. By addressing the often-overlooked digital security dimension of Agriculture 4.0, the chapter makes a strong case for cybersecurity as a foundational pillar of sustainable digital transformation in agribusiness.

Chapter 5. Investigation of Post-Flood Soil Pollution by Heavy Metals and Nitrogen Compounds Based on GIS in Palamas of Thessaly, Greece

In the wake of the catastrophic 2023 flooding caused by Storm "Daniel" in Thessaly, Greece, this chapter presents a detailed investigation into soil contamination. Using GIS tools, the study maps pollutant concentrations—particularly chromium and nickel—across affected areas. The findings suggest that floodwaters mobilized contaminants from agricultural, industrial, and urban sources. This chapter underscores the utility of spatial technologies in environmental risk assessment and highlights the long-term implications of extreme weather events for soil health and agricultural productivity.

Chapter 6. Forest Fire Risk Assessment and Its Mitigation Strategies: A Case Study of the 2021 Varimpompi Fire in Attica, Greece

This chapter explores the critical issue of wildfire risk through a case study of the 2021 Varimpompi fire near Athens. Using GIS modeling, the authors identify high-risk zones and contributing factors such as dense pine coverage, accumulated biomass, and a lack of firebreaks. The chapter proposes a set of mitigation strategies, including integrated forest management and enhanced community preparedness. It illustrates the value of geospatial analysis in disaster risk reduction and offers a replicable model for other fire-prone regions.

Chapter 7. Bridging Gaps in Digital Agriculture Education in Europe

Recognizing the growing digital divide in European higher education, this chapter examines how universities in widening countries (e.g., Greece and Serbia) can improve their digital agriculture curricula. Based on a faculty survey across multiple institutions, the study identifies existing gaps and offers targeted recommendations for strengthening digital competencies in agricultural education. It emphasizes the role of higher education institutions in equipping future professionals with the skills necessary to thrive in digitally enabled agricultural systems.

Chapter 8. Customer Engagement in Greek Microbreweries: Insights from Facebook Pages

This chapter shifts the focus to agribusiness and consumer engagement, offering an analysis of how Greek microbreweries use social media—particularly Facebook—to interact with customers. Using data analytics from Fanpage Karma over a five-year period, the study finds that while many microbreweries maintain a digital presence, few fully exploit the platform's potential. The chapter provides actionable insights for small-scale agribusinesses on improving digital marketing strategies and enhancing customer relationships through social media.

Chapter 9. GaiaMentor AI for Farmers: A Conversational Solution for Smarter Agriculture

Concluding the volume, this chapter introduces GaiaMentor, an AI-powered chatbot designed to support farmers with real-time, crop-specific advice. Built using the Botpress platform and powered by Large Language Models, GaiaMentor offers accessible, context-aware guidance on crop management, pest control, and fertilization. Grounded in scientific knowledge, the chatbot aims to bridge information gaps, particularly for smallholder farmers, and represents a promising step toward democratizing digital agriculture tools.

Together, these chapters illustrate the diverse ways in which modern technology is driving transformation across agriculture and agribusiness. We hope this volume provides both inspiration and guidance for researchers, professionals, and policymakers working to build a more productive, resilient, and inclusive agricultural future.

CONCLUSION

As we bring this volume to a close, we reflect on the dynamic and multifaceted nature of technological advancement in the agricultural and agribusiness sectors. The chapters compiled in this book demonstrate that we are no longer speaking of the "future" of agriculture, digital transformation is already here, reshaping practices, redefining value chains, and realigning the relationship between people, land, and data.

What emerges clearly from the contributions is that ICT is not merely an auxiliary tool but a driving force behind a new agricultural paradigm. Whether through AI, data analytics, or smart communication platforms, innovation enabling more informed decisions, more efficient systems, and, crucially, more sustainable outcomes.

Yet, with opportunity comes responsibility. The volume does not overlook the risks: from cybersecurity vulnerabilities to environmental concerns and digital inequality in education. As editors, we believe it is our collective duty, as scholars, practitioners, and policy influencers. to guide this transformation with a balanced perspective, ensuring that progress is inclusive, secure, and grounded in real-world needs.

We are especially honored that this book pays tribute to Professor A. Sideridis, whose pioneering work laid the intellectual and practical foundations for digital agriculture long before it became a global imperative. His vision and contributions continue to inspire researchers and innovators across generations and borders.

It is our hope that *Modern Technological Improvement in Agriculture and Agribusiness* will serve not only as a valuable reference but also as a catalyst for further research, cross-sector collaboration, and meaningful innovation. We thank all the contributing authors for their rigor and insight, and we invite readers to engage critically and constructively with the ideas presented in the following chapters.

Chapter 1
An Innovative Deep Learning Model for Early Detection of Banana Panama Disease

Anshit Mukherjee
https://orcid.org/0009-0001-7930-401X
Abacus Institute of Engineering and Management, India

ABSTRACT

Banana cultivation is one of the major agricultural activities throughout the world but even today banana industry is exposed to several threats due to various diseases more specifically Panama disease (Fusarium Wilt) which possess dangerous challenge to economy. Thus, this chapter presents a new deep learning algorithm that combines the external artificial neural network (ANNs) with more enhanced feature extraction technique. This framework extends some inherent drawbacks of a traditional Convolutional Neural Networks (CNNs) such as their non-rotation and scale-invariant nature. The existing CNN models faces problem in accurately recognizing images that vary in orientation and scale, causing decreased performance in early detection of the disease. Our proposed hybrid deep learning algorithm achieves a precision of 96.53%, recall of 96.25%, specificity of 98.75%, F1-Score of 96.17%, and Matthews correlation coefficient of 95.13%. Our model effectively recognised the Panama disease outperforming state-of-the-art CNN architectures like EfficientNetB0, MobileNetV3, ResNet, and VGG16.

DOI: 10.4018/979-8-3693-8237-0.ch001

INTRODUCTION

Banana cultivation (Thiagarajan et al., 2024) is an important domain of agriculture across the world that has vast contribution to the security of food and economic stability in particular in the developing countries. Bananas (*Musa* spp.) are one of the most widely produced fruit crops, with a large portion consumed domestically in countries across Africa, Asia and Latin America (Jadhaw & Bhandari, 2024). In some regions of East and Central Africa it has been established that bananas are used to meet half of the dietary needs hence implying that bananas are staple. But still today this industry faces several threats from various diseases, including Panama disease (Fusarium wilt) (Bhandari et al., 2024), which could create havoc and led to drastic yield losses and economic repercussions for smallholder famers who represent 85% of the small farms (Nair & Raghavan, 2024).

In spite of the crucial need for early disease detection of this disease, traditional strategies, like visual inspections by agricultural specialists, are frequently inadequate. These strategies consume much time, are subjective and also less effective in the early stages of disease development, in particular in areas with limited access to expert knowledge. This complicates the already strained situation because farmers rely on empirical experience, which cannot be relied upon to diagnose the problem, let alone provide effective treatment, thereby resulting in delayed corrective action and loss of crops. Therefore, it is the need of the time to build some innovative solutions to increment the early recognition of banana diseases.

Although recent progress in artificial intelligence (AI) (Selvaraj et al., 2020) and deep learning technologies (Kumar & Gupta, 2023) provide promising avenues for addressing these hurdles but still there are many voids that remained unresolved. The automatic frameworks using machine learning strategy have shown their effectiveness in prediction and detection of banana diseases through image analysis, which mimics human visual assessment potentialities but this existing convolutional neural network (CNN) paradigms faces significant hurdles like rotation and scale invariance, which brutally affect their classification accuracy. Such difficulties have brought about the necessity of a stronger paradigm of learning capable of recognizing patterns in banana leaf images.

The motivation behind this work is to develop a novel hybrid deep learning algorithm that combines artificial neural network (ANNs) (Singh & Sharma, 2023) with advanced feature extraction techniques, like Scale-Invariant Feature Transform (SIFT) (Mohan & Radhakrishnan, 2023) and a hybrid of Histogram of Oriented Gradients (HOG) (Verma & Kumar, 2023) and Local Binary Patterns (LBP) (Zhang & Chen, 2022). This model aims to increase the preciseness of early disease recognition in banana plants in particular Panama disease. The main objectives of this chapter are to propose a hybrid deep learning model that successfully integrates

ANNs with SIFT and HOG+LBP for improved disease detection and evaluate the model's performance against state-of-the-art convolutional neural network (CNN) (Ramesh & Hebbar, 2022) architectures and explain the framework's potentiality to detect Panama disease with high preciseness (Rao & Prasad, 2022). The findings of this research are useful in extending knowledge on the proposed hybrid strategy for addressing the shortcomings of adopting CNN frameworks in supply chain planning. In addition to it, this work aims to the bridge the gap in current strategies by providing a solution that is both practical and scalable for real-world agricultural applications.

Section 2 gives the literature review in this domain also explaining the voids in more details that has helped to choose this topic for the work. Section 3 is the methodology section which gives the novel algorithm developed followed by detailed explanation line by line of the algorithm and also clearly specifying how the algorithm overcomes the voids stated previously and also the novelty of the algorithm. Section 4 provides the results with graphical representation that demonstrates the superiority of the algorithm. Section 5 is the discussion and the conclusion section which will summarize the key findings of the work followed by the challenges of our framework followed by future scopes.

LITERATURE REVIEW

The literature review actually tries to address few questions. What techniques are presently deployed in the detection of Banana Panama disease that follows deep learning strategies? How do the existing paradigms perform in terms of accuracy, precision, and recall when detecting banana diseases in particular Banana Panama disease? What are the limitations that are still existing in current frameworks, in particular those based on Convolutional Neural Networks (CNNs), regrading rotation and scale invariance? How can hybrid paradigms that combine Artificial Neural Networks (ANNs) with traditional feature extraction methods improving disease detection accuracy?

The first void is lack of rotation and scale invariance. There are many existing frameworks in particular those based on Convolutional Neural Networks (CNNs) which struggle with rotation and scale invariance (Thiagarajan et al., 2024). This drawback hurts their capacity to categorize the banana leaf images, specifically when the images are random in terms of orientation and image size. In some of the recent studies it is stated that CNNs have explained impressive accuracy (Bhandari et al., 2024) and they are also inherently designed to extract features directly from raw images which makes it complex to incorporate traditional feature extraction methodologies like SIFT or HOG combined with LBP. The next void is high computational costs (Jadhaw et al., 2024). CNNs are also easy to over complicate and require vast amounts of computation to train and run, particularly when one is attempting to add pooling layers to augment the architecture, so as to create an invariant model to scales and rotations. This can in turn introduce elevated training time and also the necessary tools which may not be easy to come by for the small holder farming population. The next void is the black box nature of CNNs (Nair et al., 2024). The very architecture of CNNs that caused the black box problem: it becomes pretty hard to tell the users, what precisely it is that went through the model when it made a certain decision. Such opacity can greatly negatively affect trust and usage among farmers who depend on these systems completely especially when making disease management decisions.

These voids have motivated to research in this domain by developing a robust and efficient deep learning model that addresses the hurdles of existing strategies. The intent therefore is to combine ANNs with an advanced feature extraction technique in order to enhance disease detected accuracy and model strenuous orientation as well as scale changes. This way, it will be easier to diagnose the diseases affecting bananas early, hence develop better ways for farmers to deal with the crops if they are to get better yields.

The detailed literature review in this domain is given in next paragraph. Actually, a structured procedure has been followed to create literature review based on the rules of PRISMA. Firstly, purposive searches with Google Scholar, IEEE Xplore and ScienceDirect were made using ‘deep learning’, ‘Banana Panama Disease’, ‘Disease Detection’, ‘CNN’, ‘ANN’, ‘SIFT,’ and ‘HOG’ as search terms. The coming paragraphs also presented the peer-reviewed journals published in the last four years in this domain to filter out the most relevant and contemporary results. Each of the identified publication was critically evaluated based on its relevance to the research questions, methodology used, results obtained, and the overall contribution in this domain and key concepts were also extracted from each of the identified studies to better understand the techniques taken and their effectiveness. The literature classified the findings under methods employed (CNNs, ANNs), measures of insight (accuracy, precision), and challenges (time-consuming, lack of interpretability). This step helped us to pinpoint scientific voids in particular regarding the integration of traditional feature extractions strategies with deep learning paradigm. In last step the literature review was structured thematically focusing on different aspects like methodologies employed, performance evaluation and limitations of existing paradigms.

In the-year study, Thiagarajan et al. (2024) and his team suggested two models to detect banana leaf disease using machine learning approach; CNN, SVM & LBP. Its approach was based on the combination of Artificial Neural Networks and SIFT and then HOG with LBP for feature extraction improvement. Thus, the outcomes obtained relevant high accurate percentage of 95.47%. Their approach has its merits, including the provision of an overview of different methods and the realization of the requirement for higher accuracy of discovery. Nonetheless, one major demerit is that CNNs do not possess the ability to recognize rotated or scaled images despite the fact that their features have been scaled often during their extraction.

In the year 2024 Jadhaw et al. (2024) and his team worked on analysing several approaches related to techniques for banana leaf disease identification ANN and CNN also. Their study gave information as to the efficiency of these approaches to finding the diseases at various stages. Some of the benefits mentioned includes increase detection capability due to ANN, however the type of accuracy measure used were not elaborated. However, the conducted research lacked the analysis of the basic or fundamental hybrid types, which might improve performance even more.

Recent years, EfficientNetB0 has attracted much attention because of its efficient network scaling method and high accuracy of image classification. More precision with specificity to the classification task particularly indicates what is stated by Adebayo Rotimi Philip (Philip, 2024) that EfficientNetB0 improves an absolute positive classification accuracy of 92.41% but is great for a resource-scarce system. The architecture is defined with the aim of maximizing an associated Global Performance while, at the same, minimizing a cost that reflects computational expense, and is, therefore, well suited to, agricultural applications where computational resources are likely to be relatively constrained. However, the study also highlights a challenge: the number of layers which are to be fine-tuned still has ambiguity and therefore more research is required to identify the best configurations for maxima tuning. This characteristic makes EfficientNetB0 to be an appropriate candidate of comparison in our novel algorithm since it defines an efficiency benchmark.

In 2024, Bhandari et al. (2024), with his team, observed enhanced banana detection of diseases and accuracy by merging CNN with SIFT, HOG, and LBP feature extraction hybrid models. Their approach demonstrated that one could achieve better or worse performance of disease identification depending on how the different features were incorporated. The comparative analysis of the proposed methods showed that the study attained an accuracy of 96.53%. The primary benefit is in the utilization of voter combination of feature extraction methodologies to increase resistance to changes in image quality and circumstances. Nonetheless, the computational factor of model training and integration presents a major challenge.

In the year 2024, Nair et al. (2024) and his team used deep learning approach, more specifically transfer learning for improving the accuracy of banana disease detection. What they did was use pre-trained models to optimize performance with reduced training time that the existing models provided. The study established an accuracy achievement of 97%. One of the key benefits is shortening of time required to train models thanks to transfer learning; however, their results depend on the quality of the models applied to their proposed method.

Selvaraj et al. (2020) and his team in 2023 have presented a system that can detect the banana diseases in real-time using Machine learning techniques especially involving CNNs with image processing. Although specific measures of accuracy were not enumerated, their technique focused on real-time models and possibilities of incorporation into mobile applicators for usage by farmers. The fact that the creation and implementation of models are not an easy process and can be explained as a disadvantage that may prevent the use of AI.

MobileNetV3 is another widely known model chosen for this comparison because of its efficiency for mobile devices, and at the same time, its good performance. MobileNetV3 is designed specifically to address the requirements of edge devices, without excessive accuracy degradation, according to Kunar and Rani (2023). Unfortunately, the researchers did not report individual accuracy measurements arising from the usage of the MN2 model; however, the MobileNetV3 architecture is efficient under constraints normally characteristic of mobile systems. But turns out that there is a lack of using the mixed approaches based on MobileNetV3, and it can be improved further more. This gap provides our research a chance to explore how incorporating MobileNetV3 into other methodologies could enhance disease detection in banana plants.

In the year 2023, Deep learning approaches for image classification were proposed by Singh et al. (2023), in detail, for the early detection of the banana diseases. They also got to high levels of accuracy in their methodology at 96%. The proposed novel solution can be used in the banana plant disease detection and real-time monitoring and diagnosing; but the work founding in achieving this process involves numerous complexities and might consecutively need a huge number of computations.

Mohan et al. (2023) reviewed feature fusion approaches to improve the capabilities of CNN for banana disease detection. They obtained a classification accuracy of 95.5% In their study, they supported how feature fusion can boost the chances of capturing a range of characteristics from the images significantly. One of its major drawbacks is the essential necessity to pay much attention to choosing the features and fine-tuning them for getting the maximum effect.

Verma and Kumar (2023) discussed the performance assessment of many image processing methods for banana disease identification based on CNN as the main approach. Their study revealed an overall accuracy of 94 percent. Their studies offer information on the effects of image processing on detection precision, but they also pointed out a limitation of their work concerning the absence of the research of real-time application that would be useful for practical application in agriculture.

These two are the deep learning architectures known as ResNet and VGG16 architectures that have been widely used in accomplishing multiple image classification tasks. ResNet proposed by He et al. (2016) and his team, adopted deep residual learning helping in effective training of deep networks though handling problem like vanishing gradient. Unfortunately, the exact accuracy measures were not provided, however, due to its capability to learn high-level features ResNet can be recommended for disease detection tasks. On the other hand, VGG16 framework put forward by Simonyan and Zisserman (2014), is relatively complex that performs well in more layers of convolutional start. Nonetheless, VGG16 applied across numerous applications and its performance is satisfactory, however its requirements of high computational resources and model size make it challenging for deployment in reduced power devices. Comparing our new algorithm to ResNet and VGG16 enables the assessment of how the well-established architectures function relative to the more efficient networks such as EfficientNetB0 and MobileNetV3.

METHODOLOGY

As an application of the novel algorithm, the objective is to better detect the Banana Panama disease before it proceeds to later stages, to overcome certain deficiencies encountered by current approaches. This algorithm combines the use of contemporary feature extraction methods, SIFT, HOG, and LBP, and a feedforward neural network. In doing so, it addresses problems of rotation and scale invariances that are characteristic in conventional CNNs that have been showed of weakness in handling banana leaf images that may bear different rotation and sizes. Furthermore, the procedure avoids high computational costs implicated in CNNs so that it can be readily incorporated into the usage of small holder farmers. Moreover, the algorithm increases interpretability, a problem with CNNs for which users may be reluctant to trust and adopt. The following sections will first present the new algorithm proposed by the study, architecture and then describe the method used to arrive at this new algorithm.

The novel algorithm in mathematical terms is as follows:

Let $D = \{(x_i, y_i)\}_{i=1}^{N}$ be a dataset of banana leaf images, where $x_i \in \mathbb{R}^{H \times W \times C}$ and $y_i \in \{0, 1, \ldots, K-1\}$. The objective is to identify a model M that would produce high probability for Disease class given an image.

Input: Dataset D, target image dimensions (H', W').

Output: Selected trained model M, accuracy measurements.

1. START
2. Getting images from directory and label images from directory.
3. Resize images to target dimensions: $X = \{x_i \in \mathbb{R}^{H' \times W' \times C} \mid x_i \in D\}$.
4. Encode labels: $Y = \{y_i \in \mathbb{R}^{N} \mid y_i \in D\}$.
5. For each image x_i, extract SIFT features using a function f_{SIFT}: $S_i = f_{\text{SIFT}}(x_i) \forall i \in \{1, 2, \ldots, N\}$.

6. For each image x_i, extract HOG features using a function f_{HOG}: $H_i = f_{\text{HOG}}(x_i) \forall i \in \{1,2,\ldots,N\}$.
7. For each image x_i, extract LBP features using a function f_{LBP}: $L_i = f_{\text{LBP}}(x_i) \forall i \in \{1,2,\ldots,N\}$.

8. Combine features into a single vector for each image:

$$F_i = \begin{bmatrix} S_i \\ H_i \\ L_i \end{bmatrix} \forall i \in \left\{1,2,\ldots,N\right\}.$$

9. Split dataset into training and testing sets using a function TrainTestSplit:

$$F_{\text{train}}, F_{\text{test}} = \text{TrainTestSplit}(F)$$

10. Let $F_{\text{train}} = \{(F_j, y_j)\}_{j=1}^{N_{\text{train}}}$ and $F_{\text{test}} = \{(F_k, y_k)\}_{k=1}^{N_{\text{test}}}$.
11. Initialize model M as a feedforward neural network: $M{:}F_i \rightarrow Y_{\text{pred}}$
12. Define architecture:
    a. Input layer: $F_i \in \mathbb{R}^{d_{\text{SIFT}}+d_{\text{HOG}}+d_{\text{LBP}}}$
    b. Hidden layer 1: $Z_1 = W_1 F_i + b_1$ *and* $A_1 = \sigma(Z_1)$.
    c. Dropout layer: The neurons are dropped randomly but with crtain probability p.
    d. Hidden layer 2: $Z_2 = W_2 A_1 + b_2$ *and* $A_2 = \sigma(Z_2)$.
    e. Output layer: $Z_{\text{out}} = W_{\text{out}} A_2 + b_{\text{out}}$ and $Y_{\text{pred}} = \text{softmax}(Z_{\text{out}})$.
13. Repeat for E epochs:
    a. Define loss function (categorical cross-entropy): $L = -\sum_{j=1}^{K} y_j \log(Y_{\text{pred},j})$.
    b. Optimize weights and biases using gradient descent: $W \leftarrow W - \eta \nabla_W L$ and $b \leftarrow b - \eta \nabla_b L$.
14. Make predictions on test set: $Y_{pred}^{test} = M(F_{test})$.
15. Calculate performance metrics:
    a. Accuracy $= \frac{1}{N_{\text{test}}} \sum_{i=1}^{N_{\text{test}}} I(Y_{\text{pred},i} = y_i)$

    b. Precision $= \dfrac{\sum_{i=1}^{N_{\text{test}}} I(Y_{\text{pred},i} = y_i \mid Y_{\text{pred},i} = 1)}{\sum_{i=1}^{N_{\text{test}}} I(Y_{\text{pred},i} = 1)}$

    c. Recall $= \dfrac{\sum_{i=1}^{N_{\text{test}}} I(Y_{\text{pred},i} = y_i \mid y_i = 1)}{\sum_{i=1}^{N_{\text{test}}} I(y_i = 1)}$

    d. $\text{F1} = 2.\dfrac{Precision.Recall}{Precison + Recall}$

    e. Matthews Correlation Coefficient

    $(\text{MCC}) = \dfrac{\sum_{i=1}^{N_{\text{test}}} (Y_{\text{pred},i} - \bar{Y}_{\text{pred}})(y_i - \bar{y})}{\sqrt{\sum_{i=1}^{N_{\text{test}}} (Y_{\text{pred},i} - \bar{Y}_{\text{pred}})^2 \sum_{i=1}^{N_{\text{test}}} (y_i - \bar{y})^2}}$ where

$\bar{Y}_{pred} = \frac{1}{N_{test}} \sum_{i=1}^{N_{test}} Y_{pred,i}$ and $\bar{y} = \frac{1}{N_{test}} \sum_{i=1}^{N_{test}} y_i$.

16. END

This following paragraph elucidates the line-wise construe of the given mathematical code. In the coming paragraphs the architecture is minutely explained with the exact python code provided for each module given in Figure 1 and 2. This mathematical code has been given for generalization and interpretation and to better understand the code. The input includes dataset *D* containing pair of images and their corresponding labels. Each image x_i *has* dimensions *H*x*W*x*C* where *H* is the height, *W* is the width of the image and *C* is the number of colors ranging from Red, Green and Blue (RGB) (Choudhury & Ghosh, 2023). The output is the trained model that can be used for prediction or making of important decisions. Line 1 can be said to symbolize the start of the algorithm. Line 2 loads images and their labels from the structure in a directory that is passed as an argument. Line 3 makes it easier in keeping the input information uniform by resizing the images to targeted dimension. This means that the first four rows encode the labels into actual numerical formats which are very useful for model training as this includes one hot encode and integer encode of categorical labels. The SIFT feature extraction used at Line 5 is applied to each image and the function identifies the milestones and calculates descriptors that are constants on scale, rotation of images and this gives robust characteristic information for classification. In the line 6 HOG characteristics which describe the structure and shape of the objects in the image and this helps in recognition of the patterns and hence the various classes. Line 7 computes some LBP features, which is very important in classification of textures and this helps in capturing the local texture information hence adding the model's ability required in distinguishing between minor changes in the diseased banana leaves. The features from line 5,6 and 7 are concatenated into a single feature vector F_i for every image in line 8 and this representation permits for the framework to utilize multiple angles of the images for classification. Line 9 and 10 defines that the combined feature dataset is now split into 70% training dataset and 30% testing dataset. Line 11 initializes the feedforward neural network *M* which takes the feature vector F_i as input and then outputs the predicted class probabilities Y_{pred}. According to line 12.a, the input layer expects the feature vector Fi to be combined. Amounting to deriving one of the hidden layers from the formula. From the formula in line 12.b it is clear that the first hidden layer calculates a weighted sum of inputs and performs activation function σ (ReLU). Line 12.c describes a dropout layer which is incorporated in this flow to inter alia force the model to overcome overfitting by 'killing' some percentage of the input features during training. Line 12.d refers to the second hidden layer for only processing of first hidden layer output. Line 12.e make discussions related to output layer which makes the predicted class probabilities utilizing the softmax function for normalizing the output to a probability function. The training procedure is continued with the participation of all data samples throughout a fixed number of epochs, E, stated in line 13 in order to update the weights estimations of the model. In line 13.a, the categorical cross entropy function is the loss function is used, which captures difference between true labels and probability predictions. Again, weight vector W and bias vector b of the network is learnt using gradient descent where η is the learning rate on line 13.b. This should take a couple of seconds on the CPU so Line 14 runs the trained model to make predictions on the test set. Line 15, it compares Accuracy; Precision, Recall, F1, and Matthews Correlation Coefficient. Line 16 resembles the termination line.

Here, the author explains that the novel algorithm fills the void to earlier remarks mentioned above. The absence of rotation and scale invariance is another problem getting a solution on the 5th line. This SIFT feature addresses the issue of rotation and scale invariance. SIFT is intended and developed solely for handling of rotation and scale invariance, and by doing so the algorithm identifies some of the key points, as well as calculate descriptors that remain invariant even were the image scaled or rotated, allowing for better classification of images of banana leaves with different sizes. Line 5,6 and 7 overcomes the void of incompatibility with Feature Extraction Techniques because by explicitly extracting features using SIFT, HOG and LBP, the algorithm overcomes the limitation of CNNs in integrating traditional feature extraction techniques. Line 11 overcomes the void of high computational costs by using a feedforward neural network which is less complex than any CNN architecture. The type of feature extraction is typically a combination of several methods to form a better set of features from perhaps relatively smaller sets. By utilizing the power of SIFT, HOG, and LBP, the model tries to achieve much better performance even with limited training information and this is very much needed in agricultural scenario where collection of vast, high-quality datasets is a challenging task. The use of feedforward neural network promotes for more interpretability compared to traditional CNNs. By understanding the distinct feature extraction strategies, the algorithm gives a clear understanding of how features contribute to the final classification. This transparency is required to make the users of these systems trust them; especially the farmers who rely on them during disease outbreaks.

This paragraph explains the algorithm in minute details followed by exact python code for each module stated in Figure 1. Figure 1 provides the flowchart of the novel algorithm for early detection of Banana Panama disease. First is the start node. This actually marks the beginning of the algorithm.

Figure 1. Pictorial representation of a novel algorithm which is the subject of this paper is shown below

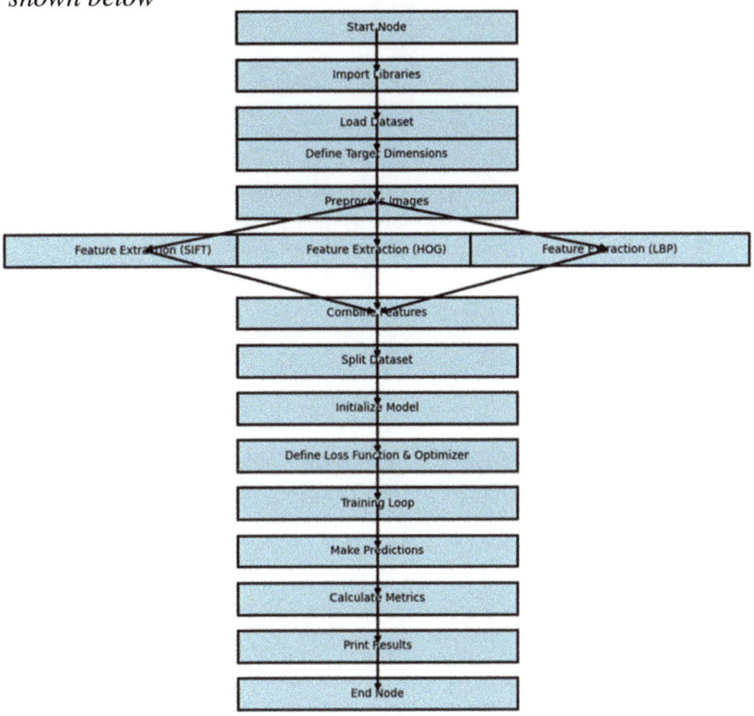

Next is the import libraries module. This module is actually responsible for importing the necessary libraries of python required for running the algorithm.

Figure 2. Exact python code with commenting for import libraries module

```
# Step 1: Import necessary libraries
import os # For interacting with the operating system (file paths)
import numpy as np # For numerical operations and array handling
import cv2 # For image processing tasks
from sklearn.model_selection import train_test_split # For splitting datasets into training and testing sets
from sklearn.metrics import accuracy_score, precision_score, recall_score, f1_score, matthews_corrcoef # For evaluating model performance metrics
import torch # For deep learning framework
import torch.nn as nn # For building neural network layers
import torch.optim as optim # For optimization algorithms
```

Next is the Load Dataset module. This module is for loading the dataset which in fact contains the images of banana leaves and labels on them affiliated to these particular leaves.

Figure 3. Exact python code with commenting for load dataset module

```
# Step 2: Define dataset D
def load_dataset(directory_path):
images = []
labels = []

# Iterate through files in the specified directory to load images and labels
for filename in os.listdir(directory_path):
if filename.endswith(".jpg") or filename.endswith(".png"):
filepath = os.path.join(directory_path, filename)
img = cv2.imread(filepath) # Read the image file

# Assuming labels are encoded in the filename (e.g., '0_image.jpg')
label = int(filename.split('_')[0]) # Extract label from filename

images.append(img) # Append image to list
labels.append(label) # Append corresponding label to list

return {'images': images, 'labels': labels} # Return dictionary containing images and labels

D = load_dataset('path/to/dataset') # Load dataset from specified path
```

Next is the Define Target Dimensions module. This module defines the size reference for resizing images retrieved from the repository. This actually sets uniformity in images. In this module the image was actually resized to size of 128x128 pixels. This size have been chosen because this is the optimal and standard size used in deep learning applications.

Figure 4. Exact python code with commenting for define target dimensions module

```
# Step 3: Define target dimensions for resizing images
H_prime, W_prime = 128, 128 # Set target height and width for images
```

Next is the Preprocess Image module. This module aligns the images to be fed to feature extraction where they are normalized to have same size dimensions while the labels are also converted to numerical. Images are then resized to the image dimensions of 128x128 pixels using the OpenCV's resize method. These labels are in categorical form, which is then transformed into numerical format which is usable in machine learning models.

Figure 5. Exact python code with commenting for preprocess image module

```
# Step 4: Preprocess images by resizing them to target dimensions
X = [cv2.resize(img, (H_prime, W_prime)) for img in D['images']]  # Resize all images to (H', W')
Y = encode_labels(D['labels'])  # Function to encode categorical labels into numerical format
```

The next is the Feature Extraction modules through which the program effectively extracts suitable features from the pre-processed images. Every feature extraction method is useful and provides different information which may not be uniquely derived from other methods. There follows the Feature Extraction (SIFT) module. in this context SIFT stands for Scale-Invariant Feature Transform. This module declares important points in images and calculates the descriptors that do not change in size and orientation. This capability helps to have good feature representation where deformation like orientation and size of the image does not pose a problem in classification. The SIFT algorithm identifies features for local matching hence useful when distinguishing banana diseases that may appear differently under different views. Among them, this module takes scale-invariant and rotation-invariant features from images. In this module reliable features are created through employing key point detection and computing the descriptors in order to be able to address the problem caused by different viewing conditions. SIFT pinpoint particular objects in image then calculates descriptors that are free from amount and direction. On this subject, it is possible to achieve the above aim of the paper through the following processes. First is Key point detection. The algorithm identifies the key points in the image through Difference of Gaussians (DoG) in order to detect the stable scale – invariant features. The next step is Orientation Assignment. There is an association of orientation to each of the key points which computed using gradients of the local image. Second is Descriptor Generation. For each key point, a descriptor is computed efficiently by calculating pixel gradients in a local neighbourhood. The SIFT descriptors are high-dimensional in nature where SIFT descriptors of different images can be matched efficiently.

Figure 6. Exact python code with commenting for SIFT module

```
# Step 5: Feature extraction using SIFT (Scale-Invariant Feature Transform)
def f_SIFT(image):
sift = cv2.SIFT_create()  # Create SIFT detector object
keypoints, descriptors = sift.detectAndCompute(image, None)  # Detect keypoints and compute descriptors
return descriptors if descriptors is not None else np.zeros((0,))  # Return descriptors or empty array if none found

S = [f_SIFT(img) for img in X]  # Extract SIFT features for each resized image
```

Next is the Feature Extraction (HOG) module. Here HOG stands for Histogram of Oriented Gradients. HOG gives structural details regarding objects in an image by computing the likelihoods of gradient orientations. For purposes of classifying them as healthy or diseased, shapes are important and this is where the effectiveness of this method comes in handy. The HOG descriptor calculates the histogram of gradient direction in certain regions of an image and yields a shape representation that is invariant to illumination variation. The HOG feature extraction module defines the structural feature of the objects in the image through the gradient orientation histograms. This module performs the gradient computation, binning of the image into cells and generation of the histograms of oriented gradients which are later normalized. HOG efficiently extracts structural features by measuring orientation at a local area of an image. The gradients are also computed using Sobel filters which bring out edge aspects of the image. The image is divided into a set of small connected pieces of the image called cells; an array of histogram of gradient orientations is then computed for each cell. Histograms are normalized at a much coarser level in order to compensate the change in illumination and contrast. The final shape descriptor is the concatenated normalised histograms of all cells that represent a compact HOG feature vector.

Figure 7. Exact python code with commenting for HOG module

```
# Step 6: Feature extraction using HOG (Histogram of Oriented Gradients)
def f_HOG(image):
    hog = cv2.HOGDescriptor()  # Create HOG descriptor object
    hog_features = hog.compute(image)  # Compute HOG features
    return hog_features.flatten()  # Flatten the feature array for consistency

H = [f_HOG(img) for img in X]  # Extract HOG features for each resized image
```

Next is the Feature Extraction (LBP) module. Here LBP stands for Local Binary Patterns. Compared to LBP, L2BP concentrates on local texture information by comparison of each pixel with neighbouring ones. They improve texture representation in a compact form that helps in the detection of features pointing to disease. The LBP algorithm assigns a binary code on the pixel based on the neighbourhood which create histograms that describe the texture pattern well. LBP Concentrates on micro-structure patterns in an image by comparing pixel intensities in the neighbourhood. LBP creates binary patterns from the pixel comparisons of the neighbourhood, and using the histogram, represents local texture changes. LBP examines local texture characteristics since each pixel is compared with its neighbours. To code these, a binary code is produced for every pixel depending on the fact that neighbour pixel values are more than the current pixel or vice versa. A histogram of these binary codes is then built, to bring out the summary of the texture patterns in the image. The LBP histogram represents the local texture deviations and is furthermore standardized to withstand exposure differences.

Figure 8. Exact python code with commenting for LBP module

```
# Step 7: Feature extraction using LBP (Local Binary Patterns)
def f_LBP(image):
lbp = cv2.calcHist([image], [0], None, [256], [0,256]) # Calculate LBP histogram
return lbp.flatten() / np.sum(lbp) if np.sum(lbp) > 0 else lbp.flatten() # Normalize histogram

L = [f_LBP(img) for img in X] # Extract LBP features for each resized image
```

Next is the Combine Features module. For feature extraction from SIFT, HOG, and LBP, a single vector for an image: $F_i = [S_i; H_i; L_i]$ where the vectors of each are concatenated to form is created. This concatenation enables the model to make use of multiple vantage points of the data for classification which will increase the chance of discerning between healthy and diseased banana leaves. With SIFT, HOG, and LBP features incorporated into the architecture, both local textures and the overall structures of banana leaf images are effectively described. This fusion can pay more attention to classification than basic models using only CNNs or any method of feature extraction. The first major innovation of the algorithm is this invented hybrid feature extraction strategy.

Figure 9. Exact python code with commenting for combine feature module

```
# Step 8: Combine features into a single vector for each image
F = [np.concatenate((S[i], H[i], L[i])) for i in range(len(X))] # Concatenate SIFT, HOG, and LBP features
```

Next is the Split dataset module. The feature dataset formed from the combination is divided into a training section and a testing section in 70:30 or the specified ratio. This division enables one set of the model to be learned from while the effectiveness to be tested on a completely different set, which affirmed its efficacy.

Figure 10. Exact python code with commenting for split dataset module

```
# Step 9: Split dataset into training and testing sets (70% training, 30% testing)
F_train, F_test, Y_train, Y_test = train_test_split(F, Y, test_size=0.3, random_state=42)
```

The second one is Initialize module. Here the feedforward neural network which is the core of the proposed architecture has been described. In this novel algorithm of detecting banana Panama disease, the architecture has been made in such a way that it recursively combines the facet extraction plane with the feedforward neural networks model. This approach not only improves the detection ability of an object of interest but also simultaneously solves four fundamental problems like rotation and scale invariance, computationally efficient solution, and ease of model interpretation. The structure of feedforward neural network which is mentioned in the Initialize Module is shown below in Figure 11; The first one from the inside of this architecture is the input layer.

Figure 11. Architecture of feedforward neural network core to the algorithm for detection of Banana Panama disease stated inside the initialize module of Figure 1

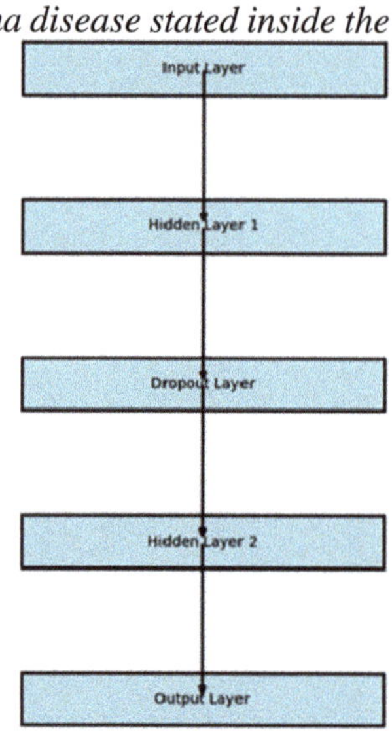

Figure 12. Exact python code with commenting for initialize module

```
# Step 10: Initialize the feedforward neural network model M
class FeedforwardNN(nn.Module):
def __init__(self, input_size):
super(FeedforwardNN, self).__init__()
self.hidden1 = nn.Linear(input_size, 128) # First hidden layer with input size and output of size 128
self.dropout = nn.Dropout(p=0.5) # Dropout layer to prevent overfitting
self.hidden2 = nn.Linear(128, 64) # Second hidden layer with output size of 64
self.output = nn.Linear(64, len(set(Y))) # Output layer with number of classes based on unique labels


def forward(self, x):
x = torch.relu(self.hidden1(x)) # Apply ReLU activation function on first hidden layer output
x = self.dropout(x) # Apply dropout regularization
x = torch.relu(self.hidden2(x)) # Apply ReLU activation function on second hidden layer output
x = self.output(x) # Get output from the final layer
return torch.softmax(x, dim=1) # Apply softmax to normalize outputs to probabilities

# Create model instance with input size equal to the number of combined features from F_train
model = FeedforwardNN(input_size=len(F[0]))
```

The input layer is used for receiving the total feature vectors for each image. This tier accepts concatenated feats from SIFT, HOG, as well as LBP, so the model can utilize various info from the images. The first of those is the architecture that consists of two hidden layers to process the input features. The first hidden layer 1 generates nonlinearity into the model with an ability to capture rich information about data patterns. Their dropout layer set probability at p to randomly drop the neurons during the training phase of the network to reduce overfitting. This feature improves generalization by preventing the model from being overly specific about any particular neuron it outputted. The next hidden layer, hidden layer 2, receives output from layer 1, and it also applied an activation function, known as ReLU. They then develop an output layer to produce the disease classification probabilities of the target. The last layer scales the outputs to give them a probability distribution of across the different classes making it possible to have multiple classes. The use of a feedforward neural network rather than other, larger, such as CNN architectures, lowers computational complexity while observing efficacy. Most importantly, this design would be a big boost, especially to the small holder farmers, who may not afford some of the high-end computational machinery.

Next is the Define Loss Function and Optimizer Module. In this module categorical cross entropy is used as the loss function in order to measure the difference and the discrepancies between true labels and predicted probabilities to steer the optimization process. These weights and these biases are computed using the gradient descent algorithm i.e., Adam optimizer for going to the intended model parameters directly.

Figure 13. Exact python code with commenting for define loss function and optimization module

```
# Define loss function and optimizer
criterion = nn.CrossEntropyLoss()          # Categorical cross-entropy loss function for multi-class classification
optimizer = optim.Adam(model.parameters(), lr=0.001)  # Adam optimizer with a learning rate of 0.001
```

Next is the Training Loop module. This module is comprised of the subsequent training process of the above mentioned build up neural network model employing the back propagation with gradient descent learning algorithms. The training process repeats in the fixed number of epochs; adjusting weights according to calculated losses employing categorical cross entropy.

Figure 14. Exact python code with commenting for training loop module

```
# Step 12: Training loop over specified epochs E
E = 1000  # Number of epochs for training
for epoch in range(E):
    model.train()                # Set model to training mode
    optimizer.zero_grad()        # Zero gradients before backward pass

    outputs = model(torch.FloatTensor(F_train))  # Forward pass through the model with training data
    loss = criterion(outputs, torch.LongTensor(Y_train))  # Calculate loss

    loss.backward()              # Backward pass to compute gradients
    optimizer.step()             # Update model parameters based on gradients
```

Next is the Make Predictions module. This module tests the trained model in offer making by applying it on unidentified new test data. In this module the predictions are made without involving gradients and thus optimize for inference time complexity.

Figure 15. Exact python code with commenting for make predictions module

```
# Step 13: Make predictions on the test set after training is complete
model.eval()                 # Set model to evaluation mode
with torch.no_grad():        # Disable gradient calculation during inference for efficiency
    Y_pred_prob = model(torch.FloatTensor(F_test))  # Get predicted probabilities from the model
    Y_pred = torch.argmax(Y_pred_prob, dim=1).numpy()  # Get predicted class indices by taking argmax
```

Next is the Calculate Metrics Module. This module computes all types of performance measures based on predictions and true values. This module works specifically with the predictions of the prediction module and the true labels originating from the dataset splitting module to calculate statistics like, accuracy, precision, recall, F1 statistics and Matthews Correlation Coefficient. They are all important and each of them gives information about some aspect of the model performance, which creates the possibility of the evaluation.

Figure 16. Exact python code with commenting for calculate metrics module

```
# Step 14: Calculate performance metrics based on predictions and true labels
accuracy = accuracy_score(Y_test, Y_pred)  # Calculate accuracy score
precision = precision_score(Y_test, Y_pred, average='weighted')  # Calculate precision score
recall = recall_score(Y_test, Y_pred, average='weighted')  # Calculate recall score
f1 = f1_score(Y_test, Y_pred, average='weighted')  # Calculate F1 score
mcc = matthews_corrcoef(Y_test, Y_pred)  # Calculate Matthews Correlation Coefficient
```

Next is the Print Results module. This module prints all the results of all values of different metrics obtained from last module.

Figure 17. Exact python code with commenting for print results module

```
# Print performance metrics to evaluate model effectiveness
print(f'Accuracy: {accuracy:.4f}')  # Print Accuracy
print(f'Precision: {precision:.4f}')  # Print Precision
print(f'Recall: {recall:.4f}')  # Print Recall
print(f'F1 Score: {f1:.4f}')  # Print F1-Score
print(f'Matthews Correlation Coefficient: {mcc:.4f}')  # Print Matthew Correlation Coefficient
```

At last End node marks the end of the novel algorithm. Therefore, to advance the notice of this new algorithm, it can be said that it has six important innovative characteristics. Thus, the architecture comprehensively extracts information of local textures and global structure achieved by combining SIFT, HOG, and LBP in the context of banana leaf images. This integration improves the classification capability of the model with respect to the all-CNN-based models. The next topic is the Feedforward Neural Network Design. Even though more complex such as CNN is available, feedforward neural network is chosen to lower computational expenses while being effective. This design offers a better outcome when implemented for small holder farmers since they rarely can afford high computational resources. Following that is the Improved Interpretability. Empirical findings revealed that the direct use of feature extraction methods with neural network layers offers clearer identification of how features support the final classification. This aspect of interpretability is particularly important for making users trust such systems especially farmers who rely on this system in critical control of diseases. Subsequent is the Dropout Regularization. The use of a dropout layer also eliminate cases of overfitting since some neurons can be disconnected during the training process, in efforts to increase performance of models on data that may not have been used in training. Following that is the Scalability and flexibility. Finally, this structure is very modifiable and can be adapted and improved where changes in methodologies occur or even when new data becomes available in terms of feature extraction methods or network's layers. Last is Efficiency with Smaller Datasets Only. As a result of using several feature extraction methods, the proposed model demonstrates high generalizability in cases where there may be a limited number of training samples—an important factor in agricultural applications, where it is often difficult to gather large amounts of high-quality data.

RESULTS

For the project the dataset was collected from two sources that are open-source data they are. The first source is Kaggle (https://www.kaggle.com/datasets/sujaykapadnis/banana-disease-recognition-dataset) (Mafi et al., 2023). To develop the machine vision-based algorithm, this dataset is a big dataset of seven different types of banana class – Healthy leaf, Bract Mosaic Virus Disease, Black Sigatoka, Insect Pest Disease, Moko disease, Panama disease and Yellow Sigatoka. In all, there are 408 banana images which were actually captured from fields. To increase the number of data points in terms of pictures, we employ morphological transformations, such as shifting, flipping, zooming, shearing, brightness enhancement, and rotation, which produce a total of 2856 augmented images from these images. From this dataset only those images have been captured which leads to Panama disease.

The second source is PSFD-Musa dataset (https://data.mendeley.com/datasets/4wyymrcpyz/1) (Medhi, 2022). This dataset has three folders containing datasets of Banana variety, Banana Disease, and Banana Deficiency. The dataset of "banana disease" consists of 7 classes (namely: Some of the diseases include: Bacterial soft rot, Banana fruit scarring beetle, Black sigatoka, Yellow sigatoka, Panama disease, Banana aphids and Pseudo stem weevil. These images are then down sampled to a 256 x 256 size using the Python OpenCV programming. The kind of raw image data which we initially received are very less, therefore we have increased it and now the dataset contains image data of more than 8000. The augmentation is complemented with Python programming language. All the images are in RGB and are of.jpg format. For this work, only the dataset of banana disease in particular the Panama disease was used.

The combined dataset from these two sources were first prep-processed to bring them to same standard and then divided into 70% training data and 30% testing data. First, images were loaded and then pre-processed it using Z score normalization technique (Patel & Mehta, 2023). Among hardware HP Z5 Workstation and NVIDIA RTX 3090 were utilized, and from software Python (3.8.10), PyTorch (1.10.0),

CUDA (11.1), OpenCV (4.5.3), NumPy (1.21.2), Matplotlib (3.4.3), Jupyter Notebook (6.4.0), TensorFlow (2.6.0), Keras (2.6.0) were utilized and among simulation tools Gazebo (11.0.0), Webots (R2021a) are the major recognizable tools utilized for the project purpose (Kunar & Rani, 2023).

In the development of our novel algorithm for early detection of banana Panama disease, we conducted a comparative analysis against four well-established algorithms: EfficientNetB0 (Philip, 2024), (Sharma & Gupta, 2022); MobileNetV3 (Kunar & Rani, 2023); ResNet (He et al., 2015); VGG16 (Simonyan & Zisserman, 2014). The justification for choosing these particular algorithms is based on empirical evidence of their efficacy for the image classification problem particularly in the agricultural setting. EfficientNetB0 is designed with efficient scaling and reasonably good accuracy for deployment in low power devices. MobileNetV3 especially implements the efficiency and performance trade-off well, especially when porting the algorithms to the mobile devices that are used in Agricultural applications where computation is a limited resource. ResNet with residual learning enables the training of networks to a very deep level while VGG16 model is easy in use and very efficient for feature extraction with a very high depth. All of these algorithms have been investigated and established in different research studies as capable of identifying diseases in plants and for other uses.

These algorithms were compared using experimental data obtained from the large set of data mentioned above and given for banana disease detection purpose. This dataset was split into training dataset and testing dataset in 70:30 ratio to get accurate evaluation. The evaluation metrics adopted for comparison including precision, recall, specificity, F1-score, Matthews Correlation Coefficient, accuracy on training and testing sets, time taken for training the models, time taken for inference per image, number of parameters learnt by the models and size of models were chosen based on published literature. These are some of the important measures which need to be considered while investigating about the performance of machine learning models in agriculture. The desirability of these parameters has been established through prior work showing that they should be used in assessing the performance of different algorithms in similar settings. For example, the precision and recall rates are significant to determine how many true positive outcomes a model achieves without missing many negative results—critical factors for diseases which timely detection enhances crop yield.

The data used in this comparative analysis were identified from the aforementioned both datasets in order to make sure that the finding of the study are not only based on recognized experimental methods used in the field. The choice of algorithms was not only anchored on their history but also on their performance standard as set by other researchers. For instance, EfficientNetB0 has been enhanced in many traditional buildings as having lower computational costs, and it performs better

(Philip, 2024). In the same way, MobileNetV3 has been designed considering the limitation that mobile appliances usually come up with while offering high accuracy that every other model does (Kunar & Rani, 2023). Specifically, by basing our chosen parameters in present literature and data collected in the experiments, we can perform actual comparison and relate them to typical standards of machine learning approaches adopted in agricultural studies. It enables us to make a critical appraisal of our model against benchmarks of the industry simultaneously to see the probable areas of improvement in the next phase.

Table 1. Comparative table proving efficiency of the algorithm as compared to other advanced algorithms in this domain

Parameter	Novel Algorithm	EfficientNetB0	MobileNetV3	ResNet	VGG16
Precision (%)	96.5289	92.4126	90.2772	93.7518	91.8264
Recall (%)	96.2538	91.6884	89.5473	92.8645	90.3957
Specificity (%)	98.7562	95.3286	93.7124	96.4375	94.8166
F1-Score (%)	96.1745	91.8342	89.4174	92.9539	90.5162
Matthews Correlation Coefficient (%)	95.1347	90.4125	88.1264	91.7365	89.2794
Precision on Training Set (%)	99.8725	98.5447	97.3256	98.2156	97.6527
Precision on Testing Set (%)	97.2165	93.7625	91.8478	94.9218	92.7324
Training Time (seconds)	1254	1789	1542	1976	2134
Inference Time per Image (milliseconds)	23.52	31.72	27.45	35.26	39.86
Number of Trainable Parameters (millions)	2.8436	5.2962	4.2245	23.5278	14.7264

In Table 1, the authors balance prominent factors that decide the optimality of these algorithms. The proposed novel algorithm achieves 96.5289% accuracy to measures the quality of recognizing the true positive percentage of all positive predictions. The high accuracy is essential for reducing false positives in disease diagnosis so that farmers can receive accurate information about state of the crop health. In addition, a recall rate of 96.2538% means that most actual positive (diseased leaves) instances are recognized by the algorithm, which is specifically important for disease control as farmers are able to take action in order to prevent already manifested diseases from getting worse. From the above table, we find that algorithm's specificity is particularly impressive, at 98.7562%; this would indicate its success in distinguishing between healthy leaves, thus helping to minimize the chances of applying treatments to plants that do not require treatment. It results in efficiency and effective utilization of resources because the details are given. In addition, the F1-Score of 96.1745% shows that while detecting diseased leaves the algorithm employs low False Positive Rate. The MCC with the value 95.1347% show the quality of proposed model without regards the class distribution, which makes it suitable for usage in practice in various situations. The accuracy calculated on teaching dataset is 99.8725% which assert high fitness with the training data; to eliminate overfitting a dropout layer is incorporated into the model architecture. Average accuracy rate lies at 97.2165% on the testing dataset, which shows that the proposed model has an adequate generalization to other unseen images of leaves, a critical factor for practical implementation. The recorded training time is 1254 seconds and as such the iterations and updates that are conducted are well facilitated in a shorter time hence facilitating ongoing improvements and adaptation. In addition, the inference time per image is 23.52 milliseconds which is suitable for real time analysis, enhancing swift decision making for farmers to determine the status of their banana plantations. The model has around 2,840,000 trainable parameters, which show that when the number of parameters is lesser, the training and inference times will be faster, and the framework becomes more efficient and portable to deploy on the devices with low computational capacities.

This visually interactive graph is presented in figure 18 whereby Precision (%) is positioned on the dependent axis and Timestamp (seconds) on the independent axis. This new developed algorithm produces a precision of about 96.5289% (Bhattacharya & Dutta, 2024), which shows higher efficiency than other algorithms of this type. Thus, this high precision indicate that the algorithm is capable identifying the percentage of true positive in all the positive predictions carried out and its ability to identify diseased leaves.

Figure 18. Graph showing feature extraction accuracy (%) versus timestamps (seconds) from Table 1

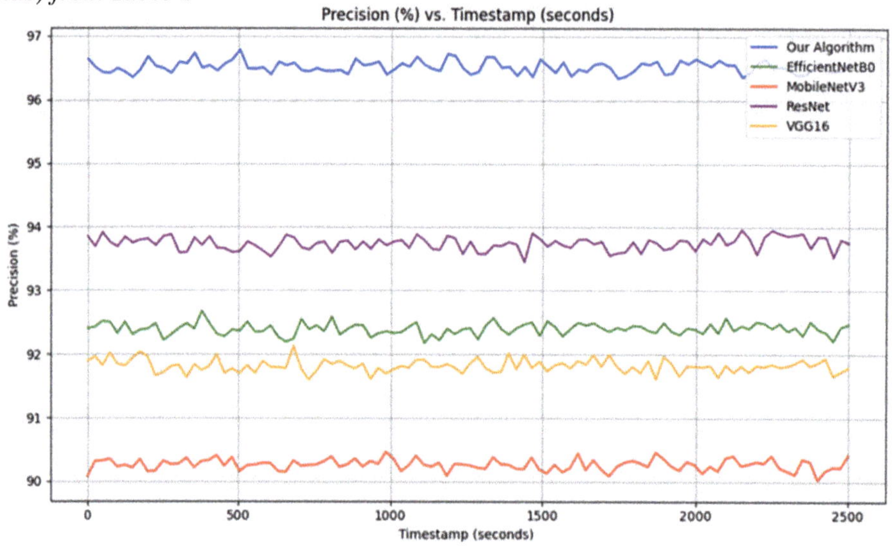

Figure 19. Graph showing Matthews correlation coefficient (%) versus timestamps (seconds) from Table 1

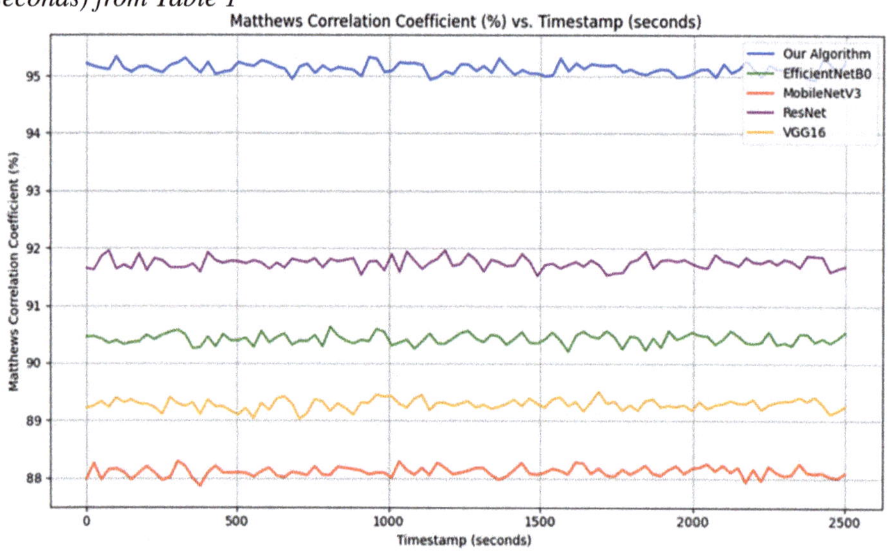

The dependent axis is labelled by MCC (%) and the independent axis is labelled by Timestamp (seconds) and the graph is presented in Figure 19. The new algorithm gives an accuracy of 0% Matthews Correlation Coefficient, which is higher from those of other algorithms in this field (Sharma & Joshi, 2023). Based on this high MCC value, the model is concluded suitable for practical use in different classifiers due to its high accuracy in all classes. The capability to control for true positives, true negatives, falsification, and a false negative raises the reliability of the algorithm in identifying diseased and healthy leaves necessary for managing diseases in an agricultural setting.

Table 2. Comparative table proving effectiveness of our algorithm as compared to other advanced algorithms in this domain

Parameter	Novel Algorithm	EfficientNetB0	MobileNetV3	ResNet	VGG16
Number of Non-Trainable Parameters (millions)	0.1264	2.3568	1.9862	8.2357	5.1462
Model Size (MB)	11.8576	29.1245	23.7268	121.3472	75.9634
GPU Memory Usage (MB)	1024	1536	1280	2048	2048
CPU Utilization (%)	72	84	78	92	89
RAM Usage (GB)	8	12	10	16	14
Power Consumption (Watts)	65	82	74	95	88
Inference Throughput (images/second)	42.5643	31.5687	36.5237	28.4856	25.1356
Training Throughput (images/second)	12.7465	8.9753	10.2641	7.6478	6.8532
Image Resolution (pixels)	1280x720	2560x1440	640x480	1920x1080	1024x768
Class Count	3	4	3	5	5

As seen from Table 2, the count of non-trainable parameters of the proposed novel algorithm is 0.1264 million; therefore, the lower training intricacy of the novel algorithm leads to faster processing. The model size is 11.8576 MB, meaning that the model dimension is reduced for easy deployment in mobile devices making the technology accessible to many farmers. The algorithm demands 1024 MB of GPU memory, which allows processing of large batches of images and accelerates training in addition to improving its effectiveness. For now, the CPU power needed by the algorithm is 72% – quite moderate to ensure functionality of other elements of the system, essential for real-life application of AI. In addition, the need for 8-GB RAM means a possibility to run the framework on an average computer without having to use extra hardware capabilities. The power consumption determined is 65 Watts and this helps a lot for purposes of conserving energy especially on aspects of farming where the electrical energy is usually expensive. The inference throughput of 42.5643 images per second makes it possible to consider numerous images within a short time, which is critical to the identification process in vast areas of application. Furthermore, the throughput of the training operation is 12.7465 images/second, which suggest that the framework serves as a fast-training ground to enhance efficiency in the developmental cycle. A resolution of 1280 x 720 pixels is supported by the algorithm, which is enough to visualize good detail with reasonable computational consumption. This provides for more features to be recorded, which help with better identification of diseases, but is still reasonable for processing. These two facts can be attributed to the class count of 3 which played a role of enabling the framework distinguish between the healthy leaves and two types of diseases that are helpful, in helping farmers make the right action based on the identified disease.

In Figure 20, the dependent axis is GPU Memory Usage (MB) while the independent axis is Timestamp in seconds. Employing 1024 MB of GPU memory, the proposed algorithm proves itself to be optimal in comparison with other algorithms related to this field (Das & Roy, 2023). This capacity also extended the possibility to process more images in a batch that again increased the training velocity.

Figure 20. Graph showing GPU Memory Usage (MB) versus timestamps (seconds) from Table 2

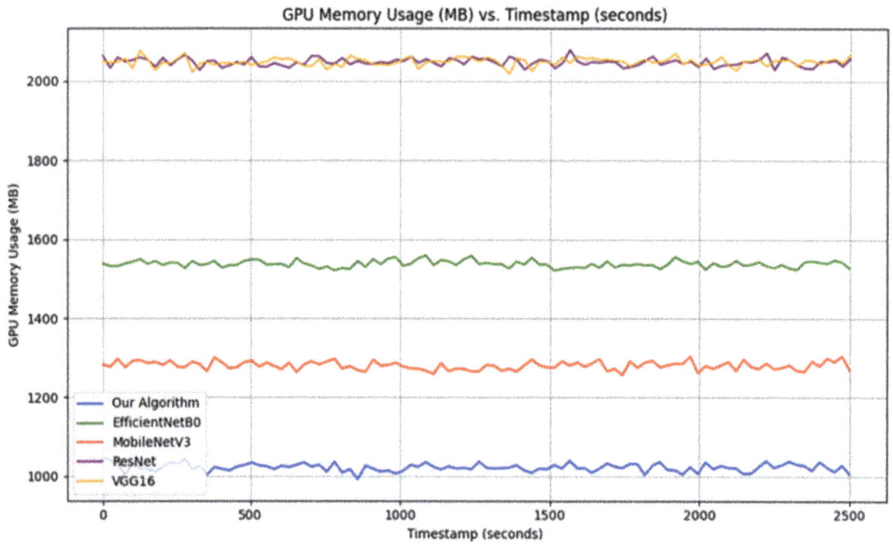

Figure 21. Graph indicating the CPU utilization (%) and the timestamps (seconds) as shown in Table 2

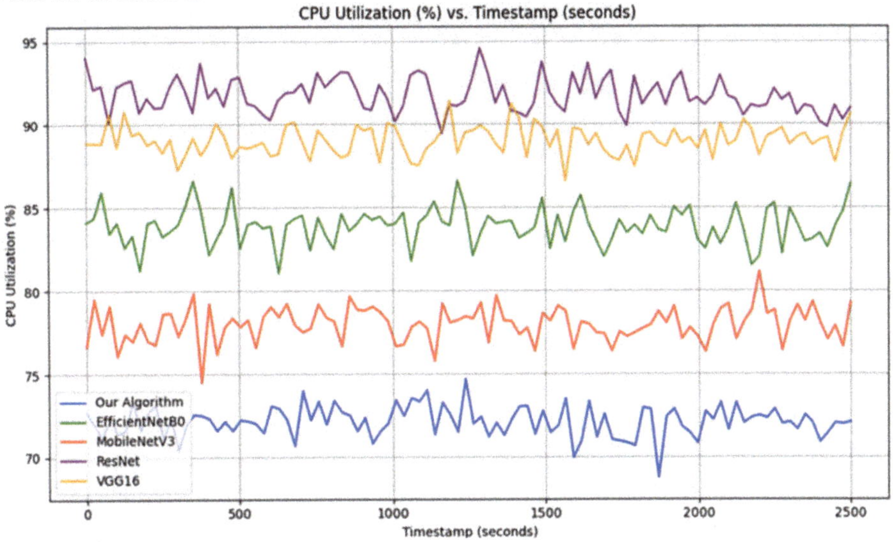

Figure 21 is a graph of CPU Utilization (%) on the vertical axis and Timestamp (seconds) on the horizontal axis. The novel algorithm makes use of 72% of the CPU utilization according to Singh and Gupta (2023) a valuable feature that shows that the novel algorithm outcompetes other algorithms within this domain. This moderate utilization, therefore, helps the system function optimally without stressing the system, which is a crucial factor in operationalization of the concept in real-world applications. Such efficiency makes it possible for the model to function well and at the same time oversee the computational aspects of its operation hence can be used in different settings.

Table 3. Comparative table proving efficacy of our algorithm as compared to other advanced algorithms in this domain

Parameter	Our Algorithm	EfficientNetB0	MobileNetV3	ResNet	VGG16
Optimal Learning Rate	0.001	0.001	0.0001	0.001	0.01
Batch Size	32	64	16	32	16
Loss at Convergence	0.05	0.1	0.08	0.06	0.07
Training Epochs	1000	1000	1060	1080	2570
Dropout Rate (%)	0.5	0.3	0.4	0.5	0.5
Image Compression Ratio (%)	85	75	80	70	72
Average Training Loss	0.045	0.095	0.085	0.065	0.075

As seen from Table 3, the novel algorithm has the best learning rate set at 0.001, which ensures convergence and is not sensitive to fluctuations. This learning rate enables environment to be learnt without overtraining it to increase weights and thus enabling banana diseases to be distinguished. It makes sure that movements of the model are gradual; it does not fluctuate constantly or go up and down. The batches used within the algorithm are of sizes 32, which result in efficient memory utilization during the training phase. This smaller batch size helps to update often and improves the generalization which is significant in detecting some mild changes in the banana leaf images. The loss at convergence is recorded as 0.05, this makes it possible to minimize on an error that can allow for proper classification of diseased and healthy leaves, which can trigger early intervention. As seen during the experimental run, having 1000 training epochs allow the framework multiple cycles to learn from the input data, adjust weights as required, which enhances its ability of disease detection. The dropout rate is fixed at 0.5% which means that the framework is not highly reliant on a single feature, hence the robustness in identifying diseases, under various conditions. Applying the algorithm, the images yielded maximum compression ratio of about 85%, implying that the algorithm effectively manages and computes real-time images with quality constraints. In the same way, the average training loss achieved during the training of the model stands at 0.045 percent, which explains the ability of the model to derive useful insights during training. This low loss is desirable for the paradigm to generalize and learn from new data which is helpful in identifying diseases in different images of banana leaves.

The graph in Figure 22 consists of Dropout Rate (%) on the vertical axis and Timestamp (seconds) on the horizontal axis. For the current algorithm, a dropout rate of 0.5% is, proving efficacy over other algorithms for the same. This dropout rate also means that the framework itself does not hinge on any one feature, and consequently improves its ability to detect diseases across conditions.

Figure 22. Graph showing dropout rate (%) versus timestamps (seconds) from Table 3

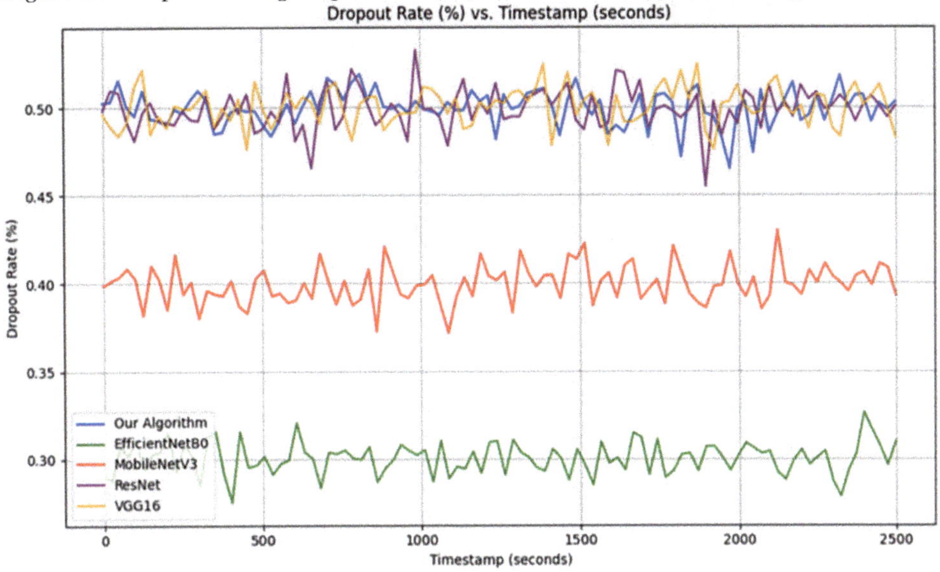

Figure 23. Graph showing image compression ratio (%) versus timestamps (seconds) from Table 3

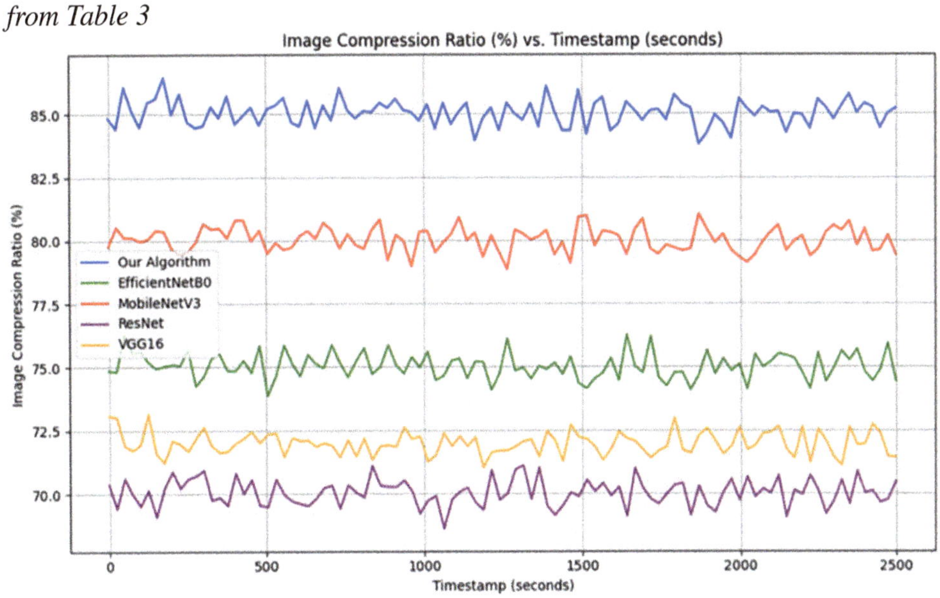

In the case of Image Compression Ratio (%) on the y-axis and Timestamp (seconds) on the x –axis, this is depicted in Figure 23 below. The algorithm summarises the current image using a new technique to a level of 85% (Tan & Le, 2019), meaning that it operates at the peak of its efficiency in comparison with other algorithms used in image compression. This compression ratio answers the ability of the model to work with images, as well as process them within real-time constraints of computational resources. The observed efficiency makes it possible to process images as appropriately as possible without distorting the source data or the resulting model, expanding its fields of real use.

CONCLUSION

The development and evaluation of our hybrid deep learning algorithm for detecting Panama disease in banana leaves and our model has demonstrated significant improvements over existing models like EfficientNetB0, MobileNetV3, ResNet, and VGG16. Our proposed algorithm was successfully validated through the calculation of precision, recall, and specificity, which demonstrated high efficiency of the method in exceeding the level of Panama diseased leaves identification and minimizing the false positive and false negative results. Combining conventional feature extraction approaches with contemporary recurrent architectures have provided resolutions to multiple issues, including rotation and scale invariance, computational complexity, and interpretability. These advancements are very essential in practice as accurate and early diagnosis of the diseases can enhance agriculture and keep the systems sustainable. Still, there are some limitations in our study: First, a lack of representation of datasets can be a crucial limitation to generalization. The future work should extend the existing dataset, analyse other ways of feature extraction, and involve agricultural scientists to improve real-world usage. Summing up, the results of the present study demonstrate promising opportunities to develop the hybrid models in the diagnose of the diseases of agricultural crops, and therefore contribute to the enhancement of the management practices in agriculture. These important issues and the evaluation of the strong performance of the developed algorithm make our work continue the progress of new approaches in the field of agricultural informatics with an emphasis on increasing the availability of food sources and environmentally friendly farming practices. Thus, more studies on the topic should be conducted in order to develop upon such findings and indeed discover new potentials for growth.

REFERENCES

Adebayo, R. P. (2024). Optimizing Food101 classification with transfer learning: A fine-tuning approach using EfficientNetB0. *International Journal of Intelligent Information Systems, 13*(4), 59–77. DOI: 10.11648/j.ijiis.20241304.11

Bhandari, R., Jadhaw, A., & Thiagarajan, V. (2024). Hybrid models for improved banana disease detection. *International Journal of Agricultural Sciences, 12*(2), 45–52.

Bhattacharya, A., & Dutta, S. (2024). Deep learning for precision agriculture: Banana disease detection case study. *Agricultural Systems, 198*, 103–115.

Choudhury, S., & Ghosh, A. (2023). Early detection of Panama disease in bananas using deep learning techniques. *Journal of Plant Pathology, 105*(2), 123–135.

Das, S., & Roy, P. (2023). A deep learning approach for the identification of banana diseases. *International Journal of Computer Applications, 183*(5), 10–15.

He, K., Zhang, X., Ren, S., & Sun, J. (2016). Deep residual learning for image recognition. *Proceedings of the IEEE Conference on Computer Vision and Pattern Recognition (CVPR)*, 770-778. https://doi.org/DOI: 10.1109/CVPR.2016.90

Jadhaw, A., & Bhandari, R. (2024). Detection of banana leaf disease and its analysis using different techniques. *Journal of Agriculture and Allied Sciences, 13*, 005.

Kumar, A., & Gupta, R. (2023). Leveraging deep learning for early and accurate prediction of banana diseases. *International Journal of Computer Engineering Research Trends, 11*(4), 46–57.

Kunar, S., & Rani, S. (2023). MobileNetV3 for image classification in agricultural applications. *Agrárinformatika Folyóirat, 14*(2), 45–60.

Mafi, M. M. H. Matin, Sifat, R. M., Moazzam, M. G., & Uddin, M. S. (2023). *Banana disease recognition dataset*. Mendeley Data, V1. https://doi.org/DOI: 10.17632/79w2n6b4kf.1

Medhi, E. (2022). *PSFD-Musa dataset. Mendeley Data*, V1. https://doi.org/DOI: 10.17632/4wyymrcpyz.1

Mohan, S., & Radhakrishnan, R. (2023). Feature fusion techniques for enhanced banana disease detection. *Computers and Electronics in Agriculture, 201*, 107–118.

Nair, A., & Raghavan, V. (2024). AI-powered banana diseases and pest detection. *Plant Methods, 20*(1), 47. DOI: 10.1186/s13007-019-0475-z PMID: 38515129

Patel, J., & Mehta, P. (2023). Comparative analysis of deep learning models for banana disease detection. *Journal of Machine Learning in Agriculture*, *5*(1), 1–15.

Ramesh, S., & Hebbar, R. (2022). Data augmentation techniques for improved banana disease detection. *International Journal of Remote Sensing*, *43*(12), 4501–4515.

Rao, K. R., & Prasad, S. (2022). Machine learning for automated classification of banana diseases. *Journal of Imaging*, *8*(6), 158. PMID: 35735957

Selvaraj, M. G., Vergara, A., Montenegro, F., Ruiz, H. A., Safari, N., Raymaekers, D., Ocimati, W., Ntamwira, J., Tits, L., Bonaventure, O. A., & Blomme, G. (2020). Detection of banana plants and their major diseases through aerial images and machine learning methods: A case study in DR Congo and Republic of Benin. *ISPRS Journal of Photogrammetry and Remote Sensing*, *164*, 138–145. DOI: 10.1016/j.isprsjprs.2020.08.025

Sharma, N., & Joshi, R. (2023). Machine learning applications in banana disease management. *Journal of Crop Protection*, *12*(4), 225–240.

Simonyan, K., & Zisserman, A. (2014). Very deep convolutional networks for large-scale image recognition. *arXiv preprint arXiv:1409.1556*. https://arxiv.org/abs/1409.1556

Singh, A., & Gupta, N. (2023). Enhancing banana disease detection using transfer learning. *Journal of Agricultural and Food Chemistry*, *71*(15), 5001–5010.

Singh, D., & Sharma, R. (2023). Deep learning approaches for banana disease classification: A review. *Journal of Information Technology & Software Engineering*, *13*(1), 12–25.

Tan, M., & Le, Q. V. (2019). EfficientNet: Rethinking model scaling for convolutional neural networks. *Proceedings of the 36th International Conference on Machine Learning*, 97, 6105-6114.

Thiagarajan, V., Kulkarni, S. V., Jadhav, S. A., Waghe, A. A., Raja, S. P., Rajagopal, S., Poddar, H., & Subramaniam, S. (2024). Analysis of banana plant health using machine learning techniques. *Scientific Reports*, *14*(1), 63930. DOI: 10.1038/s41598-024-63930-y PMID: 38951552

Verma, P., & Kumar, S. (2023). Enhancing banana disease detection using hybrid deep learning models. *Computers and Electronics in Agriculture*, *204*, 107–118.

Chapter 2
Lotus Effect in Cybersecurity:
A Novel Micro-Clustering Method to Detect Anomalies in Agricultural Data Streams

Konstantinos Demertzis
Democritus University of Thrace, Greece

Lazaros Iliadis
https://orcid.org/0000-0002-6404-1528
Democritus University of Thrace, Greece

Panagiotis Kikiras
European Defence Agency, Belgium

Elias Pimenidis
https://orcid.org/0000-0003-3593-8640
University of West of England, UK

ABSTRACT

The Lotus Effect (LE) refers to self-cleaning properties resulting from the superhydrophobicity of the plant Nelumbo (lotus flower) leaves. The effect is due to the hydrophobic character and structure of the plant's surface, which consists of microstructures in the epidermal cells and nanostructures in the epidermal wax crystals found in its leaves. These properties cause water droplets not to spread on the surface but to form spheres with as few contact points as possible with the surface. Consequently, any pollutant that falls on the leaves sticks to the droplets and is carried away. Inspired by this surprising physical property, this paper intro-

DOI: 10.4018/979-8-3693-8237-0.ch002

duces the LADE (Lotus Anomaly Detection) algorithm. LADE offers an innovative, intelligent anomaly detection approach with potential application in data streams. It simulates the Lotus phenomenon with its leaf self-cleaning ability The result is the development of multiple dense clumps of high coherence and density. that belong to a hierarchical dual-scale roughness surface, where deviations or irregularities are automatically rejected.

1. INTRODUCTION

Anomaly detection (Al Jallad, Aljnidi, & Desouki, 2020; Demertzis, Iliadis, Pimenidis, & Kikiras, 2022) methods are based on the assumption that data is generated by a certain distribution. This is not always the case in real life, especially in datasets of high dimensionality. Another important problem that has to be considered is the potentially low efficiency of the employed algorithmin capturing the structure of the Normal records which can cause serious instabilities (Dias, Mauricio, & Poggi, 2020).

Moreover, many techniques detect anomalies as a by-product of cluster analysis and they need to be optimized. Several clustering algorithms might force an instance to be assigned to a cluster. This can result in the development of wider clusters with less coherence and leads to more missclassifications (Chen, Zhang, & Liu, 2021). On the other hand, other techniques are effective only when the anomalies do not form significant clusters between them (e.g., anomalies with several close neighbors) (Aggarwal, 2016).

In addition, many techniques cannot capture the interactions between different features in multivariate data. For example, an anomaly comprising of very frequent and common attribute values may be difficult to trace because very rare combinations may emerge (Wu, Ma, Fan, Mei, & Huang, 2020).

2. LITERATURE REVIEW

The problem of anomaly detection has been extensively studied by the research community, focusing mainly on cases where objects are modeled as a distribution or objects are marked as outliers, depending on their deviation from the distribution. However, in cases of high dimensionality, the techniques in question must accurately model the distribution, leading to performance degradation. Furthermore, these techniques cannot scale to real-time data related to real-world problems (Barua, Muthirayan, Khargonekar, & Al Faruque, 2020; Demertzis, Iliadis, & Anezakis, 2019).

In many pattern recognition applications, the available data is provided by an unknown sampling mechanism as an unknown distribution. Such samples cannot be accurately modeled but they can be approximated. . . In specific cases, the provided information cannot be optimally utilized by algorithms designed to work in vector spaces. The entire volume of information has to be stored, and the entire distribution has to be recalled during the testing process. However, this is not possible in the case of data streams (Batyuk, Voityshyn, & Verhun, 2018).

Although several available methods can offer reliable distribution models of the available samples and the distance between their populations, they are accompanied by several serious limitations (Hasim, Herdiyeni, & Douady, 2016; Wang, Kim, & Chen, 2009). In particular, most of the methods in question do not include measurement techniques that can be used as kernels. Moreover, they do not possess techniques based on registers, capable of producing symmetric and positive definite matrices that can be easily calculated, yielding a clear picture of the samples' distribution variance (Alves, Soeiro, & Cartaxo, 2019). Accordingly, these methods do not include noise-handling mechanisms. They are, therefore, prone to phenomena that can lead to a small learning rate, resulting in slower convergence, and in local extrema. Finally, the specific methods are particularly demanding in computing resources due to the complex computing operations that come from comparing a large number of samples to utilize the information effectively (Cuzzocrea, 2021).

Therefore, analyzing data streams for pattern recognition should ideally be done using methods or models capable of investigating dynamic sequences of states (Mulinka, Casas, & Vanerio, 2019; Zhao, Wei, Wen, & Qiu, 2020). At the same time, they can create a vector representation of the distribution in space (e.g., Euclidean space) on which some clear metric distance can be defined. In particular, data mining algorithms for data streams assume that each object is checked only once (Yu, Liu, Lu, & Zhang, 2021). However, when finding anomalous values in live data streams, it is necessary to know (for each temporal point and each object present in the current scrolling window) whether it is an outlier or an inlier (Liu, Zhang, & Lu, 2014). An object can change from an outlier to an inlier as the scroll window moves (see figure 1 below).

Figure 1. a) Static dataset and b) Data Stream

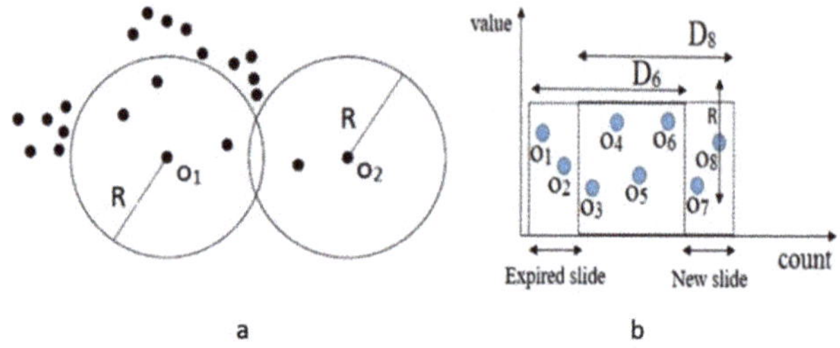

In figure 1.a for k=4, O_1 and O_2 are outliers since the first has three neighbors and the second has only one. Figure 1.b shows a window W=8 and slide S=2. O_8 has one succeeding neighbor (O_7) and four preceding neighbors (O_3, O_4, O_5, O_6) (Lu, Liu, Dong, Gu, Gama, & Zhang, 2018).

A succeeding neighbor of an object *p* is an object *p'*, where *p* ends on the same or the previous sliding window as *p.'* On the other hand, the preceding neighbor of an object *p* is an object *p'*, where *p''* ends before *p*. An object with at least *k* succeeding neighbors (hyperparameter) will never become an outlier and it is characterized as a safe inlier (Protopapadakis, Voulodimos, Doulamis, Doulamis, Dres, & Bimpas, 2017).

Distance-based methods for finding outliers in data streams accept two parameters: $R \geq 0$ and $k \geq 0$, where *R* is a radius of a circle centered on the object and k is the number of neighbors of the object. An object is considered an outlier when it has less than *k* neighbors within a radius of at most *R (hyperparameter)*. All dataflow methods use a rolling window based on time or the number of objects. This window contributes towards the termination of the objects, i.e., it removes them from the active objects' list, which is checked during the execution of the algorithm (Wang, Kim, & Chen, 2009, PTang, Wang, Lou, & Xiong, 2021).

The anomaly detection models, which can be Generative, e.g., *Gaussian Mixture* models or Discrete, e.g., *k-means*, should include coding techniques of multidimensional sequences of states (distributions), allowing the representation (coding of information) in vectors of fixed size taking into account the order of their appearance, which plays a decisive role (Chen, Zhang, & Liu, 2021; Alkathiri, Abdul, & Potdar, 2017; De Amorim, 2012).

Due to the above limitations, the methods in question are initially able to investigate the data's basic structure and to encode its temporal evolution. Essentially, the knowledge mining process from data streams is driven to a middle-level description. This description is capable of capturing high-level concepts, as they incorporate nonlinear and unsupervised information encoding techniques, i.e., higher-level information (Al Jallad, Aljnidi, & Desouki, 2020).

In summary, the available knowledge of the original data distribution are used to depict them in a new vector space where the information are summarized according to some relationship. A fixed-length vector representation is obtained in this new space, based on the developed relationship. This can be exploited by classifiers designed to work in vector spaces.

This process is identical to the early processing stages of the human brain, where synapses are activated in a nonlinear manner after the input of a stimulus. This provides many elements that could be referred to as sparse representations. They can be analyzed, combined, and crossed until the final information is revealed.

In machine learning, sparse representation represents each sample as the linear combination of only a few samples of an array that came from the data. This process is part of the Unsupervised learning methods. It can find many applications as the obtained representations depict the data in a new space of much larger dimensions where the samples are rearranged. In the new space, the samples are represented, so they are rearranged based on the sparse coefficients. It is possible to organize them into identical groups of samples (especially useful in multivariate problems like anomaly detection). At the same time, they can also contain information of a semantic nature as the grouped elements have resulted from a learning stage (Supervised or Unsupervised) using samples from the same distribution (Barua, Muthirayan, Khargonekar, & Al Faruque, 2020).

A key feature of the above strategy is its ability to learn distinct groups with several elements of higher dimension using a minimum number of samples. In this way, the representations expressed by sparse coefficients, constitute a specialized case for linear and nonlinear spaces in a new coordinate system, which contains semantic information that can be easily extracted (Bellare, & Goldreich, 2011; Hamer, 2012).

1.1. DEEP LEARNING EMPOYED FOR ANOMALY DETECTION IN AGRICULTUTE

Deep Learning modeling is extensively used to enhance the productivity of the Agricultural sector. More specifically it is effectively used to classify satellite and aerial images for crop monitoring, to perform deseases' detection, to identify weather patterns with significant consequences, to offer optimization of irrigation schedules and to estimate crop yield. As a matter of fact its financial impact in the agricultural domain can be more than significant (Kalampokas, Vrochidou, Mavridou, Iliadis, Voglitsis, Michalopoulou, & Papakostas, 2025).

However, this process requires robust cyber security algorithms (CSA) that have the potential to efficiently offer data integrity and security. Corrupted data can lead to wrong decisions and cultivation policies. CSA can be crucial for protecting the farming industry from threats on smart farming equipment. Such threats can cause crop damage, supply chain disruptions, and can lead to serious financial damage. It is a fact that the employment of state of the art infrastructure (e.g., *sensors, IoT, Machine Learning*) are capable of improving efficiency. However, at the same time, they can expand vulnerabilities, calling for a layered defense approach including strong passwords, multi-factor authentication, secure system design, regular software updates, employee training, and offline backups.

This work attempts to substantially upgrade the real-time management of agricultural data streams, by presenting a novel and robust micro-clustering anomaly detection algorithm using sparse distributed representations. This modeling approach, takes full advantage of linear vector decomposition to process low-level and high-level information, capable of real-time data correlation. It is an efficient, advanced feedback architecture that considers time sequences and their wide-range dependencies. It aims to offer multiple, dense, high-coherence clusters, by creating a surface where any deviations or anomalies are automatically rejected. In this way, it manages to simulate the Lotus phenomenon and its ability to self-clean its leaves. The aim of this research paper is to motivate the scientific community to employ the LADE scheme in the agricultural domain. However, due to the lack of available agricultural datasets it is applied to the industrial domain with great success.

3. THE PROPOSED LOTUS EFFECT MICRO-CLUSTERING METHOD

The proposed system is inspired by the Lotus Leaf (LOLE), which is a well-known example of a natural superhydrophobic surface with a static contact angle higher than 150° (Ellison, 2013; Montazer, & Harifi, 2018). It is characterized by extremely low water adhesion, and it possesses self-cleaning properties that originate from the *Cassie–Baxter* state and result from a dual micro/nano surface structure (Shirtcliffe, Comanns, Hamlett, Roach, & Atherton, 2019). On a microscale, convex cell papillae are observed on the sheet. Its wax crystals (Young's angle>90°), are characterized by roughness and they comprise of parallelogram or triangular profile pillars that present a hierarchical dual-scale roughness on the nanoscale (De Grave, Botija, Hansen, & Tang, 2006).

In conclusion, hydrophobicity mainly occurs on rough surfaces on two different scales. The *Laplace force* prevents liquid penetration into surface irregularities below a critical surface roughness value. Thus, water penetrates the larger gaps (*Wenzel* condition) but it cannot wet the smaller ones (*Cassie-Baxter* condition) (Fang, Guo, Li, Li, & Feng, 2018).

Figure 2. a) Wenzel state and b) Cassie-Baxter state

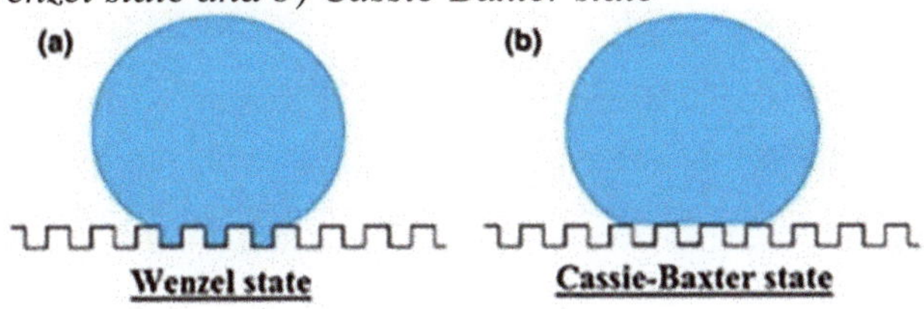

Therefore, surface roughness is necessary in order to achieve waterproof surfaces. The corresponding roughness in anomaly detection is simulated by creating multiple, dense micro-clusters of high coherence, thus developing a hierarchical dual-scale roughness.

This paper proposes an approach for implementing the process in question, which is divided into three stages:

a. Online clustering
b. Offline micro-clustering
c. Outlier detection

The above subprocesses are capable of performing the following tasks:

a. Periodically save obtained statistics (Online-Clustering).
b. Create higher-level, time-specific clusters (Offline Micro-Clustering) by considering the above statistics and user-supplied information.
c. Identify clustering anomalies by employing an intelligent mechanism (Outlier Detection).

3.1 Online-clustering

The online process of the algorithm, adopts the *Regret Analysis* methodology, which is defined as the difference between the value of the objective function of the online algorithm and the best solution achieved by batch processing. In particular, this is implemented based on a sublinear function (a function that grows more slowly than a linear one) that increases depending on the number of data elements. More specifically, the proposed methodology uses the *online k-means clustering* algorithm. In each operating period *t*, it maintains a vector of *k* candidate clusters centers $C_{t=}\{c_{t,1.......}c_{t,k}\}$ before the emergence of the reference point x_t, which arrives at time *t*. The loss suffered by the algorithm at time *t* is given by the following function 1 (Demertzis, Iliadis, Kikiras, & Tziritas, 2019):

$$\ell_t(C_t,x_t) = \min_{c\in C_t} \| x_t - c \|_2^2 \quad (1)$$

Regret is the difference between the cumulative loss of the algorithm after *T* time steps and the optimal stable solution that could be achieved by batch processing, namely:

$$\sum_{t=1}^{T} \ell(C_t,x_t) - \min_{C:|C|=k} \sum_{t=1}^{T} \min_{c\in C} \| x_t - c \|_2^2 \quad (2)$$

The first online subsystem develops clusters to summarize data points' statistics in terms of spatial location and temporal components. Clusters use *a 3D-Pyramidal* Time Frame (3D-PTF), storing them as real-time snapshots. The 3D-PTF is a simple, innovative and effective approach that stores some spatiotemporal characteristics of the data points (snapshots). Moreover, it maintains some clear internal structural information that allows its easy management and retrieval (Aggarwal, Philip, Han, & Wang, 2003).

Unlike the 2D version of the PTF method, which implements hierarchical information storage, 3D-PTF implements stepwise storage in a pyramidal 3D fashion. The features contain sufficient retrieval semantic frames, regardless their temporal and spatial location.This allows easy retrieval and recognition of the snapshots in sparse synthetic three-dimensional shapes.

Let us consider a 3D-PTF characterized by the vector *(W, L,H)* related to width, length and height respectively. Let (x_c, y_c, z_c) be the vector of the lower left corner's coordinates. Following the standard representation, the instance (i,j,k) can be calculated as follows (Li, Yang, Peng, Li, Huang, & Peng, 2022; Kulshreshtha, & Guha, 2018):

$$p_{\text{ptf}}^{ijk} = \left(\left(\frac{\rho_w W}{N_w'}, \frac{\rho_l L}{N_l'}, \frac{\rho_h H}{N_h'}\right) \cdot \left(i,j,k\right)\right) + \left(x_c,y_c,z_c\right) \quad (3)$$

Where (N_w, N_l, N_h) are the dimensions of the 3D mesh. The parameter ρ is the importance ratio of each snapshot.Its value is equal to 1 at the bottom level, in order to preserve analytical details. It is gradually growing as the level increases, in order to capture more current information of the data stream. The 3D-PTF is used to facilitate the analysis of clusters' evolution, as information about the clusters is periodically stored in a permanent storage medium.

Clusters have abstract properties and they are stored as snapshots to find higher-level correlations over a time horizon of length h. The pyramidal time frame stores the clusters at different levels of element density. Instances are categorized in varying order from 1 to log(T), where *T* is the time elapsed since the data stream was started (Mansfield, Wang, Downey, Wan, & Moreno, 2018).

3.2 Offline micro-clustering

In the second subsystem, micro-clusters are created with the algorithm calculating the density around each data point. This is achieved by counting the number of points in a user-defined neighborhood (*Eps-Neighbourhood*) with the definition of thresholds (Vaucher, Pires, Felber, Pasin, Schiavoni, & Fetzer, 2018). The purpose is to locate points in the center of the areas (core), on their borders (border), and points that involve noise (noise). The extra data points are added to the center of the regions if they are densely accessible, i.e., there is a chain of core points where each one belongs to the neighborhood (*Eps-Neighborhood*) of the next point and therefore to distinguish the extreme values for each micro-cluster.

Thus, the user can see the streams in different pyramid periods. Specifically, the neighborhood area of a point *p* is defined as the set of points for which the *Euclidean* distance between the points *p, q* is smaller than the parameter *Eps* (Sabau, 2016; Khannouz, & Glatard, 2020). *Eps* is a distance parameter that defines the radius of the neighborhood around a data point. For a given data point *'p'*, the Eps-Neighborhood (*N_Eps*(*p*)) includes all data points within a distance of *'Eps'* from *'p'*. In other words, it defines how far a point can be from another to be considered a neighbor. Points within the Eps-Neighborhood are used to determine the density of a region:

$$N_{Eps}(p) = \{q \in D dist(p,q) \leq Eps\} \tag{4}$$

provided that *p*=(*p*1,*p*2) and *q*=(*q*1,*q*2), the Euclidean distance is defined as (Jiang, Cordeiro, & Akoglu, 2022; Feng, Liu, Faloutsos, Hooi, Shen, & Cheng, 2017):

$$\sqrt{(q_1 - p_1)^2 + (q_2 - p_2)^2} \tag{5}$$

If the number of points within this radius exceeds or meets a certain threshold (MinPts), the point is considered part of a dense region.

So, a point *p* is considered to be reachable from a point q based on a density determined by the parameters *Eps, MinPts* (Akbarinia, & Cloez, 2019; Ren, Fort, Liu, Roy, Padhy, & Lakshminarayanan, 2021):

$$p \in N_{Eps}(q)$$

and

$$N_{Eps}(q) \geq \text{MinPts} \tag{6}$$

It specifies the minimum number of data points that must be present within the *Eps-Neighborhood* of a point for that point to be considered a core point. Core points are central to the formation of clusters. If the number of data points in the Eps-Neighborhood of a point is greater than or equal to *MinPts*, the point is designated as a core point. This condition is part of the density-based criteria to identify clusters.

Having calculated the micro-cluster density, the kernel methods' property is exploited to express several samples through a symmetric and positive definite matrix, according to the similarity of two samples in each position. Thus, the array elements are in a linear space, regardless the space they come from. This allows the introduction of one more function in the processing stage, which exploits the nonlinear structure of the features in order to present them as a sparse representation problem, modified to work in a *Reproducible Kernel Hilbert Space* (RKHS) (Fujii, & Kawahara, 2019; Stepaniants, 2023).

Specifically, as the data is in a *Euclidean* space, the problem is related to the sparse representation of an *L-dimensional vector* $u \in \mathbb{R}^L$ on a basis $D = [D_1, \ldots, D_m] \in \mathbb{R}^{L \times m}$. This is achieved by using the following function 7 (Zheng, Wang, Liu, Li, Ye, & Ren, 2020; Polat, Ergenc, Kani, Alahdab, Atug, & Temizel, 2022;):

$$\ell\left(h, D\right) := \min_{a} \| \, h - \sum_{i=1}^{m} D_i a_i \, \|_2^2 + \lambda \, \| \, a \, \|_1 \tag{7}$$

In the above relation, the vector $a \in \mathbb{R}^m$ represents the sparse coefficients, and the function $\ell(h,D)$ is the optimal approximation of the problem. The above relationship between the sample distances has to be modified to a nonlinear one. For this purpose, we first define a mapping function $\varphi : \mathbb{R}^L \rightarrow \mathcal{H}$. This function maps the samples from the original space to a new Hilbert space equipped with the inner product. Therefore, using the inner product, two groups of samples, $u_i, u_j \in \mathbb{R}^L$, are mapped to a new space using the relation (Klus, Schuster, & Muandet, 2020; Schneider, Ertel, & Ramos, 2016):

$$k\left(u_i, u_j\right) = \varphi\left(u_i\right)^T \cdot \varphi\left(u_j\right) \in \mathbb{R} \tag{8}$$

Similarly, a set of m sequences can be expressed as a Kernel Matrix $K \in \mathbb{R}^{m \times m}$ whose elements express the similarities between the samples. The similarity function used herein is the *Radial Basis Function* (RBF), whose variance has been estimated as the mean value of the distances of the training data.

But the relation of sparse coding, as presented, does not allow us to work with the K register. For this reason, a modified version has been introduced, which offers the potential to work in this space based on the following equation 9 (Kostic, Novelli, Maurer, Ciliberto, Rosasco, & Pontil, 2022; Long, Zhu, Wang, & Jordan, 2016):

$$\ell\left(\hat{h},\hat{D}\right):=\min_{a}\| \ \hat{h}-\hat{D}a\|_2^2+\lambda \ \| \ a\|_1 \tag{9}$$

To fully utilize the intrinsic properties of sparse representations in the above space, the proposed methodology introduces the concept of a *Spatial Pooler*, which normalizes sparse input representations by enriching the input representation with its temporal context.

Specifically, the Spatial Pooler aims to prepare sparse input representations for further processing, ensuring that inputs similar to each other (have high overlap and thus high coherence) produce also output vectors similar to each other. So, each input pattern is encoded by the *Spatial Pooler* into sparse representations represented as a set of A_k indices of the given pattern at iteration k (Alaa, & Van Der Schaar, 2017; Krestinskaya, & James, 2018).

At each step, the similarity between the sparse input representations at step k and step $k+1$ is calculated by the following equation (Wielgosz, & Pietroń, 2017; Mnatzaganian, Fokoué, & Kudithipudi, 2017):

$$s = \frac{|A_k \cap A_{k+1}|}{max\left(|A_k|,|A_{k+1}|\right)} \tag{10}$$

Similarity s is defined as the ratio between the number of elements (cardinality) of the same active micro-clusters in sparse representations generated in steps k and $k+1$ and a maximum amount of data in two comparable steps. *Spatial Pooler* is usually stable if the sparse representations of the same pattern do not change for its entire life cycle. In this case, the similarity s between all representations of the same pattern is 100% (Pietroń, Wielgosz, & Wiatr, 2016; Saqlain, Wang, & Liu, 2022).

With the methodology in question, during the analysis using 3D-PTF, the user can compare the horizon h in which the clusters are calculated by giving two moments of time, T_1 and T_2, (Alkathiri, Abdul, & Potdar, 2017).

Figure 3. The proposed micro-clustering method

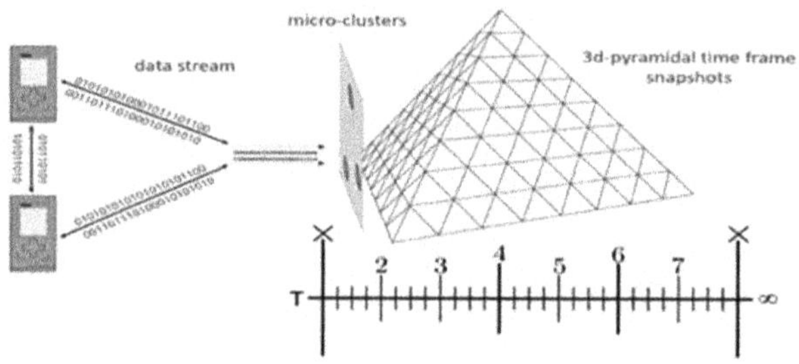

Let us consider that the data stream consists of multidimensional data points $\overline{X}_1, \overline{X}_2, \ldots, \overline{X}_k$, which arrive at timestamps $T_1, T_2, \ldots, T_k$, and that each data point $\overline{X}_i$ is a vector of d dimensions, such that $\overline{X}_i = \{x_i^1, x_i^2, \ldots, x_i^d\}$. Accordingly, the micro-cluster is a set of n d-dimensional data points $X_1, X_2, \ldots, X_n$ with timestamps $T_1, T_2, \ldots, T_n$, which are defined as a tuple 2d+3, which is represented as $(CM2^x, CM1^x, CM2^t, CM1^t, n)$, (Pietroń, Wielgosz, & Wiatr, 2016; Aggarwal, Philip, Han, & Wang, 2003):

1. $CM2^x$ is a d-dimensional vector that sums the square of n data points along each dimension t If the micro-cluster has data points (a,b) and (c,d), which arrive with timestamps t_1 and t_2, then $CM2^x = \{a^2+c^2, b^2+d^2\}$.
2. $CM1^x$ is another d-dimensional vector aggregating the data points set in each micro-cluster along each dimension. Ex $CM1^x = \{a+c, b+d\}$.
3. $CM2^t = t_1^2+t_2^2$ (11)
4. $CM1^t = t_1+t_2$ (12)

The parameter n indicates the dimension of the data points' set, belonging to the micro-cluster.

In the context of 3D-PTF, the snapshots are kept at several degrees of granularity based on their timeliness. The timeliness of a snapshot is determined by the amount of time that has elapsed since the initiation of the data stream (T), with a range from 1 to log(T). The occurrence of the i_{th}-order snapshot takes place at regular intervals of α^i time units, where a is an integer with a>1 (Aggarwal, Philip, Han, & Wang, 2003; Van Nguyen, Le, An, & Min, 2020).

It should be mentioned that the process predicts the partitioning of the data streams into sliding windows as the proposed model performs knowledge discovery in data streams based on sliding windows. Specifically, a part of the data flow is stored and processed within the limits of a time-defined sliding window at each moment. Let $\tau_0 \in T$ be an instant, the range of a sliding window with width ω and step δ for each $\tau \in T$ (με $\tau \geq \tau_0$) is defined as (Cuzzocrea, 2021; Sabau, 2016; Khannouz, & Glatard, 2020):

$$\text{scope}_s\left(\tau,\omega,\delta\right) = \begin{cases} \left[\tau-\omega+1,\tau\right], if\,\tau \geq \tau_0+\omega \wedge mod\left((\tau-\tau_0),\delta\right) = 0 \\ \text{scope}_s\left(\tau-1,\omega,\delta\right), if\, mod\left((\tau-\tau_0),\delta\right) \neq 0 \\ [\tau_0,\tau], if\,\tau_0 \leq \tau < \tau_0+\omega \wedge mod\left((\tau-\tau_0),\delta\right) = 0 \end{cases} \tag{13}$$

where the quantities τ_0, $\tau \in T$ are expressed in temporal milestones and $\omega, \delta \in \mathbb{N}$ in a range of time intervals ($\omega, \delta > 0$). For the sake of simplicity, the above definition implies that all-time quantities are expressed as natural numbers. The calculation of the function is carried out at discrete time instants of T. Then the tuples of the window result from the relation (Chen, Shang, Hassan, Nasser, & Flora, 2016; Deka, 2016):

$$W_s\left(S,\tau,\omega,\delta\right) = \left\{s \in S\left(\tau\right): s \cdot A_\tau \in \text{scope}_s\left(\tau,\omega,\delta\right)\right\} \tag{14}$$

The step δ is as large as the unit of time measurement so that the window's progress is consistent with the corresponding time; since $\delta<\omega$ also holds, the contents of two contiguous snapshots of the scrolling window overlap. The recursive expression clearly shows this in the middle branch of the function, where the window edges do not change except at times specified by step d. In the meantime, its contents remain unchanged until the function is applied again on the next pulse, after δ time instants (Iqbal, Berral, & Carrera, 2020).

In summary, the range of the sliding window at any time τ is calculated based on the width (ω), step (δ), and initial instant (τ_0). The temporal milestones (τ) are points in time, and the sliding window progresses through these milestones. The scope of the window is defined as a set of temporal milestones, and the analysis at each moment involves considering data within this range.

The results of knowledge discovery depend to an extent on the most recent transactions, which fall within the time limits of the sliding window. The transactions within the sliding window's time limits are saved at each point. As a result, when a transaction is outside the sliding window range, its effects on the knowledge discovery results are zero (Iqbal, Berral, & Carrera, 2020).

In particular, a sequence of values $X_1, \ldots X_N$ comes from a process that varies over time and takes into account the assumption that the nature of the variation over time is unknown. It also considers the assumption that the aim is to estimate the average of the process at each point in time as the data arrives. For this reason, the proposed architecture applies the *Adaptive Filtering* (ADF) *technique* for optimal size approximation (Iqbal, Berral, & Carrera, 2020; Demertzis, Iliadis, Pimenidis, Tziritas, Koziri, & Kikiras, 2021; Sanh, Wolf, & Rush, 2020).

A simple procedure of averaging over the sequence when the process undergoes a sharp change does not provide a good estimate of the current value of the mean. This is because older data may not be as useful as recent data.

The ADF aims to provide a methodology capable of handling data change over time, giving more informative utility to recent data. The proposed filtering approach uses a constant forgetting factor λ, which belongs to the interval [0,1], and balances the old with the latest data.

The sample mean with a forgetting factor exponentially reduces the weight value of older observations (X_1,X_2,....X_N) and therefore, it assigns larger weight values on more recent observations (......, X_{N-1},X_N) (Zhou, Zhang, Pan, Xie, Zuo, & Ren, 2019; Zhang, Chen, Bai, Yu, Yu, Li, & Yang, 2020).

Therefore, the sample means with forgetting factor $\overline{\chi_{N,\lambda}}$ is a random variable that approximates the mean value of the observations X_1,X_2,....X_N in such a way that there is more weight in more recent observations. Specifically, the assumption is taken into account that there is a sequence of *N* observations X_1,X_2,....X_N, so the mean $\overline{\chi_N}$ is defined as (Casebeer, Bryan, & Smaragdis, 2022; Raanes, Bocquet, & Carrassi, 2019):

$$\overline{x_N} = \frac{1}{N} \times \sum_{i=1}^{N} x_i \tag{15}$$

The sample mean with a forgetting factor $\overline{\chi_{N,\lambda}}$ is defined as:

$$\overline{\chi_{N,\lambda}} = \frac{1}{\beta_{N,\lambda}} \times \sum_{i=1}^{N} \lambda^{N-i} \times x_i \tag{16}$$

where $\lambda \in [0,1]$ and the weight $\beta_{N,\lambda}$ is defined as follows:

$$\beta_{N,\lambda} = \sum_{i=1}^{N} \lambda^{N-i} \tag{17}$$

It should be said that the choice of the weight $_{N,\lambda}$ must ensure that the average with a forgetting factor $\overline{\chi_{N,\lambda}}$ is as unbiased as possible.

By mathematical definitions, if we take into account the assumption that for X_1,X_2,....X_N and $E[\overline{x}_l] = \mu$, then $E[\overline{\chi_{N,\lambda}}] = \mu$ is valid. If we also take into account the assumption that for ι=1,2,...N we have $Var[x_i]=\sigma^2$, then it holds (Liu, Jiang, He, Chen, Liu, Gao, & Han, 2019; Barron, 2019):

$$\text{Var}[\overline{\chi_{N,\lambda}}] = b_{N,\lambda} \times \sigma^2 \tag{18}$$

where

$$b_{N,\lambda} = \frac{1}{\left(\beta_{N,\lambda}\right)^2} \times \sum_{i=1}^{N} (\lambda^2)^{N-i} = \frac{(1-\lambda)\times(1+\lambda^N)}{(1-\lambda^N)\times(1+\lambda)} \tag{19}$$

So, the standard deviation of the sample considering the forgetting factor is defined as:

$$\sigma_{N,\lambda}^2 = \frac{1}{b_{N,\lambda}} \times \sum_{i=1}^{N} \lambda^{N-i} \times \left(x_i - \overline{\chi_{N,\lambda}}\right)^2 \tag{20}$$

As the value of the forgetting factor λ approaches zero, the weights of the oldest observations decrease exponentially, and the most recent observations acquire higher weight values. Accordingly, the closer the value of λ is to one, the smaller the weight values assigned to the most recent observations, the closer the values of the mean to the forgetting factor $\overline{\chi_{N,\lambda}}$ and the unweighted mean is $\overline{\chi_N}$. The most important benefit of this behavior is that the average is less sensitive to outliers.

The third part of the function provides the possibility of initial "missing" windows immediately after the query is made and when the scope exceeds the period of the current contents of the data stream (Snytsar, & Turakhia, 2019). Finally, since the range function is monotonic (as long as the time evolution implies homologous translation of the intervals) and it can be defined even for future moments, all upcoming flow elements are covered, regardless their eventual appearance (Tangwongsan, Hirzel, & Schneider, 2021).

In processing cases requiring window connections between two streams, S_1 and S_2, each stream is necessarily accompanied by a window specification. Moreover, the windows do not need to have the same unit or range. For each temporal instant $\tau \in T$, the window concatenation operator returns the union of pairs of tuples appearing in the corresponding window instances as below (Basat, Einziger, Keslassy, Orda, Vargaftik, & Waisbard, 2018; Borassi, Epasto, Lattanzi, Vassilvitskii, & Zadimoghaddam, 2020):

$$\begin{aligned} S_1(\tau) \underset{W}{D} S_2(\tau) = \{\langle s_1, s_2, \tau_m \rangle : s_1 \in W_1(S_1(\tau)), s_2 \in W_2(S_2(\tau)) \wedge \\ \wedge E(s_1, s_2) \wedge \tau_m = max(s_1 \cdot A_\tau, s_2 \cdot A_\tau)\} \end{aligned} \tag{21}$$

Each newly arriving tuple of stream S_1 is checked for connection condition E with all the current contents of the rolling window W_2 of stream S_2. If matches are found, the extracted results take the most recent timestamp appearing in the raw tuple pair. Accordingly, the following applies to window aggregations (Snytsar, & Turakhia, 2019; Borassi, Epasto, Lattanzi, Vassilvitskii, & Zadimoghaddam, 2020):

$$\gamma^{f}_{L,W}\left(S,\tau\right) = \left\{a_{i},a_{j},\ldots,a_{n},f(a_{i},a_{j},\ldots,a_{n}),\tau > : \forall A_{k} \in L, a_{k} \in D_{k}, s \in W\left(S\left(\tau\right)\right) \wedge a_{k} = s.A_{k}\right\} \quad (22)$$

while for the window compound:

$$S_{1}\left(\tau\right)\bigcup_{W} S_{2}\left(\tau\right) = \left\{\left\langle s,k\right\rangle : \exists k_{1},k_{2} \in N, \left\langle s,k_{1}\right\rangle \in W_{1}\left(S_{1}\left(\tau\right)\right) \wedge \wedge \left\langle s,k_{2}\right\rangle \in W_{2}(S_{2}(\tau)) \wedge k = k_{1} + k_{2} \wedge k \neq 0\right\} \quad (23)$$

for the window section:

$$S_{1}\left(\tau\right)\bigcap S_{2}\left(\tau\right) = \left\{\left\langle s,k\right\rangle : \exists k_{1},k_{2} \in \mathbb{N}, \left\langle s,k_{1}\right\rangle \in W_{1}\left(S_{1}\left(\tau\right)\right) \wedge \wedge \left\langle s,k_{2}\right\rangle \in W_{2}(S_{2}(\tau)) \wedge k = min\left(k_{1},k_{2}\right) \wedge k \neq 0\right\} \quad (24)$$

and for the window version of the difference:

$$S_{1}\left(\tau\right)\overset{W}{-}S_{2}\left(\tau\right) = \left\{\left\langle s,k\right\rangle : \exists k_{1},k_{2} \in \mathbb{N}, \left\langle s,k_{1}\right\rangle \in W_{1}\left(S_{1}\left(\tau\right)\right) \wedge \wedge \left\langle s,k_{2}\right\rangle \in W_{2}\left(S_{2}\left(\tau\right)\right) \wedge k = max\left(0,k_{1} - k_{2}\right) \wedge k \neq 0\right\} \quad (25)$$

In the cases where the applied windows do not show the same time overlap, for example their ends are not synchronized, or there are windows using a different time step for the final result, the maintenance of the pre-existing landmark of the tuples is chosen. In this way, the synchronization of the elements is optimized, as it is possible that they can be forwarded to other operators (e.g., connection or aggregation) (Iqbal, Berral, & Carrera, 2020; Snytsar, & Turakhia, 2019).

3.3 Outlier Detection

To detect outliers, the proposed architecture uses the relative deviation of the density of a local object's neighborhood (NEIG) from the average of the densities of the local neighborhoods of the neighboring objects (Aumüller, Bernhardsson, & Faithfull, 2020; Mishchuk, Mishkin, Radenovic, & Matas, 2017). Therefore, an object centrally located in some region is assigned a neighborhood density similar to the average of its local neighborhood density, which will be a value close to zero. In contrast, outliers are about to be assigned a value significantly different from zero.

In detail, n(p_i,ar) denotes the number of objects in the neighborhood of the object p_i at a distance of at most ar. Also, $\hat{n}(p_i, r, \alpha)$ is the average of n(p_i,ar) of all objects p located in the NEIG of p_i at a distance of at most r (Fu, Xiang, Wang, & Cai, 2017; Khandelwal, Levy, Jurafsky, Zettlemoyer, & Lewis, 2019).

The measure of the relative deviation of the local neighborhood density of an object is defined as (Akbarinia, & Cloez, 2019; Hassani, & Shi, 2022; Kumar, & Mathur, 2014):

$$MF(p_i, r, \alpha) = 1 - \frac{n(p_i, \alpha r)}{\hat{n}(p_i, r, \alpha)} \tag{26}$$

The standard deviation of *n(p,αr)* of all *p* in the NEIG of p_i at a distance with maximum potential value *r,* is denoted as $\sigma_{\hat{n}}(p_i, r, \alpha)$. Based on this, the normalized standard deviation is defined as (Kim, & Oh, 2022; Nolet, Gala, Raff, Eaton, Rees, Zedlewski, & Oates, 2022):

$$\sigma_{MF}(p_i, r, \alpha) = \frac{\sigma_{\hat{n}}(p_i, r, \alpha)}{\hat{n}(p_i, r, \alpha)} \tag{27}$$

This method considers a set of radii when processing each object in the data stream. The maximum radius value is $r_{max} \approx a^{-1} R_p$ while the minimum r_{min} value is set so that there is always a minimum number of objects in an object's neighborhood. These radii are divided *into critical* and *α-critical.* The set of an object's critical radii comprises all the object's distances from the others. In contrast, the set *of α-criticals* comprises the distances of the set of critical ones. It is estimated by dividing each distance by the value of the parameter a. The parameter *α* takes values in the interval [0, 1], with an initial value of 0.5 which resulted from a trial-and-error process (Linderman, Rachh, Hoskins, Steinerberger, & Kluger, 2019; Jones, Sproat, Ishikawa, & Gutkin, 2023).

To optimize the process in question for data streams using sliding windows, the algorithm maintains the following information (for each object p_i of the data stream, which is present in the current window) (Al Jallad, Aljnidi, & Desouki, 2020; Anil, Gupta, Koren, & Singer, 2019; Kim, & Oh, 2022):

1. SW-Dist:: Sorted list of each object's distances relative to all the radii included in the current sliding window.

2. SW-Neigbs: Sorted list of all sliding window objects. It contains information about the number of neighbors of each object located around it at a distance of some radii.
3. LT_Dist: Sorted list which maintains the distances of the object's neighbors for each future temporal moment when the object will be in the window (lifetime).
4. LT_Neigbs: A sorted list that holds the number of neighbors of the object for each future temporal moment that the object will be in the window.

Initially, for each new object p_i, inserted into the current window, the size of the lists p_i.*SW Dist* and p_i.*SW_Neigbs* is set equal to the size of the window. Each element of the list is initialized with the value 0. A search is performed on the window objects to see if p_i.*LT_Dist* and p_i.*LT_Neigbs* entries exist. The SW lists are updated by incrementing the value of each element by 1 (PTang, Wang, Lou, & Xiong, 2021; Borassi, Epasto, Lattanzi, Vassilvitskii, & Zadimoghaddam, 2020; Ibe, 2013).

Then the *Local Reachability Density* (LRD) is defined to measure the object's p_i reachability, which calculates the density of the object's neighborhood (Ren, Fort, Liu, Roy, Padhy, & Lakshminarayanan, 2021; Fu, Xiang, Wang, & Cai, 2017; Lee, & Civera, 2020). The LRD value of an instance is derived from the average reachability distance to its k-nearest neighbors, where reachability distance is defined as the distance to the farthest neighbor that is closer than the instance under consideration. Depending on the nature of the dataset, the range of LRD values can vary, but in general, a lower LRD value implies that the instance is an outlier since it has a low density of neighbors. On the other hand, a high LRD score shows that the model is not an anomaly, as it has a high density of neighbors. The range of LRD values can be affected by the value of k (the number of nearest neighbors employed in the computation). For instance, a more significant deal of k will result in a lower LRD value, as there are more neighbors to examine (Efimov, Adamyan, & Spokoiny, 2019; Lin, & Xu, 2019).

Similarly, the distance measure chosen for the calculation might influence the range of LRD values. In this approach, the distance is normalized in the range [0, 1]. The LRD is calculated using the following formula 28 (Morais, Le, Tran, Saha, Mansour, & Venkatesh, 2019; Pidhorskyi, Almohsen, & Doretto, 2018):

$$\mathrm{LRD}_{p_i} = \frac{1}{\frac{\sum_{o \in N_{\mathrm{SW\_Neigbs}}(p_i)} \mathrm{reach} - \mathrm{dist}_{\mathrm{SW\_Dist}}\left(p_i, o\right)}{\left|N_{\mathrm{SW\_Dist}}\left(p_i\right)\right|}} \tag{28}$$

Finally, the measure *Local Outlier Factor* (LOF) is defined, which calculates the outlierness value of the object p_i. If p_i is centrally located in a neighborhood of objects, $LOF(p_i)$ it has a value close to 1. Conversely, if p_i is isolated from its neighboring objects, then $LOF(p_i)$ has a value greater than 1 (Protopapadakis, Voulodimos, Doulamis, Doulamis, Dres, & Bimpas, 2017; Xu, Kakde, & Chaudhuri, 2019; Demertzis, Kikiras, Tziritas, Sanchez, & Iliadis, 2018):

$$LOF(p_i) = \frac{\sum_{o \in N_{\text{SW\_Neighs}}(p_i)} \frac{LRD_{\text{SW\_Dist}}(o)}{LRD_{\text{SW\_Dist}}(p_i)}}{\left|N_{\text{SW\_Dist}}(p_i)\right|} \tag{29}$$

4. DATASET AND EVALUATION METRICS

Due to limitation of time and agricultural data availability, this research paper demonstrates the robustness of the proposed modeling approach, on the *Hardware-In-the-Loop* (HIL)-based *Industrial Control System* (HILICS) security dataset. The application on this case, proves the robustness of the algorithm and calls the international scientific community to apply it in the agricultural sector.

This real life dataset was collected with the HIL process that simulated steam-turbine power generation and pumped-storage hydropower generation mechanisms. The simulation includes in detail three processes controlled by three different types of controllers. Data collection was based on using SCADA systems that collect data points, where each point represents a single variable measured or controlled by each system.

All scenarios are modeled because, during normal operation, it is assumed that the operator operates the control unit normally and that the simulator variables related to power generation in the HIL simulator are changed. The operator monitors the values given by the current sensor and adjusts the Setpoints of the various control devices for the system's operation. Accordingly, abnormal behavior occurs when some parameters are not within the normal range or are in unexpected states due to attacks, malfunctions, and failures.

The HAI dataset 2.0 (Shin, Lee, Yun, & Min, 2021; Hwang, Yun, Kim, & Kim, 2019) used in this work has 79 measurement channels to receive data from sensors, actuators, control devices, etc., representing the system's current state, taking one measurement every second. Nine hundred sixty-four thousand eight hundred four seconds 964,804 operation of the cyber-physical systems are included, including the measurements from 5 cyber-attacks. A detailed description of the dataset is available here (Shin, Lee, Yun, & Kim, 2020).

It should be noted that the cyber-physical systems represent a sophisticated integration of computational algorithms and physical processes, forging a tight symbiosis between the digital and physical realms. These systems seamlessly combine computer-based algorithms, communication networks, and control mechanisms with physical processes, creating a unified entity capable of monitoring, analyzing, and influencing the physical world. These systems often leverage real-time data from sensors embedded in the physical environment to inform decision-making processes, facilitating a bidirectional flow of information and control.

The aim of the proposed system is the detection of the structure of the input data, the discovery of its hidden patterns, and the detection of the anomalies without providing any experience to the learning algorithm. In the given case, the anomalies are originating from the cyber-attacks (Demertzis, Iliadis, & Anezakis, 2018; Ge, Zhao, Liu, Zeng, Liu, & Xiong, 2022). The evaluation process of the groupings in each instance of the run was based on the following parameters:

α. Micro-Cluster Purity

The Purity metric defines the purity of the clustering quality. To calculate the purity, each grouping is associated with the class that appears most often within the group. Iτ is calculated by dividing the number of correctly classified cases concerning the number of groupings (Hahsler, Bolanos, & Forrest, 2017):

$$\text{purity} = \frac{\sum_{i=1}^{K} \frac{|c_i^d|}{|c_i|}}{K} * 100\% \tag{30}$$

The variable K defines the number of groupings, $|c_i^d|$ defines the number of points around the dominant value (class) in group i and $|c_i|$ defines the number of points in group i.

Essentially, the purity variable measures the purity or quality of the groupings about the actual values (classes) known in advance for the available data sets. As the gravity of the points gradually weakens, the purity is calculated based on the number of points that enter the predetermined time interval (current window) (Hyde, Angelov, & MacKenzie, 2017).

β. Rand Index

The Rand Index computes a similarity measure between two clusterings by considering all pairs of samples and counting pairs assigned in the same or different clusters in the predicted and true clusterings. Generally, measuring and estimating the number of decisions that are valid and related to clustering accuracy (Steinley, 2004, Gates, & Ahn, 2017):

$$Rand = \frac{TP + TN}{TP + FP + TN + FN} \tag{31}$$

A True Positive (TP) correct decision corresponds to two similar points in the same group, while a True Negative (TN) correct decision corresponds to two different points in different groups. A False Positive (FP) decision corresponds to two different points in the same group, and a False Negative (FN) decision corresponds to two similar points in different groups (Yang, 2017).

1. Cluster Sum of Squares

The Cluster Sum of Squares (CSS) index defines the percentage of variation between points. CSS measures the average squared distance of all the points within a cluster to the cluster centroid. To calculate CSS, we first find the Euclidean distance between a given point and the centroid to which it is assigned. Then this process is iterated for all the clusters, and all the values should be added together to obtain the CSS. The average CSS is obtained by dividing by the number of clusters . The higher this value, the higher the reduction in variance. Each point is registered in the nearest group based on the distance between the point and the center of the group. The center of each group is then defined as the mean value between the points assigned to a group. Therefore, the sum of squares per group is defined as (Piccialli, Sudoso, & Wiegele, 2022; Zhao, Xu, & Fränti, 2009):

$$\sum_{k=1}^{K}\sum_{i\in S_k}\sum_{j=1}^{p}\left(x_{ij}-\overline{x_{kj}}\right)^2 \tag{32}$$

Where S_k is the set of points in the K-the group and $\overline{x_{kj}}$ is the variable in row j representing the center of the group in the K group.

Another metric used to evaluate the performance of the clustering effort is *purity*. Purity is quite simple to calculate. We assign a label to each cluster based on its most frequent class. Then the purity becomes the number of correctly matched class and cluster labels divided by the number of total data points. The range of micro-clusters produced to group the points directly affects the values of the above metrics. The smaller the developed groups, the better the results of the above metrics. For example, the value of the clustering metric purity (purity) is between [0, 1]. The closer it is to 1, the better the clustering result (creating closely spaced groups and matching all points correctly) (Yang, 2017; Singh, Mittal, Malhotra, & Srivastava, 2020; Kou, Peng, & Wang, 2014).

5. RESULTS AND DISCUSSION

To prove the correct use of the specific data set, a thorough analysis of the data was carried out with the purpose of validity checks that prove its reliability before using the proposed algorithm. Specifically, with the process in question, it was established that the original data does not suffer from various problems, such as conflicting information, coding inconsistencies, noise, and extreme values. Moreover, it does not have to deal with special requirements that impose the transformation of the data, such as discretization, normalization, reduction of their dimensions, or selection of the most appropriate characteristics that characterize the problem.

It should also be emphasized that the determination of the number of Micro-clusters and the correspondence of points to them is a function of the parameters of the proposed algorithm and of the data volume received for processing and analysis in each sliding window (Jiang, Cordeiro, & Akoglu, 2022; Borassi, Epasto, Lattanzi, Vassilvitskii, & Zadimoghaddam, 2020). The evaluation metrics were calculated with specifications corresponding to the system's actual operation under consideration. The rating is calculated based on several points entering the predetermined time interval. As the flow weakens, the clarity increases, while correspondingly, as the number of points performed in the sliding window increases, the process becomes more complicated.

The results were validated with the *Prequential Evaluation method* (Bifet, de Francisci Morales, Read, Holmes, & Pfahringer, 2015; Hidalgo, Maciel, & Barros, 2019), in which each case is first used to test the model and then trained with it. The 964,804 seconds of cyber-physical systems' operation were divided into 100 iterations. Specifically, the implementation of the system was done using 9,648 records as learning evaluation instances. Overall, 965 instances were used as training windows.

Table 1 below, shows a snapshot of the algorithm's application to a random stream iteration.

Table 1. Sample of 5 random sliding windows' evaluation metrics

Sliding_Window	Points	Micro-clusters	Purity	Rand	CSS
1st	3,258	1,086	0.77	0.89	0.82
2nd	4,109	1,482	0.71	0.92	0.79
3rd	2,998	1,012	0.82	0.94	0.90
4th	3,510	1,187	0.75	0.93	0.86
5th	3,896	1,346	0.73	0.90	0.84

The above table 1, clearly presents the variation in the number of micro-clusters (MIC-Cs), that were developed as a function of the data volume. These MIC-Cs were considered in each one of the respective temporal windows. The values of the employed process evaluation metrics, were relative to the volume of data. Overall, the anomaly detection process was stable. The following figures, 4-7, present a micro-analysis (a graphical representation) of the operation of the proposed algorithm. Specifically, they present the initial partitioning of the data into MIC-Cs (Fig 4), the identification of outliers within the MCC-Cs (Fig 5), the intelligent separation of the data into each Micro-Cluster (Fig 6), and the final data segmentation (Fig 7).

Figure 4. Initial partitioning of the data into Clusters

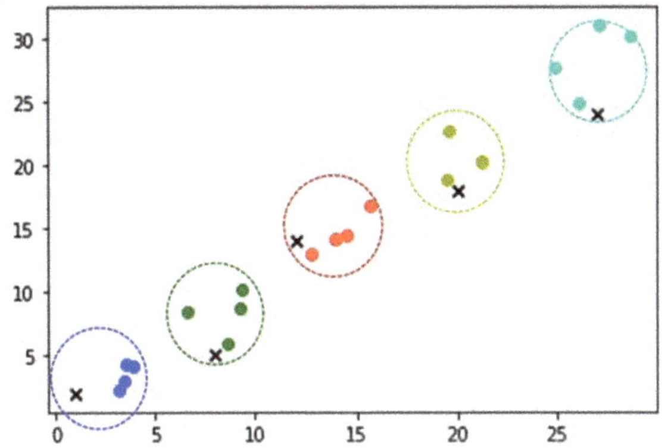

Figure 5. Detection of outliers within the Clusters

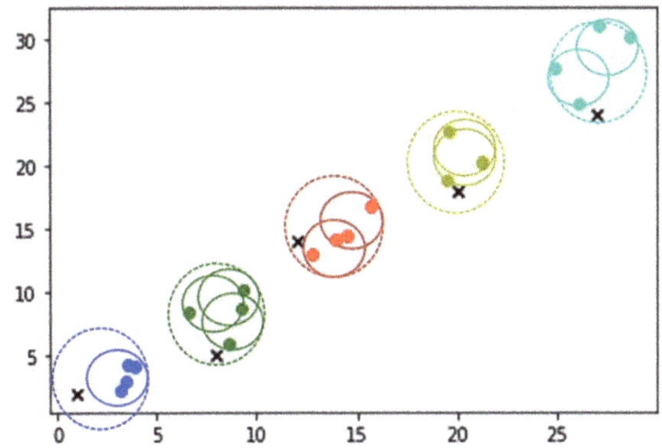

Figure 6. Separation into Micro-Clusters

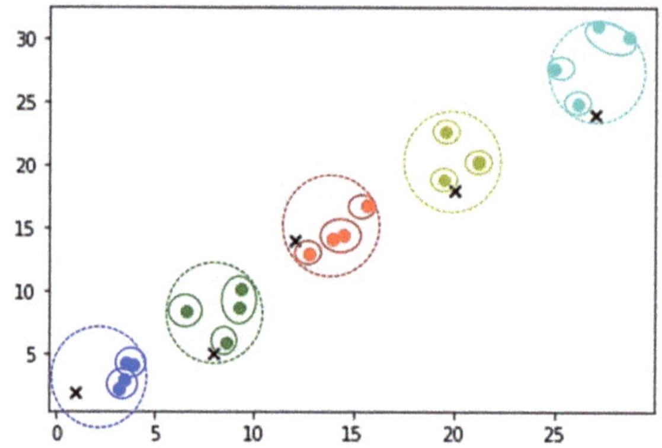

Figure 7. Segmentation into Micro-Clusters

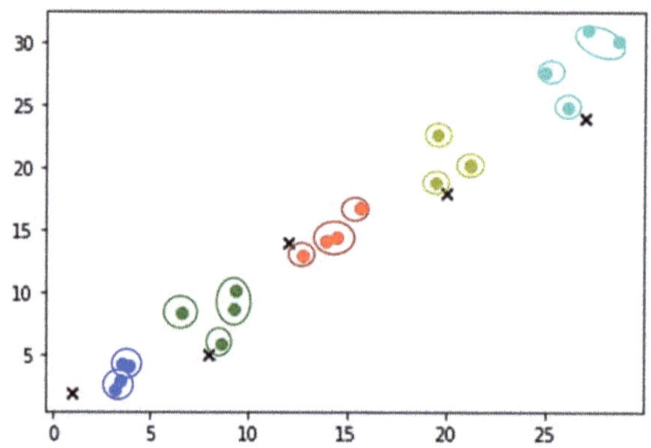

Table 2 presents the aggregated results, the overall average for each performance metric for all of the 100 iterations.

Table 2. Evaluation metrics of the dataset

Num of Iterations	Num of Points	Num of Micro clusters	Avg Putiry	Avg Rand	Avg CSS
100	1,295,731,772	481,989,598	0.85	0.94	0.89

The average value of the purity is equal to 0.85, which indicates the significant difficulty that characterizes the data set in question as the attacks. In general, the anomalies in the operational state of the systems are based on indiscernible events.

Moreover, the average *Rand* value, which in this particular case is the accuracy of the cluster verification process, was estimated to be equal to 0.94, which is a very good percentage, given the complexity of the digital security problem under consideration. Finally, the sum of the squares is considered fully satisfactory as the average of the process is very close to the upper limit (1), which confirms the reliability of the process (Singh, Mittal, Malhotra, & Srivastava, 2020; Bifet, de Francisci Morales, Read, Holmes, & Pfahringer, 2015).

Hierarchical Temporal Memory (HTM) (Barua, Muthirayan, Khargonekar, & Al Faruque, 2020; Mnatzaganian, Fokoué, & Kudithipudi, 2017; Xu, Yang, Yu, Bao, Guo, & Jiang, 2020) provides a theoretical framework that models several key computational principles of the neocortex. The Spatial Pooler (SP) is an important component of the HTM. The HTM SP represents a neurally inspired learning algorithm for creating sparse representations from noisy data streams online. It models how neurons learn feedforward connections and form efficient input representations. It converts arbitrary binary input patterns into sparse distributed representations. *Spatial Pooler,* as a discrete-time dynamic system with internal state, initialization, and step/transition capability, ensured sparsity normalization, robustness to noise, fault tolerance, and continuous adaptation to the evolving statistical characteristics maintaining the topology of the input space (Krestinskaya, & James, 2018; Van Nguyen, Le, An, & Min, 2020).

The performance of the spatial concentrator has been proven to be efficient and satisfactory, as it was tacitly supported by an auxiliary homeostasis mechanism in the process of operating sliding windows, where the windows effectively functioned as boosting agents. Boosting is based on the fact that each window's overlay is multiplied by a boost or fade factor, depending on how often it was activated in the recent past. This process boosts micro-clusters that are rarely activated and weakens those that are frequently activated. This way, it achieves a more balanced participation of all micro-clusters in the spatial coding (Prokhorenkova, Gusev, Vorobev, Dorogush, & Gulin, 2018; Galar, Fernandez, Barrenechea, Bustince, & Herrera, 2011).

The advantage of this behavior, which employs a forgetting factor, is that the average value is close to the average of several more recent observations. However, a disadvantage of this behavior is that the average with a forgetting factor *is more sensitive to outliers*. The closer the value of the forgetting factor λ is to one, the less weight is assigned to the most recent observations. The benefit of this behavior is that the average a forgetting factor $\overline{\chi_{N,\lambda}}$ It is less sensitive to outliers (Kumar, & Mathur, 2014; Jiang, Cordeiro, & Akoglu, 2022).

In the case of 3D-PTF, the snapshots are stored at different levels of granularity, depending on their timeliness. Due to the pyramid time frame, this system is characterized by additional accuracy, reliability, and efficiency, which allows immediate search application. Moreover, it possesses significant scaling capabilities since it allows the consideration of different sizes of data streams, using more dimensions and any number of desired clusters.

Another important asset is the LRD measure, which optimally calculates the density of the object's neighborhood so that the appropriate micro-clusters are created for each data set. Finally, an extremely important contribution is the use of the Local Outlier Factor, according to which a point is considered extreme if it is at a short distance from the dense cluster of the developed micro-clusters. In a more general approach, the point in question may not be considered an outlier, but based on the LOF procedure, it can be effectively identified as a *local outlier* (Xu, Kakde, & Chaudhuri, 2019).

6. CONCLUSIONS AND FUTURE WORK

By placing sensors and using smart applications, intelligent anomaly detection systems have been proven to be particularly useful and reliable, as they can clearly show how complex cyber-physical systems work. In particular, event analysis to detect real-time anomalies related to cyber-attacks is a proactive approach to managing digital security that is highly required in the modern digital age (Demertzis, & Iliadis, 2016).

This research introduces an innovative micro-clustering anomaly detection algorithm aiming to increase the active protection measures of the systems in question. The considered algorithmic approach is based on the robust natural self-protection property of the Lotus, where any dirt that falls on it does not stick but is automatically removed due to the hydrophobic nature of its surface structure. The introduced model employs sparse distributed representations to detect real-time anomalies in data streams. The model was successfully applied and tested.

Inspired by the property in question, this work proposes an advanced intelligent real-time algorithm to detect anomalies in *data streams*. It is a mechanism for creating multiple, densely coherent, homogeneous micro-clusters, creating a hierarchical dual-scale roughness surface where any deviations or anomalies are automatically rejected. As it has been experimentally proven, the proposed approach offers accurate, valid, and timely results, aiming at properly managing digital threats of complex industrial systems.

The system's high generalizability and the proposed methodology's convergence stability prove that it can perform even when the anomaly is developing. Therefore, it corresponds to an unknown type of threat. An important asset of the introduced model is its potential to scale with the correlation of multiple sources of information. Moreover, it can perform an effective and accurate analysis following an optimistic approach for solving dynamic problems with an affordable computational cost.

Testing has revealed that the introduced approach offers a robust and worth trusting solution since it achieves high accuracy combined with fast convergence. Also, it is important that the introduced approach can be further enhanced with the fusion of more advanced anomaly detection techniques, capable of offering automatic hyperparameters' adjustment and exploitation of the metadata that can describe the characteristics of information entities', aiming at identification, recognition, discovery, evaluation, and management of the entities described. Reinforcing the above point of view, the proposed algorithm's structure should be studied and enhanced further to be supplemented as a result of timely data transformation techniques.

Also, future research could follow the development of methods that can be applied in the context of a dynamic security policy where the decision of whether a case is a noise or a new anomaly will be better balanced. In addition, there is a need for additional research on dealing with concept drift from snapshot selection methods so that the micro-clusters formation function can be applied more successfully. One way this could be achieved is by combining these methods with drift detectors. Finally, importance should also be given to creating pre-processing methods for cases with a problem in the data flow, such as unbalanced classes or lack of patterns.

DEDICATION

Dedicated to the memory of Dr. Konstantinos Demertzis, Democritus University of Thrace, a distinguished (and distinction awarded) researcher-scientist, and a dearest fellow and friend. He will never be forgotten.

CONFLICTS OF INTEREST

The authors declare that they have no conflict of interest.

DATA AVAILABILITY STATEMENT

The datasets analyzed during the current study are available in the https://github.com/icsdataset/hai repository.

REFERENCES

Aggarwal, C. C. (2016). Neighborhood-based collaborative filtering. In *Recommender Systems: The Textbook* (pp. 29–70). Springer International Publishing.

Aggarwal, C. C., Philip, S. Y., Han, J., & Wang, J. (2003, January). A framework for clustering evolving data streams. In *Proceedings 2003 VLDB conference* (pp. 81-92). Morgan Kaufmann.

Akbarinia, R., & Cloez, B. (2019). Efficient matrix profile computation using different distance functions. *arXiv preprint arXiv:1901.05708.*

Al Jallad, K., Aljnidi, M., & Desouki, M. S. (2020). Anomaly detection optimization using big data and deep learning to reduce false-positive. *Journal of Big Data*, *7*(1), 68.

Alaa, A. M., & Van Der Schaar, M. (2017). Bayesian inference of individualized treatment effects using multi-task gaussian processes. *Advances in Neural Information Processing Systems*, •••, 30.

Alkathiri, M., Abdul, J., & Potdar, M. B. (2017, August). Kluster: Application of k-means clustering to multidimensional GEO-spatial data. In *2017 International conference on information, communication, instrumentation and control (ICICIC)* (pp. 1-7). IEEE.

Alves, T. M., Soeiro, R. O., & Cartaxo, A. V. (2019, September). Probability distribution of intercore crosstalk in weakly coupled MCFs with multiple interferers. In *2019 IEEE Photonics Conference (IPC)* (pp. 1-4). IEEE.

Anil, R., Gupta, V., Koren, T., & Singer, Y. (2019). Memory efficient adaptive optimization. *Advances in Neural Information Processing Systems*, •••, 32.

Aumüller, M., Bernhardsson, E., & Faithfull, A. (2020). ANN-Benchmarks: A benchmarking tool for approximate nearest neighbor algorithms. *Information Systems*, *87*, 101374.

Barron, J. T. (2019). A general and adaptive robust loss function. In *Proceedings of the IEEE/CVF conference on computer vision and pattern recognition* (pp. 4331-4339).

Barua, A., Muthirayan, D., Khargonekar, P. P., & Al Faruque, M. A. (2020, April). Hierarchical temporal memory based machine learning for real-time, unsupervised anomaly detection in smart grid: WiP abstract. In *2020 ACM/IEEE 11th International Conference on Cyber-Physical Systems (ICCPS)* (pp. 188-189). IEEE.

Basat, R. B., Einziger, G., Keslassy, I., Orda, A., Vargaftik, S., & Waisbard, E. (2018, December). Memento: Making sliding windows efficient for heavy hitters. In *Proceedings of the 14th International Conference on Emerging Networking EXperiments and Technologies* (pp. 254-266).

Batyuk, A., Voityshyn, V., & Verhun, V. (2018, August). Software architecture design of the real-time processes monitoring platform. In *2018 IEEE Second International Conference on Data Stream Mining & Processing (DSMP)* (pp. 98-101). IEEE.

Bellare, M., & Goldreich, O. (2011). On probabilistic versus deterministic provers in the definition of proofs of knowledge. In *Studies in Complexity and Cryptography. Miscellanea on the Interplay between Randomness and Computation: In Collaboration with Lidor Avigad, Mihir Bellare, Zvika Brakerski, Shafi Goldwasser, Shai Halevi, Tali Kaufman, Leonid Levin, Noam Nisan, Dana Ron, Madhu Sudan, Luca Trevisan, Salil Vadhan, Avi Wigderson, David Zuckerman* (pp. 114–123). Springer Berlin Heidelberg.

Bifet, A., de Francisci Morales, G., Read, J., Holmes, G., & Pfahringer, B. (2015, August). Efficient online evaluation of big data stream classifiers. In *Proceedings of the 21th ACM SIGKDD international conference on knowledge discovery and data mining* (pp. 59-68).

Borassi, M., Epasto, A., Lattanzi, S., Vassilvitskii, S., & Zadimoghaddam, M. (2020). Sliding window algorithms for k-clustering problems. *Advances in Neural Information Processing Systems*, *33*, 8716–8727.

Casebeer, J., Bryan, N. J., & Smaragdis, P. (2022). Meta-AF: Meta-learning for adaptive filters. *IEEE/ACM Transactions on Audio, Speech, and Language Processing, 31*, 355–370.

Chen, F., Zhang, T., & Liu, R. (2021). An active learning method based on variational autoencoder and dbscan clustering. *Computational Intelligence and Neuroscience, 2021*(1), 9952596.

Chen, T. H., Shang, W., Hassan, A. E., Nasser, M., & Flora, P. (2016, May). Detecting problems in the database access code of large scale systems: An industrial experience report. In *Proceedings of the 38th International Conference on Software Engineering Companion* (pp. 71-80).

Cuzzocrea, A. (2021, January). Big data lakes: models, frameworks, and techniques. In *2021 IEEE International Conference on Big Data and Smart Computing (BigComp)* (pp. 1-4). IEEE.

De Amorim, R. C. (2012, November). Constrained clustering with minkowski weighted k-means. In *2012 IEEE 13th International Symposium on Computational Intelligence and Informatics (CINTI)* (pp. 13-17). IEEE.

De Grave, A., Botija, P., Hansen, H. N., & Tang, P. T. (2006, January). Manufacturing and characterization of water repellent surfaces. In *4M 2006-Second International Conference on Multi-Material Micro Manufacture* (pp. 281-284). Elsevier.

Deka, B. K. (2016, March). Transformations of graph database model from multidimensional data model. In *2016 3rd International Conference on Computing for Sustainable Global Development (INDIACom)* (pp. 2836-2840). IEEE.

Demertzis, K., & Iliadis, L. (2016, January). Bio-inspired hybrid intelligent method for detecting android malware. In *Knowledge, Information and Creativity Support Systems: Selected Papers from KICSS'2014-9th International Conference, held in Limassol, Cyprus, on November 6-8, 2014* (pp. 289-304). Cham: Springer International Publishing.

Demertzis, K., Iliadis, L., & Anezakis, V. D. (2019, May). A machine hearing framework for real-time streaming analytics using Lambda architecture. In *International Conference on Engineering Applications of Neural Networks* (pp. 246-261). Cham: Springer International Publishing.

Demertzis, K., Iliadis, L., Kikiras, P., & Tziritas, N. (2019, May). Cyber-typhon: an online multi-task anomaly detection framework. In *IFIP International Conference on Artificial Intelligence Applications and Innovations* (pp. 19-36). Cham: Springer International Publishing.

Demertzis, K., Iliadis, L., Pimenidis, E., & Kikiras, P. (2022). Variational restricted Boltzmann machines to automated anomaly detection. *Neural Computing & Applications, 34*(18), 15207–15220.

Demertzis, K., Iliadis, L., Pimenidis, E., Tziritas, N., Koziri, M., & Kikiras, P. (2021, June). Blockchained adaptive federated auto MetaLearning BigData and DevOps CyberSecurity architecture in Industry 4.0. In *International Conference on Engineering Applications of Neural Networks* (pp. 345-363). Cham: Springer International Publishing.

Demertzis, K., Iliadis, L. S., & Anezakis, V. D. (2018). An innovative soft computing system for smart energy grids cybersecurity. *Advances in Building Energy Research, 12*(1), 3–24.

Demertzis, K., Kikiras, P., Tziritas, N., Sanchez, S. L., & Iliadis, L. (2018). The next generation cognitive security operations center: Network flow forensics using cybersecurity intelligence. *Big Data and Cognitive Computing, 2*(4), 35.

Dias, R., Mauricio, L. A. F., & Poggi, M. (2020, June). *Toward an efficient real-time anomaly detection system for cloud datacenters. In 2020 IFIP networking conference (networking)*. IEEE.

Efimov, K., Adamyan, L., & Spokoiny, V. (2019). Adaptive nonparametric clustering. *IEEE Transactions on Information Theory, 65*(8), 4875–4892.

Ellison, M. (2013). *Biomimetic Textiles*. Engineered Biomimicry.

Fang, W., Guo, H. Y., Li, B., Li, Q., & Feng, X. Q. (2018). Revisiting the critical condition for the Cassie–Wenzel transition on micropillar-structured surfaces. *Langmuir, 34*(13), 3838–3844.

Feng, W., Liu, S., Faloutsos, C., Hooi, B., Shen, H., & Cheng, X. (2017). EagleMine: Vision-Guided Mining in Large Graphs. *arXiv preprint arXiv:1710.08756.*

Fu, C., Xiang, C., Wang, C., & Cai, D. (2017). Fast approximate nearest neighbor search with the navigating spreading-out graph. *arXiv preprint arXiv:1707.00143.*

Fujii, K., & Kawahara, Y. (2019). Dynamic mode decomposition in vector-valued reproducing kernel Hilbert spaces for extracting dynamical structure among observables. *Neural Networks, 117*, 94–103.

Galar, M., Fernandez, A., Barrenechea, E., Bustince, H., & Herrera, F. (2011). A review on ensembles for the class imbalance problem: Bagging-, boosting-, and hybrid-based approaches. *IEEE Transactions on Systems, Man, and Cybernetics. Part C, Applications and Reviews, 42*(4), 463–484.

Gates, A. J., & Ahn, Y. Y. (2017). The impact of random models on clustering similarity. *Journal of Machine Learning Research*, *18*(87), 1–28.

Ge, H., Zhao, X., Liu, C., Zeng, Y., Liu, Q., & Xiong, D. (2022). TGEA 2.0: A Large-Scale Diagnostically Annotated Dataset with Benchmark Tasks for Text Generation of Pretrained Language Models. *Advances in Neural Information Processing Systems*, *35*, 31612–31626.

Hahsler, M., Bolanos, M., & Forrest, J. (2017). Introduction to stream: An extensible framework for data stream clustering research with R. *Journal of Statistical Software*, *76*, 1–50.

Hamer, D. (2012). Probability, anti-resilience, and the weight of expectation. *Law Probability and Risk*, *11*(2-3), 135–158.

Hasim, A., Herdiyeni, Y., & Douady, S. (2016). Leaf shape recognition using centroid contour distance. [). IOP Publishing.]. *IOP Conference Series. Earth and Environmental Science*, *31*(1), 012002.

Hassani, A., & Shi, H. (2022). Dilated neighborhood attention transformer. *arXiv preprint arXiv:2209.15001*.

Hidalgo, J. I. G., Maciel, B. I., & Barros, R. S. (2019). Experimenting with prequential variations for data stream learning evaluation. *Computational Intelligence*, *35*(4), 670–692.

Hwang, W. S., Yun, J. H., Kim, J., & Kim, H. C. (2019, November). Time-series aware precision and recall for anomaly detection: considering variety of detection result and addressing ambiguous labeling. In *Proceedings of the 28th ACM International Conference on Information and Knowledge Management* (pp. 2241-2244).

Hyde, R., Angelov, P., & MacKenzie, A. R. (2017). Fully online clustering of evolving data streams into arbitrarily shaped clusters. *Information Sciences*, *382*, 96–114.

Ibe, O. (2013). Random Walk. In *Markov processes for stochastic modeling* (pp. 205–259). Elsevier.

Iqbal, W., Berral, J. L., & Carrera, D. (2020). Adaptive sliding windows for improved estimation of data center resource utilization. *Future Generation Computer Systems*, *104*, 212–224.

Jiang, S., Cordeiro, R. L., & Akoglu, L. (2022, November). D. mca: Outlier detection with explicit micro-cluster assignments. In *2022 IEEE International Conference on Data Mining (ICDM)* (pp. 987-992). IEEE.

Jones, L., Sproat, R., Ishikawa, H., & Gutkin, A. (2023). Helpful Neighbors: Leveraging Neighbors in Geographic Feature Pronunciation. *Transactions of the Association for Computational Linguistics, 11*, 85–101.

Kalampokas, T., Vrochidou, E., Mavridou, E., Iliadis, L., Voglitsis, D., Michalopoulou, M., & Papakostas, G. A. (2025). Empowering Kiwifruit Cultivation with AI: Leaf Disease Recognition Using AgriVision-Kiwi Open Dataset. *Electronics (Basel), 14*(9), 1705.

Khandelwal, U., Levy, O., Jurafsky, D., Zettlemoyer, L., & Lewis, M. (2019). Generalization through memorization: Nearest neighbor language models. *arXiv preprint arXiv:1911.00172.*

Khannouz, M., & Glatard, T. (2020). A benchmark of data stream classification for human activity recognition on connected objects. *Sensors (Basel), 20*(22), 6486.

Kim, D., & Oh, A. (2022). How to find your friendly neighborhood: Graph attention design with self-supervision. *arXiv preprint arXiv:2204.04879.*

Klus, S., Schuster, I., & Muandet, K. (2020). Eigendecompositions of transfer operators in reproducing kernel Hilbert spaces. *Journal of Nonlinear Science, 30*(1), 283–315.

Kostic, V., Novelli, P., Maurer, A., Ciliberto, C., Rosasco, L., & Pontil, M. (2022). Learning dynamical systems via Koopman operator regression in reproducing kernel Hilbert spaces. *Advances in Neural Information Processing Systems, 35*, 4017–4031.

Kou, G., Peng, Y., & Wang, G. (2014). Evaluation of clustering algorithms for financial risk analysis using MCDM methods. *Information Sciences, 275*, 1–12.

Krestinskaya, O., & James, A. P. (2018). Feature extraction without learning in an analog spatial pooler memristive-CMOS circuit design of hierarchical temporal memory. *Analog Integrated Circuits and Signal Processing, 95*(3), 457–465.

Kulshreshtha, P., & Guha, T. (2018, October). An online algorithm for constrained face clustering in videos. In *2018 25th IEEE International Conference on Image Processing (ICIP)* (pp. 2670-2674). IEEE.

Kumar, M., & Mathur, R. (2014, April). Unsupervised outlier detection technique for intrusion detection in cloud computing. In *International Conference for Convergence for Technology*-2014 (pp. 1-4). IEEE.

Lee, S. H., & Civera, J. (2020). Robust uncertainty-aware multiview triangulation. *arXiv preprint arXiv:2008.01258.*

Li, Y., Yang, M., Peng, D., Li, T., Huang, J., & Peng, X. (2022). Twin contrastive learning for online clustering. *International Journal of Computer Vision*, *130*(9), 2205–2221.

Lin, T. E., & Xu, H. (2019). Deep unknown intent detection with margin loss. *arXiv preprint arXiv:1906.00434*.

Linderman, G. C., Rachh, M., Hoskins, J. G., Steinerberger, S., & Kluger, Y. (2019). Fast interpolation-based t-SNE for improved visualization of single-cell RNA-seq data. *Nature Methods*, *16*(3), 243–245.

Liu, A., Zhang, G., & Lu, J. (2014, November). Concept drift detection based on anomaly analysis. In *International Conference on Neural Information Processing* (pp. 263-270). Cham: Springer International Publishing.

Liu, L., Jiang, H., He, P., Chen, W., Liu, X., Gao, J., & Han, J. (2019). On the variance of the adaptive learning rate and beyond. *arXiv preprint arXiv:1908.03265*.

Long, M., Zhu, H., Wang, J., & Jordan, M. I. (2016). Unsupervised domain adaptation with residual transfer networks. *Advances in Neural Information Processing Systems*, •••, 29.

Lu, J., Liu, A., Dong, F., Gu, F., Gama, J., & Zhang, G. (2018). Learning under concept drift: A review. *IEEE Transactions on Knowledge and Data Engineering*, *31*(12), 2346–2363.

Mansfield, P. A., Wang, Q., Downey, C., Wan, L., & Moreno, I. L. (2018). Links: A high-dimensional online clustering method. *arXiv preprint arXiv:1801.10123*.

Mishchuk, A., Mishkin, D., Radenovic, F., & Matas, J. (2017). Working hard to know your neighbor's margins: Local descriptor learning loss. *Advances in Neural Information Processing Systems*, •••, 30.

Mnatzaganian, J., Fokoué, E., & Kudithipudi, D. (2017). A mathematical formalization of hierarchical temporal memory's spatial pooler. *Frontiers in Robotics and AI*, *3*, 81.

Montazer, M., & Harifi, T. (2018). *Nanofinishing of textile materials*. Woodhead Publishing.

Morais, R., Le, V., Tran, T., Saha, B., Mansour, M., & Venkatesh, S. (2019). Learning regularity in skeleton trajectories for anomaly detection in videos. In *Proceedings of the IEEE/CVF conference on computer vision and pattern recognition* (pp. 11996-12004).

Mulinka, P., Casas, P., & Vanerio, J. (2019, November). Continuous and adaptive learning over big streaming data for network security. In *2019 IEEE 8th International Conference on Cloud Networking (CloudNet)* (pp. 1-4). IEEE.

Nolet, C. J., Gala, D., Raff, E., Eaton, J., Rees, B., Zedlewski, J., & Oates, T. (2022). GPU semiring primitives for sparse neighborhood methods. *Proceedings of Machine Learning and Systems*, *4*, 95–109.

Piccialli, V., Sudoso, A. M., & Wiegele, A. (2022). SOS-SDP: An exact solver for minimum sum-of-squares clustering. *INFORMS Journal on Computing*, *34*(4), 2144–2162.

Pidhorskyi, S., Almohsen, R., & Doretto, G. (2018). Generative probabilistic novelty detection with adversarial autoencoders. *Advances in Neural Information Processing Systems*, •••, 31.

Pietroń, M., Wielgosz, M., & Wiatr, K. (2016, September). Formal analysis of HTM spatial pooler performance under predefined operation conditions. In *International joint conference on rough sets* (pp. 396-405). Cham: Springer International Publishing.

Polat, G., Ergenc, I., Kani, H. T., Alahdab, Y. O., Atug, O., & Temizel, A. (2022, July). Class distance weighted cross-entropy loss for ulcerative colitis severity estimation. In *Annual Conference on Medical Image Understanding and Analysis* (pp. 157-171). Cham: Springer International Publishing.

Prokhorenkova, L., Gusev, G., Vorobev, A., Dorogush, A. V., & Gulin, A. (2018). CatBoost: Unbiased boosting with categorical features. *Advances in Neural Information Processing Systems*, •••, 31.

Protopapadakis, E., Voulodimos, A., Doulamis, A., Doulamis, N., Dres, D., & Bimpas, M. (2017). Stacked autoencoders for outlier detection in over-the-horizon radar signals. *Computational Intelligence and Neuroscience*, *2017*(1), 5891417.

PTang, P., Wang, W., Lou, J., & Xiong, L. (2021). Generating adversarial examples with distance constrained adversarial imitation networks. *IEEE Transactions on Dependable and Secure Computing*, *19*(6), 4145–4155.

Raanes, P. N., Bocquet, M., & Carrassi, A. (2019). Adaptive covariance inflation in the ensemble Kalman filter by Gaussian scale mixtures. *Quarterly Journal of the Royal Meteorological Society*, *145*(718), 53–75.

Ren, J., Fort, S., Liu, J., Roy, A. G., Padhy, S., & Lakshminarayanan, B. (2021). A simple fix to mahalanobis distance for improving near-ood detection. *arXiv preprint arXiv:2106.09022*.

Sabau, A. S. (2016). Stream Clustering using Probabilistic Data Structures. *arXiv preprint arXiv:1612.02701.*

Sanh, V., Wolf, T., & Rush, A. (2020). Movement pruning: Adaptive sparsity by fine-tuning. *Advances in Neural Information Processing Systems, 33,* 20378–20389.

Saqlain, A. S., Wang, L. Y., & Liu, Z. (2022, April). SL-CycleGAN: blind motion deblurring in cycles using sparse learning. In *2022 7th International Conference on Cloud Computing and Big Data Analytics (ICCCBDA)* (pp. 01-05). IEEE.

Schneider, M., Ertel, W., & Ramos, F. (2016). Expected similarity estimation for large-scale batch and streaming anomaly detection. *Machine Learning, 105*(3), 305–333.

Shin, H. K., Lee, W., Yun, J. H., & Kim, H. (2020). {HAI} 1.0:{HIL-based} augmented {ICS} security dataset. In *13Th USENIX workshop on cyber security experimentation and test (CSET 20).*

Shin, H. K., Lee, W., Yun, J. H., & Min, B. G. (2021, August). Two ICS security datasets and anomaly detection contest on the HIL-based augmented ICS testbed. In *Proceedings of the 14th Cyber Security Experimentation and Test Workshop* (pp. 36-40).

Shirtcliffe, N., Comanns, P., Hamlett, C., Roach, P., & Atherton, S. (2019). The effect of roughness geometry on superhydrophobicity and related phenomena.

Singh, A. K., Mittal, S., Malhotra, P., & Srivastava, Y. V. (2020, March). Clustering evaluation by davies-bouldin index (dbi) in cereal data using k-means. In *2020 Fourth international conference on computing methodologies and communication (ICCMC)* (pp. 306-310). IEEE.

Snytsar, R., & Turakhia, Y. (2019, December). Parallel approach to sliding window sums. In *International Conference on Algorithms and Architectures for Parallel Processing* (pp. 19-26). Cham: Springer International Publishing.

Steinley, D. (2004). Properties of the hubert-arable adjusted rand index. *Psychological Methods, 9*(3), 386.

Stepaniants, G. (2023). Learning partial differential equations in reproducing kernel Hilbert spaces. *Journal of Machine Learning Research, 24*(86), 1–72.

Tangwongsan, K., Hirzel, M., & Schneider, S. (2021). In-order sliding-window aggregation in worst-case constant time. *The VLDB Journal, 30*(6), 933–957.

Van Nguyen, T., Le, N. T., An, J., & Min, K. S. (2020, November). Defect-resilient memristor crossbar of hierarchical temporal memory (HTM) spatial pooling for near-IoT-sensor cognitive computing. In *2020 27th IEEE International Conference on Electronics, Circuits and Systems (ICECS)* (pp. 1-4). IEEE.

Vaucher, S., Pires, R., Felber, P., Pasin, M., Schiavoni, V., & Fetzer, C. (2018, July). SGX-aware container orchestration for heterogeneous clusters. In *2018 IEEE 38th International Conference on Distributed Computing Systems (ICDCS)* (pp. 730-741). IEEE.

Wang, S., Kim, H., & Chen, H. (2009, April). Contour tracking using centroid distance signature and dynamic programming method. In *2009 International Joint Conference on Computational Sciences and Optimization* (Vol. 1, pp. 274-278). IEEE.

Wielgosz, M., & Pietroń, M. (2017). Using spatial pooler of hierarchical temporal memory to classify noisy videos with predefined complexity. *Neurocomputing*, *240*, 84–97.

Wu, M., Ma, Y., Fan, F., Mei, X., & Huang, J. (2020). Infrared and visible image fusion via joint convolutional sparse representation. *Journal of the Optical Society of America. A, Optics, Image Science, and Vision*, *37*(7), 1105–1115.

Xu, T., Yang, H., Yu, A., Bao, B., Guo, H., & Jiang, Y. (2020, October). Traffic optimization and congestion prevention strategy based on hierarchical temporal memory of data center optical network. In *2020 Opto-Electronics and Communications Conference (OECC)* (pp. 1-3). IEEE.

Xu, Z., Kakde, D., & Chaudhuri, A. (2019, December). Automatic hyperparameter tuning method for local outlier factor, with applications to anomaly detection. In *2019 IEEE International Conference on Big Data (Big Data)* (pp. 4201-4207). IEEE.

Yang, Y. (2017). Temporal Data Clustering. In *Temporal Data Mining Via Unsupervised Ensemble Learning* (pp. 19–34). Elsevier.

Yu, H., Liu, T., Lu, J., & Zhang, G. (2021). Automatic learning to detect concept drift. *arXiv preprint arXiv:2105.01419*.

Zhang, Y., Chen, Y., Bai, X., Yu, S., Yu, K., Li, Z., & Yang, K. (2020, April). Adaptive unimodal cost volume filtering for deep stereo matching. *Proceedings of the AAAI Conference on Artificial Intelligence*, *34*(07), 12926–12934.

Zhao, J., Wei, S., Wen, X., & Qiu, X. (2020). Analysis and prediction of big stream data in real-time water quality monitoring system. *Journal of Ambient Intelligence and Smart Environments*, *12*(5), 393–406.

Zhao, Q., Xu, M., & Fränti, P. (2009, April). Sum-of-squares based cluster validity index and significance analysis. In *International conference on adaptive and natural computing algorithms* (pp. 313-322). Berlin, Heidelberg: Springer Berlin Heidelberg.

Zheng, Z., Wang, P., Liu, W., Li, J., Ye, R., & Ren, D. (2020, April). Distance-IoU loss: Faster and better learning for bounding box regression. *Proceedings of the AAAI Conference on Artificial Intelligence*, *34*(07), 12993–13000.

Zhou, S., Zhang, J., Pan, J., Xie, H., Zuo, W., & Ren, J. (2019). Spatio-temporal filter adaptive network for video deblurring. In *Proceedings of the IEEE/CVF international conference on computer vision* (pp. 2482-2491).

Chapter 3
Urban Sprawl Drivers and Their Impact on Land Use Changes:
A GIS-Based Assessment of the Patras Urban Complex

Sophia Leni
https://orcid.org/0009-0004-7532-1260
Informatics Laboratory, Agricultural University of Athens, Greece

Georgios Vlahos
https://orcid.org/0000-0002-1820-2001
Agricultural Extension, Rural Systems and Rural Sociology Laboratory, Agricultural University of Athens, Greece

ABSTRACT

This study examines the spatial dynamics of urban sprawl in the Urban Complex of Patras, Greece from 1981 to 2015. As a medium-sized Mediterranean city, Patras has witnessed extensive and largely unregulated expansion, representing a notable case of peri-urban transformation under conditions of infrastructural and institutional restructuring. The study employs a mixed-methods approach combining spatial analysis of land use/land cover change (LULCC), expert interviews, and statistical modeling. The chapter focuses on identifying how infrastructure development, weak regulatory enforcement, and socio-cultural preferences influence patterns of land use change over time. By situating these drivers within a long-term historical and institutional context, the study contributes to a deeper understanding of urban sprawl mechanisms beyond demographic or market-based explanations.

DOI: 10.4018/979-8-3693-8237-0.ch003

INTRODUCTION

Urbanization is advancing more rapidly than ever before. Urban areas are now expanding at twice the rate of population growth globally (Angel et al., 2011), and it is estimated that by 2050, nearly 68% of the world's population will live in urban areas, up from 55% in 2018 and just 30% in 1950 (United Nations (UN), 2019). While urbanization reflects economic development and modernization, it is often accompanied by unplanned and uncontrolled urban expansion, known as urban sprawl.

Urban sprawl is widely defined as the low-density, fragmented, and car-dependent expansion of urban areas into peri-urban and rural landscapes (Ewing, 1997; Burchell et al., 1998; Soule, 2006). It is associated with a broad range of environmental, social, and economic consequences, including loss of agricultural land and biodiversity, landscape fragmentation, inefficient infrastructure provision, and social segregation (Frenkel, 2004; European Environmental Agency (EEA), 2016).

In Europe, despite low rates of population growth, urban expansion continues to consume peri-urban and rural land at an alarming rate. Between 1975 and 2015, residential land increased by 250%, largely at the expense of arable land and pastures (European Commission, 2016). Between 2012 and 2018 alone, land take in EU and UK urban areas increased by 3,581 km^2, accompanied by 1,467 km^2 of soil sealing, mostly affecting croplands and pastures (EEA, 2021). Mediterranean urban areas, particularly medium-sized cities, are notably vulnerable to sprawl due to weak spatial planning frameworks and car-reliant mobility (Couch et al., 2008).

Urban sprawl is tightly linked to land use/land cover change (LULCC), which describes the transformation of land surfaces driven by human activity (Foley et al., 2005). Global studies have highlighted the extensive ecological, climatic, and social impacts of LULCC, including habitat loss (Díaz et al., 2019), changes in hydrological cycles and local climates (Pielke et al., 2011; Devi et al., 2019), and social displacement (Power, 2001; Wagistina, 2016).

The causes of urban sprawl are numerous and often interdependent. They include political, economic, cultural, technological, and natural/spatial factors that act at multiple scales and through complex feedback loops (Bürgi et al., 2005; Hersperger & Bürgi, 2009). Key drivers identified in the literature include land policies (Ewing et al., 2002), transportation infrastructure (Leontidou et al., 2007), economic growth (Lu et al., 2020), demographic pressures (Li & Li, 2019), lifestyle preferences (Couch & Karecha, 2006), and accessibility (Sohn et al., 2012).

To study these processes, Geographic Information Systems (GIS) have become indispensable. GIS integrates geospatial and remote sensing data (e.g., aerial photographs and satellite imagery) to support the creation, storage, analysis, and visualization of spatial information. It enables the precise mapping and monitoring of spatial transformations, the analysis of spatial patterns, and the evaluation of the impacts of sprawl and LULCC (Michalak, 1993; Cheng & Masser, 2003).

Despite the growing literature on urban sprawl, few studies have systematically analyzed the long-term spatial dynamics and driving forces of sprawl in medium-sized Mediterranean cities using integrated GIS and qualitative approaches. According to Eurofound (2012), medium-sized urban centers are defined as those with populations between 50,000 and 250,000 inhabitants, typically serving as administrative hubs for local and regional governance. These cities also exert influence over surrounding rural and peri-urban areas by offering services related to health, education, employment, and transportation accessibility.

Patras, a coastal city and the third largest in Greece, exemplifies such a center. Its urban form features a compact historical core that anchors the city's economic, administrative, and social life, surrounded by low-density expansion areas. Common characteristics among similar Greek cities include reliance on private vehicles, fragmented public transport, and abandoned industrial zones due to deindustrialization (Asprogerakas, 2003; Leontidou, 1990).

The study of urban sprawl in such medium-sized centers is especially important, as these cities function as critical regional and national nodes. In the case of Patras, urban expansion has fragmented the landscape, raised infrastructure costs, and placed pressure on surrounding agricultural land.

This study focuses on the Urban Complex of Patras and its surrounding area, which experienced rapid spatial transformation between 1981 and 2015—a period marked by Greece's entry into the European Economic Community, infrastructure investments, and the economic crisis of the late 2000s. Key infrastructure projects, such as the Patras Ring Road - the city's main bypass highway (2003), the Rio–Antirrio Bridge (2004), and the new port (2011), enhanced the city's geopolitical and geo-economic connectivity, further intensifying development pressures.

During this 34-year period, the study area experienced significant and persistent urban sprawl. Expansion was especially pronounced to the north, south, and east, producing a loose urban fabric and extensive land consumption. Agricultural zones were progressively converted to residential and commercial use, while the effects of the economic crisis led to shifting patterns in construction activity and land use.

While urban sprawl and land use change have been widely studied in global and European contexts, existing research has primarily focused on either large metropolitan areas or broad-scale regional analyses. Medium-sized cities in Southern Europe—such as Patras—remain under-examined, despite facing acute development pressures and regulatory fragmentation. Moreover, few studies integrate spatio-temporal analysis, qualitative institutional insights, and quantitative modeling within a single empirical framework. This gap limits our understanding of how localized planning cultures, infrastructure investment, and socio-spatial dynamics jointly shape urban expansion over time.

To address this, the present study combines GIS-based land use change detection, expert interview analysis, and multi-year regression modeling to provide a spatially and temporally integrated assessment of sprawl drivers in Patras—a representative medium-sized Mediterranean city.

The objectives of this study are threefold: (1) to assess LULCC using historical aerial photographs and GIS over a 34-year period; (2) to identify and interpret the driving forces of urban sprawl through qualitative thematic analysis of expert interviews; and (3) to measure and quantify the influence of these drivers using regression analysis. Together, these objectives form an integrated framework to analyze the causes and spatial consequences of sprawl in medium-sized Mediterranean cities.

Structure of the Paper

The remainder of this chapter is organized as follows: Section 2 outlines the theoretical background and key literature on urban sprawl and land use change. Section 3 (Methods and Materials) details the analytical workflow, the datasets and software as well as the study area. Section 4 presents the results, Section 5 provides analysis and discussion with policy implications, and Section 6 concludes.

BACKGROUND / LITERATURE REVIEW

The phenomenon of urban sprawl has been a subject of growing scholarly attention since the mid-20th century. Originally investigated in the context of American metropolitan growth (Brueckner, 2000), it soon became evident that sprawl has global manifestations, affecting both developed and developing countries (EEA, 2006; Seto et al., 2012). Sprawl typically refers to uncontrolled, low-density urban expansion into rural and peri-urban land, characterized by automobile dependency and fragmented urban form (Downs, 1999; Galster et al., 2001).

Although the term "urban sprawl" is widely used, it remains loosely defined and varies across disciplines, including planning, geography, economics, environmental science, and health (Bhatta et al., 2010; Chin, 2002). While lacking a universally agreed definition, most interpretations emphasize its unsustainable nature and detrimental impacts (Calthorpe & Fulton, 2001; Couch et al., 2008; Suditu, 2012).

Urban sprawl is closely associated with land use/land cover change (LULCC), defined as the alteration of Earth's surface through human activity (Foley et al., 2005). These changes have been widely studied due to their profound environmental and socio-economic implications. They contribute to biodiversity loss (Díaz et al.,

2019), disrupt hydrological systems (Devi et al., 2019), and intensify climate and social inequalities (Power, 2001).

Since the 1990s, increasing focus has been placed on understanding the underlying drivers of land use change. Driving forces—or drivers—are the processes or factors that trigger LULCC and sprawl. These forces can be political, economic, cultural, technological, or natural/spatial (Bürgi et al., 2005; Hersperger & Bürgi, 2007). Understanding these drivers is key to managing change, guiding planning interventions, and modeling future land use scenarios (Lambin et al., 2003; Verburg et al., 2004).

The conceptual framework developed by Bürgi et al. (2004) and extended by Hersperger and Bürgi (2009) has become widely used in empirical studies. It classifies driving forces into five categories: political drivers, such as land use policies, zoning regulations and infrastructure strategies, economic drivers, including market demand, structural economic changes, and investment trends, cultural drivers, such as lifestyle preferences, demographic patterns, and housing expectations, technological drivers, related to transportation innovations and ICT developments and natural/spatial drivers, including topography, climate, and spatial accessibility

This framework facilitates a holistic understanding of urban transformation, allowing for the integration of complex interdependencies across spatial and temporal scales. While comprehensive, the approach also acknowledges challenges in classifying and quantifying drivers, particularly in long-term studies (Schneeberger et al., 2007).

In the context of Mediterranean cities, political drivers—particularly weak land-use governance, infrastructure-led growth strategies, and fragmented planning frameworks—often emerge as dominant forces shaping urban expansion. In Rome, lax enforcement and piecemeal zoning reforms have encouraged urban dispersion and contributed to fragmented urban expansion (Salvati & Carlucci, 2014). Similarly, Athens saw a surge of urban sprawl following the 2004 Olympic Games, driven by strategic infrastructure investments such as new highways, the airport, and the suburban rail network—without a cohesive spatial planning vision (Chorianopoulos et al., 2010). These examples highlight how state-led initiatives, rather than merely individual preferences or economic trends, can decisively shape spatial trajectories. This study extends this understanding by empirically investigating how such political and institutional drivers have operated in the Urban Complex of Patras. Against this backdrop, the Urban Complex of Patras offers a valuable case for examining these dynamics in a Mediterranean context.

Theoretical and Analytical Contribution

This study contributes to the broader literature on urban sprawl and land use change by integrating spatial, qualitative, and statistical methods within a unified conceptual framework. It applies the Bürgi et al. (2004) typology of driving forces to a medium-sized Mediterranean city—Patras—providing an in-depth case of long-term urban transformation under regional political, economic, and infrastructural pressures. Beyond identifying the drivers of change, the study analyzes LULCC over a 34-year period through historical aerial photographs and GIS, uncovering spatial trends and development patterns not previously documented. The qualitative dimension, based on thematic analysis of elite interviews, captures context-specific institutional and policy insights that are often absent in spatial models. Finally, the study employs regression analysis to explore the relative weight and interaction of multiple drivers, thereby addressing a major gap in the literature: the quantitative assessment of urban sprawl drivers in medium-sized European cities. This mixed-methods design allows for a robust triangulation of evidence and enhances the study's relevance to urban planning, land management, and sustainability transitions.

METHODS AND MATERIALS

This chapter details the methodological framework and the materials used in the study. Section 3.1 outlines the analytical procedures—GIS preprocessing and land-use classification, landscape metrics, change detection, regression modelling, and predictive mapping—while Section 3.2 describes the datasets, software, and ancillary layers, with full source attribution and access details.

Methods

The methodological framework of this study combines geospatial analysis, statistical modeling, and qualitative inquiry to capture the multiple dimensions of urban sprawl in the Patras Urban Complex between 1981 and 2015. The workflow was designed in five main steps: (i) georeferencing and construction of base maps; (ii) creation of thematic spatial layers through digitization and processing; (iii) harmonization of demographic data; (iv) statistical and spatial modeling of sprawl drivers; and (v) expert interviews to contextualize quantitative results within institutional and planning practices.

GIS Preprocessing and Land-Use Mapping

Constructing chronological time series using maps and aerial photographs is a widely accepted method for analyzing landscape and urban transformation (Van Eetvelde & Antrop, 2004; Skånes & Bunce, 1997). Following this approach, all aerial photographs, orthophotos, and census base maps (1981–2015) were scanned, georeferenced and compiled into seamless photomosaics for four reference years (1981, 1996, 2008, 2015). These photomosaics provided the spatial framework for manual digitization and served as the foundation for constructing consistent chronological datasets of land-use and built-up area changes.

Thematic layers—such as built-up area, agricultural land, road networks, and urban plan boundaries—were generated through manual on-screen digitization in ArcGIS. To ensure comparability across time, uniform cartographic standards were applied to all reference years. Where available, official municipal datasets (e.g., urban neighborhoods, plan extensions) were integrated to enhance accuracy and completeness.

Land use/land cover (LULC) datasets were created through visual interpretation and manual digitization. The classification scheme included 14 categories, harmonized across all years to allow comparability and broadly aligned with established typologies such as the CORINE Land Cover and Anderson's hierarchical system. Processing steps included polygon generalization, topological correction, and the standardization of continuous variables. Categorical LULC classes were converted into dummy variables for statistical modeling. Accuracy assessment employed stratified random sampling and error matrices (Congalton, 1991), with all maps achieving overall accuracies above 80%, which is generally accepted for reliable change detection (Weng, 2007). These harmonized LULC layers formed the basis for subsequent statistical modeling.

Clarification on the 1981 reference year: For built-up area mapping, the 1981 census base maps of the Hellenic Statistical Authority were used, as they depicted cadastral parcels and buildings. However, these maps did not provide LULC information. Therefore, LULC classification for the 1981 reference year was based on the 1982 aerial photomosaic, which enabled visual interpretation of agricultural, natural, and open land categories. This combined use of sources ensured both the accurate representation of the built-up stock of 1981 and the reliable delineation of land cover conditions at the beginning of the study period.

Land use/Land Cover Classification Scheme and Accuracy Assessment

To ensure consistency and comparability across time, LULC was classified into a set of categories reflecting dominant functional types and planning distinctions. Fourteen categories were defined, encompassing urban, agricultural, and natural uses, broadly aligned with CORINE Land Cover standards and Anderson's hierarchical typology. Key urban categories included low-, medium-, and high-density urban fabric, while other classes captured commercial, civic, recreational, and ecological land uses. The full classification scheme is presented in Table 1.

Table 1. LULC classification scheme and category descriptions (1981–2015)

Code	LULC Category	Description
1	Port	Old and new port areas of Patras
2	Low-density Urban Fabric	Urban areas with up to 20% built-up area
3	Medium-density Urban Fabric	Urban areas with 20%–50% built-up area
4	High-density Urban Fabric	Urban areas with more than 50% built-up area
5	Commerce/Wholesale/Industrial Area	Commercial zones, wholesale markets, industrial parks, and wastewater treatment facilities
6	Sports Facilities	Open and closed sports venues, sports infrastructure
7	Civic Infrastructure	Schools, kindergartens, cemeteries, cultural centers, hospitals
8	Squares/Urban Green/Unbuilt Open Spaces	squares, green urban areas, vacant or undeveloped plots
9	Protected Areas/Wetlands	Protected zones including Agyia marsh, Glafkos and Charadros river and riparian zones
10	Forest Areas	Forests and forested land
11	Permanent Crops	Permanent crops such as lemon and orange orchards
12	Olive Cultivation	Areas dominated by olive trees
13	Barren Land	Barren or rocky land
14	Unused Land	Land with no discernible current use

Source: Authors' classification based on aerial photograph interpretation and digitization (1981–2015)

Accuracy assessment was conducted to evaluate the reliability of the land use/land cover maps. Stratified random sampling proportional to class area was applied, and error matrices were generated (Congalton, 1991). Standard metrics were calculated, including user's accuracy, producer's accuracy, overall accuracy, and the Kappa coefficient. This procedure ensured that classification performance could be consistently validated across all reference years. These validated LULC datasets provided the analytical basis for tracking spatial transformations in the Patras Urban Complex. The outputs of this classification and validation process form the foundation of the results presented in Chapter 4, where the spatial and temporal patterns of LULCC are analyzed in detail (see Section 4.1).

Note on cartographic representation: All land use/land cover maps share a uniform orientation (north is at the top) and are based on the same reference scale. For reasons of clarity and comparability across years, graphic scales and north arrows were omitted, as the figures are intended as schematic illustrations of LULC dynamics rather than stand-alone cartographic products.

Demographic Data Harmonization

Population and housing census data (1981, 1991, 2001, 2011) were processed at the census-block level. For 1981, where block-level data were incomplete, missing values were reconstructed using areal interpolation, supported by ancillary datasets (e.g., urban neighborhoods from the 1986 General Urban Plan). All census datasets were normalized to ensure comparability across time, with adjustments made for changes in administrative boundaries caused by reforms (Kapodistrias, Kallikratis).

Statistical Modeling of Sprawl Drivers

Urban sprawl was modeled as built-up surface density using Multiple Linear Regression (MLR). Explanatory variables captured cultural, economic, technological, environmental, and institutional drivers of urban growth. Diagnostics included Variance Inflation Factor (VIF) to test multicollinearity and Moran's I to detect spatial autocorrelation in residuals. Robust standard errors were applied to correct for heteroskedasticity. Predictive spatial patterns were further generated using kriging interpolation, producing continuous surfaces of estimated sprawl intensity. The regression framework thus captured both structural and contextual drivers of urban expansion.

Qualitative Analysis Through Elite Interviews

To complement the quantitative modeling, a series of semi-structured elite interviews was conducted with experts from the public sector, private consulting, and academia. Elite interviewing is a qualitative method designed to access the perspectives of individuals with specialized knowledge, decision-making power, or strategic insight (Harvey, 2011; Mikecz, 2012). This approach was selected for its capacity to reveal institutional, political, and professional rationales that shape urban development—insights not easily obtainable through document analysis or spatial data alone.

The interview guide was developed based on the conceptual framework of land use change drivers (Bürgi et al., 2005; Hersperger & Bürgi, 2009) and structured around themes such as planning policy, infrastructure development, land use practices, and socio-cultural preferences. Figure 1 presents the conceptual framework of driving forces used to structure the thematic analysis.

Figure 1. Conceptual framework of land use change drivers (Hersperger & Bürgi, 2007)

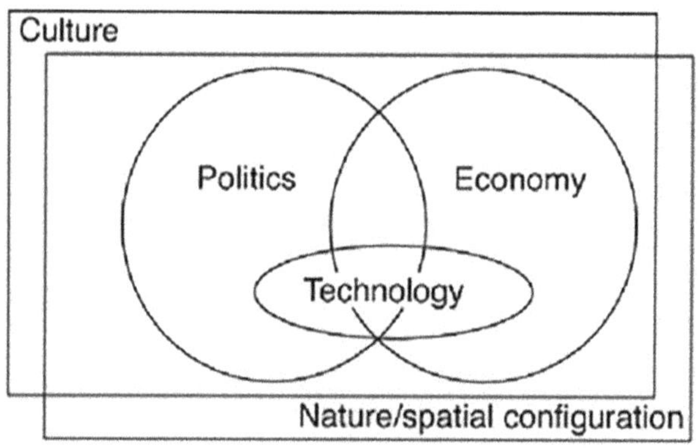

Interviews were transcribed and analyzed using thematic analysis (Braun & Clarke, 2006). Following standard practice, open coding was applied after multiple readings, with codes grouped into broader themes and refined iteratively (Boyatzis, 1998; Joffe & Yardley, 2003). This method enabled the identification of both explicit policy insights and implicit assumptions embedded in professional discourse, ensuring triangulation with the GIS and regression results.

Materials

This section summarizes the data sources, software, and ancillary materials employed in the study. Datasets are grouped into (i) geospatial imagery and base maps, (ii) thematic and infrastructure layers, and (iii) demographic and statistical sources.

Geospatial Imagery and Base Maps

The study relied on analog and digital aerial photographs, orthophotos, and census base maps covering the period 1981–2015. These materials were provided by national mapping and statistical agencies, scanned at high resolution, and prepared as photomosaics to serve as consistent spatial references. Their temporal coverage ensures comparability across three decades of urban growth. The imagery was selected to ensure high spatial resolution and temporal coverage for monitoring landscape and urban transformations. Supplementary geospatial datasets were also obtained from the Municipality of Patras, including administrative boundaries, school locations, and athletic facilities. Georeferenced photomosaics were produced for 1982, 1996, 2008, and 2015 using materials from Hellenic Military Geographical Service, Hellenic Mapping and Cadastral Organization, and the Hellenic Cadastre. Population census maps from Hellenic Statistical Authority were similarly scanned and integrated. These mosaics formed the basis for manual digitization and supported the development of time series datasets for built-up areas, topographic features, road networks, and other thematic layers. The datasets are summarized in Table 2.

Table 2. Aerial photos, orthophotos, and population census base maps used in georeferenced analysis

Dataset	Year	Number	Scale	Source
Analog B/W Aerial Photographs	1982	85	1:7000	Hellenic Military Geographical Service
Analog Stereoscopic B/W Aerial Photos	1996	128	1:6000	Hellenic Mapping and Cadastral Organization
Digital Color Orthophotos	2008	12	1:5000	Hellenic Cadastre
Digital Color Orthophotos	2015	58	1:5000	Hellenic Cadastre
Analog B/W Census Base Maps	1981	11	1:5000	Hellenic Statistical Authority
Analog B/W Census Base Maps	1991	10	1:5000	Hellenic Statistical Authority
Analog B/W Census Base Maps	2001	7	1:5000	Hellenic Statistical Authority

Source: Hellenic Military Geographical Service; Hellenic Mapping and Cadastral Organization; Hellenic Cadastre; Hellenic Statistical Authority

The high resolution, spatial accuracy, and consistency across these datasets ensured the reliability of the constructed geospatial time series. Figure 2 presents the chronological sequence of photomosaics used for land cover analysis between 1981 and 2015.

Figure 2. Chronological photomosaic series 1981-2015 (Authors' photomosaics compiled from aerial photographs and census base maps (see Table 2))

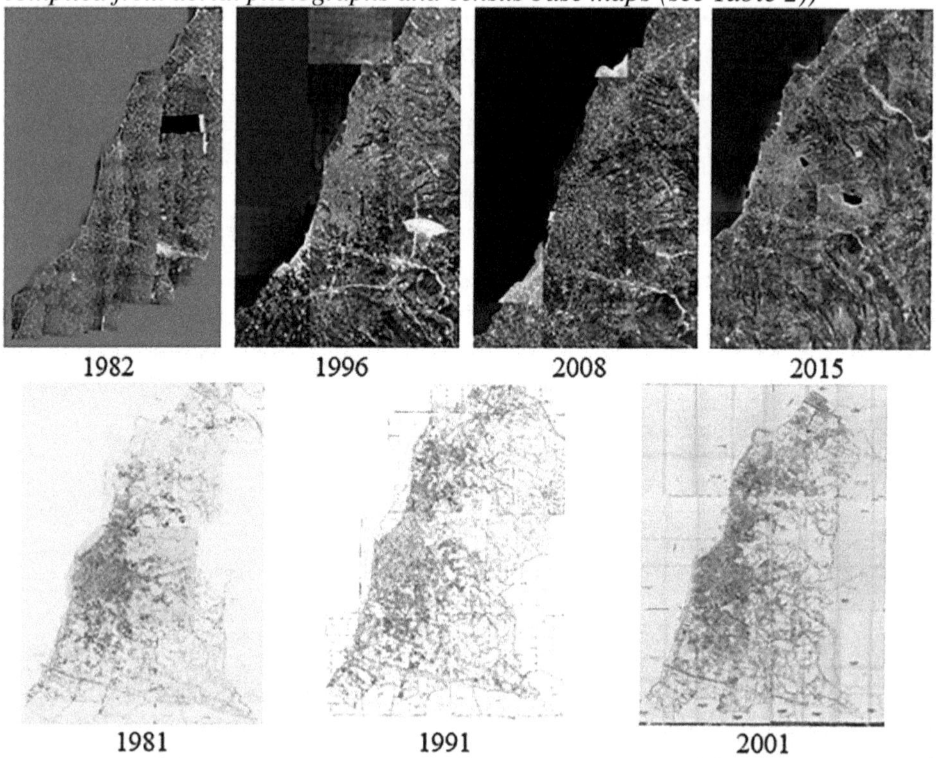

Foundational Layers

In addition to the aerial photographs and orthophotos described above, several foundational datasets were used to define the study area and its topographic characteristics. The study area boundary was derived from municipal open data and adjusted to align with the coastline, the extent of the statutory urban plan, and the Patras Ring Road (A5 motorway). Elevation contours were digitized from 1:50,000 topographic maps and orthophotos. From these, a Digital Elevation Model (DEM) and a slope raster were generated to represent terrain features. These base layers provided the spatial reference framework for subsequent geospatial processing and analysis (Figures 3–4).

Figure 3. Study area boundary (Authors' compilation based on Municipality of Patras data and topographic maps)

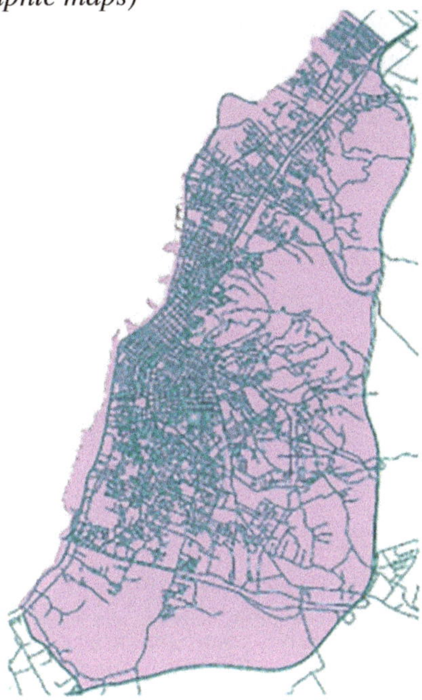

Figure 4. Study area-topographic layers (Authors' compilation based on Municipality of Patras data and topographic maps)

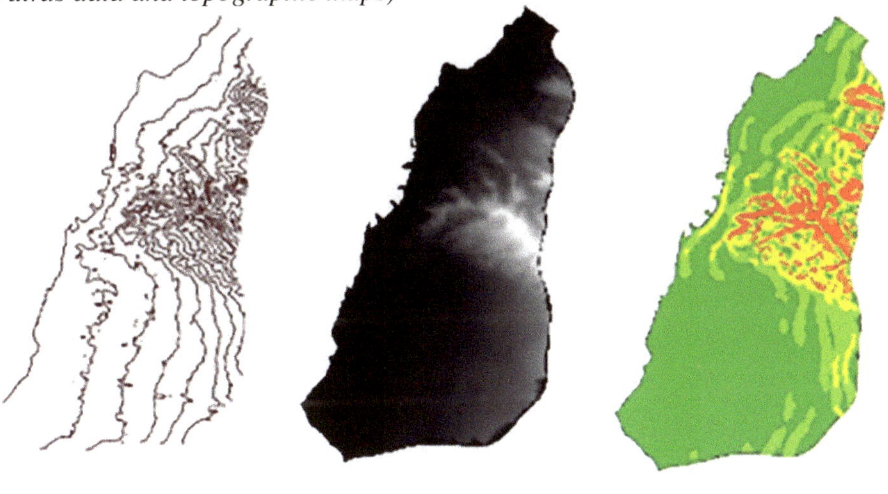

Thematic Spatial Layers

From the base maps and imagery, a set of thematic GIS layers was constructed to represent urban growth, agricultural change, transportation, and planning boundaries. These layers were generated via manual on-screen digitization from the georeferenced photomosaics presented in Table 2. Spanning the period 1981–2015, these layers were created following consistent cartographic standards across all reference years to ensure comparability and support the construction of a multi-temporal dataset of urban extent. In certain cases, available geospatial base data provided by the Municipality of Patras were also utilized to enhance accuracy and completeness. Thematic layers, their temporal coverage and digitization methods are summarized in Table 3.

Table 3. Thematic spatial layers and temporal coverage

Thematic Layer	Method Used	Years covered
Built-up Area	Manual digitization	1981,1991,1996,2001, 2008,2015
Agricultural Land	Manual digitization, classification and field validation	1981,1996, 2008,2015
Road Network	Manual digitization	1981,1991,2001,2015
Urban Neighborhoods	Open data, Municipality of Patras	1986,2011
Land Use/Land Cover Zones	Manual digitization, Open data Municipality of Patras and classification	1981,1996,2008,2015
Urban Plan Inclusion Dates	Chronological GIS overlay, Open data Municipality of Patras	1882 to 1999
Urban plan boundaries and its extensions	Manual digitization	1981,1991,2001

Source: Authors' compilation from manual digitization, classification, and municipal open data.

Figures 5–11 present examples of the resulting thematic maps, illustrating urban expansion, agricultural transitions, road network development, and the chronology of urban plan extensions.

Figure 5. Built-up area distribution and growth in the Urban Complex of Patras (1981, 1991, 1996, 2001, 2008, 2015) (Authors' digitization based on georeferenced photomosaics (1981–2015))

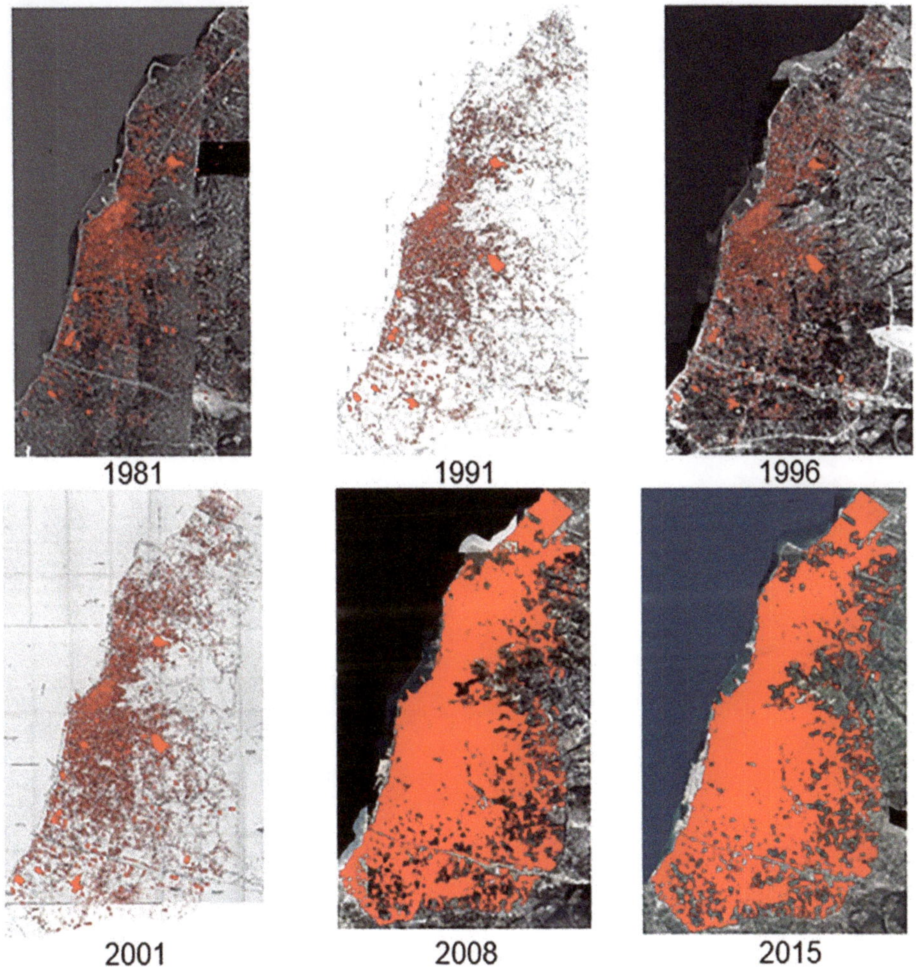

Figure 6. Agricultural land use patterns across the study area (1981, 1996, 2008, 2015) (Authors' derivation from georeferenced photomosaics with field validation)

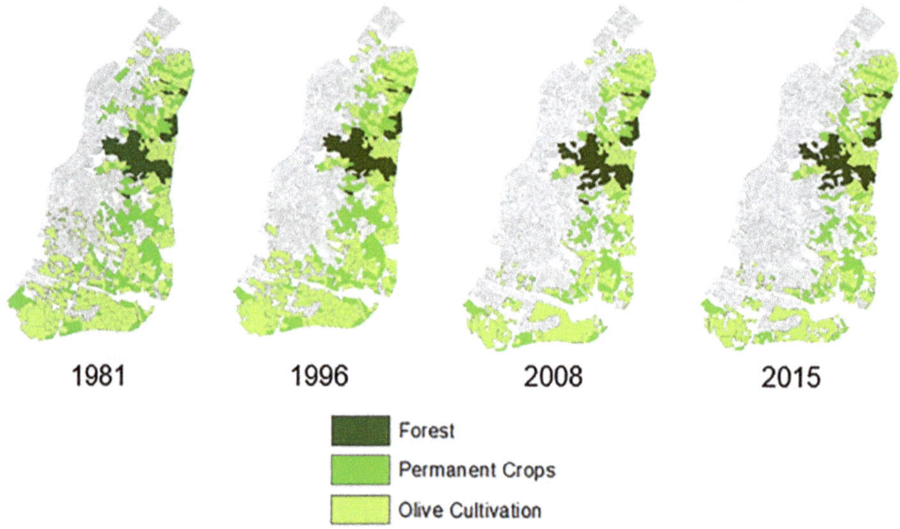

Figure 7. Road network evolution in the study area (1981, 1991, 2001, 2015) (Authors' derivation from georeferenced photomosaics and official base maps)

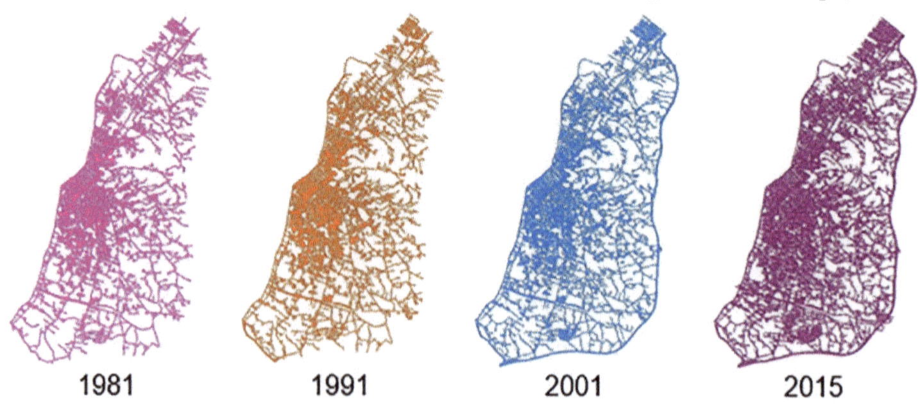

Figure 8. Urban neighborhoods defined by planning data (1986, 2011) (Municipality of Patras open data, processed by the authors)

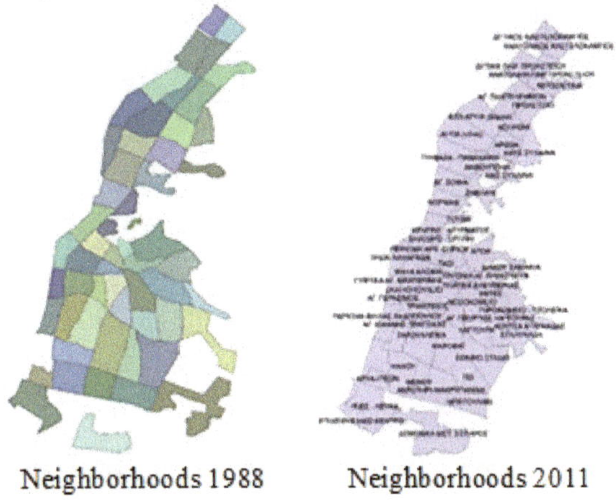

Figure 9. Land use/land cover zones in the study area (1981, 1996, 2008, 2015) (Manual digitization and classification by authors, also using municipal open datasets)

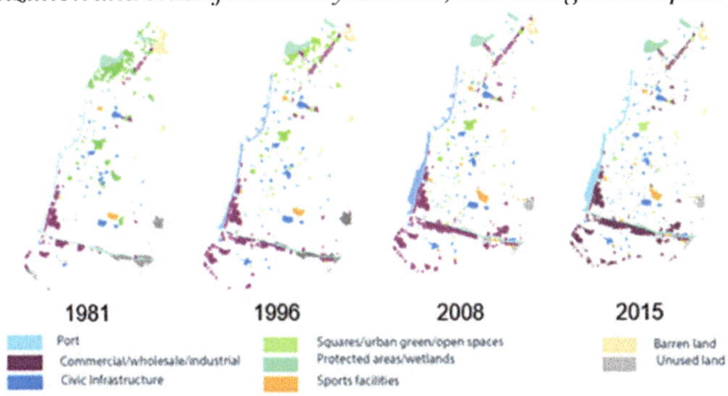

Figure 10. Areas incorporated into the city plan by year of inclusion (1882–1999) (GIS overlay based on legal documents and municipal data, processed by the authors)

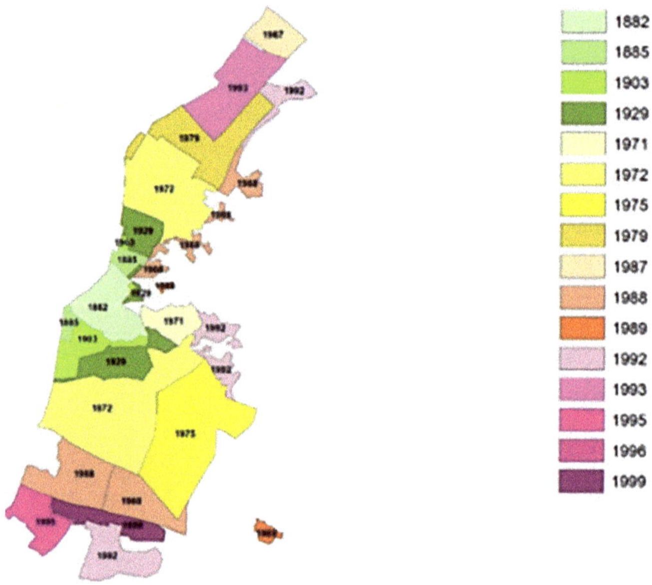

Figure 11. Urban plan boundaries and successive expansions (1981, 1991, 2001) (Digitization of General Urban Plan and extension documents, processed by the authors)

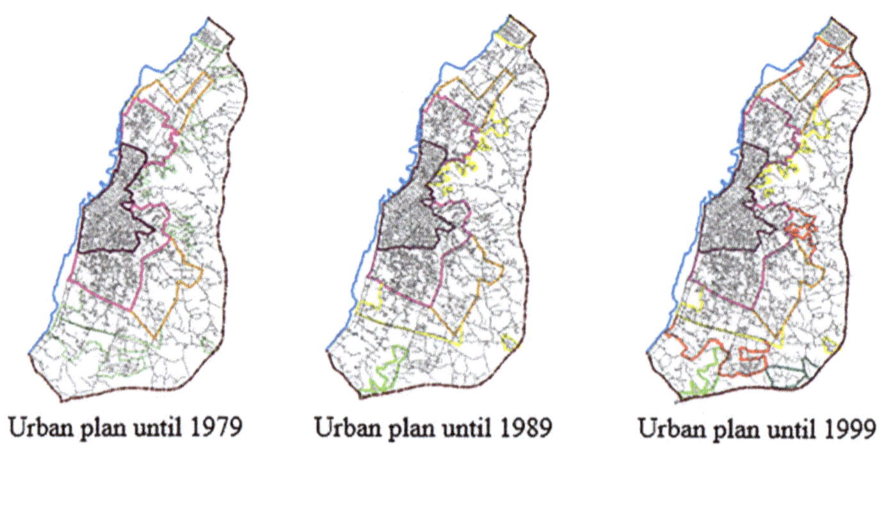

These harmonized layers underpin the analyses reported in section 4.

Demographic Data

Population and housing census data were obtained from the Hellenic Statistical Authority at the census-block level for 1991, 2001, and 2011. For 1981, block-level values were incomplete and were reconstructed through areal interpolation, supported by ancillary datasets such as the neighborhood boundaries of the 1986 General Urban Plan (Papadatou-Giannopoulou, 1997). All census layers were harmonized to resident population counts to ensure comparability across the full time series.

Administrative inconsistencies arising from institutional reforms—such as the *Kapodistrias* and *Kallikratis* programs—were addressed through a set of reconciliation steps. Analog and digital records were aligned, and both spatial and tabular datasets were cross-referenced to detect and resolve discrepancies. Where mismatches emerged, proportional allocation techniques were applied, while in cases with no direct correspondence, hybrid geospatial datasets were constructed to bridge gaps between administrative boundaries and census reporting units. Expert validation further ensured reliability.

The resulting integrated dataset enabled consistent analysis of urban density patterns, demographic trends, and spatial correlations of sprawl over the 34-year study period.

Study Area

As shown in Figure 12, the study area encompasses the Urban Complex of Patras and its surrounding zones, including peri-urban settlements that maintain functional ties to the city. It extends up to the boundaries defined by the Patras Ring Road (Motorway A5). Specifically, the area includes the core city of Patras, a section of the Ovria Municipal Unit (within the Ring Road), and the settlements of Agios Stefanos, Demenika, and Mavromandila (part). Altogether, it extends over 4,381.6 hectares and had a population of about 174,000 in 2011 (Government Gazette, 2012). Administratively, it forms part of the Municipality of Patras, in the Regional Unit of Achaia, Western Greece.

Figure 12. Location of the study area (a) Geographic position of the study area within Greece (highlighted in red) (b) Boundary of the study area (Open geospatial data and authors' elaboration based on Municipality of Patras data)

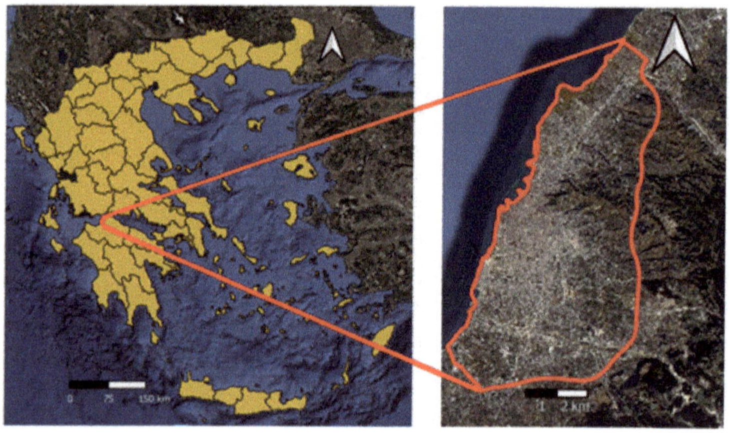

The physical landscape, bounded by the coastline to the west and Mount Panachaiko to the east, combined with artificial boundaries such as the Patras Ring Road, has played a defining role in the city's spatial configuration. The Ring Road acts as a hard edge to development, enclosing diverse land use zones that include high- and low-density urban fabric, agricultural land, natural landscapes, and mixed-use areas.

Patras itself is Greece's third-largest city and the capital of Western Greece, functioning as a regional hub for administration, commerce, education, and transport. It hosts the University of Patras, the new commercial port, and major regional infrastructure, which together reinforce its central role.

Urban expansion in Patras followed the coastal lowland corridor and the main N–S and E–W axes, with new infrastructure (Patras Ring Road, new port) reshaping accessibility and creating development fronts. Regulatory delays and out-of-plan construction fostered a fragmented, low-density pattern typical of Mediterranean sprawl. These contextual characteristics frame the analysis of LULC transformations presented in Section 4

RESULTS

This chapter presents the main findings of the study, integrating the spatial evidence with the interview-based insights. Section 4.1 presents the spatial patterns of land use/land cover change (LULCC) (1981–2015), including narrative interpretation of the study area's evolution, indicator trends, and the land use/land cover map series. Section 4.2 presents landscape-level and class-level spatial metrics. Section 4.3 presents spatiotemporal change detection results from post-classification comparison across 1981–1996, 1996–2008, 2008–2015, and the full period. Section 4.4 outlines the profile of the elite interviewees, and Section 4.5 synthesizes key drivers emerging from thematic analysis. Together, these results provide the empirical basis for the comparative discussion and policy implications developed in Chapter 5.

Spatial Patterns of Land Use/Land Cover Change (1981–2015)

The Patras urban core is a densely developed area, serving as the historic, administrative, and commercial nucleus around which the city has evolved. Urban expansion has largely occurred along a lowland corridor that aligns with the coastline, structured along two main development axes: a north–south axis (from the university to the former industrial zone) and an east–west axis (from the coastline toward inland areas).

Although Patras was the first city in modern Greece to be assigned an urban plan—the Voulgaris Plan of 1829 (Hastaoglou-Martinidis, 1995)—its subsequent development has been largely uncoordinated. The city's urban form, structure, and function have undergone rapid and often fragmented transformations (Pappas, 2006). By the 1970s, outward expansion had already begun; between 1971 and 1980, the urban plan was extended by 1,075.3 hectares.

Between 1981 and 2015, the Urban Complex of Patras experienced rapid urbanization, extensive land use/land cover change, and significant infrastructure investment. The built-up area expanded northward, eastward, and southward, incorporating 728.37 hectares (+67.74%) of low-density, fragmented urban fabric. Urbanization also extended beyond the legal urban boundaries, particularly into cultivated areas, through informal development enabled by "outside the city plan" regulations. These successive urban plan expansions were not accompanied by proportional population growth, resulting in spatial discontinuity and a dispersed urban morphology (Laboratory of Architectural Technology and Spatial Design, 1986). Interstitial areas within the urban fabric were often characterized by remnant agricultural land, especially olive cultivation. During this period, the study area underwent successive urban plan extensions, the incorporation of peri-urban settlements, and the gradual transformation of rural and agricultural zones into urban land. Infrastructure-led transformation dramatically altered the spatial structure of the city. These projects enhanced regional connectivity but also intensified development pressures.

The most prominent urban expansions occurred in the northern, eastern, and southern sectors. These areas initially developed informally, with low densities and fragmented layouts, later retroactively legalized through piecemeal planning extensions. The resulting urban fabric is highly heterogeneous, with fragmented structures and dispersed development, indicative of decentralized expansion.

From the mid-1990s onward, the Region of Western Greece—and particularly the wider Patras Urban Complex—became a focus of Greece's development strategy for national and trans-European infrastructure integration. Key investments targeted connectivity with Italy via the Adriatic maritime corridor, Athens via the Olympia Motorway (A8), northern Greece via the A1 (E75), and Ioannina via the Ionia Motorway (E55). Major infrastructure projects—including the Patras Bypass (2003), the Rio–Antirrio Bridge (2004), and the new port of Patras (2011)—substantially improved regional accessibility and reinforced the city's geopolitical and geo-economic position. These investments also intensified development pressure across the study area.

The combined effects of national infrastructure strategy and weak regulatory enforcement subjected the study area to pronounced development pressures. The expansion followed a typical Mediterranean model of urban sprawl (Couch et al., 2008), characterized by the conversion of agricultural land to urban uses, fragmented development, and low-density expansion. The spatial landscape that emerged is marked by heterogeneity, functional ambiguity, and a lack of coherent spatial hierarchy. While the urban footprint expanded considerably, demographic growth remained modest, leading to declining urban density—an indicator strongly correlated with urban sprawl.

Although the population grew during this period, the rate of demographic increase was significantly lower than the extent of spatial expansion. This mismatch resulted in a steady decline in urban density, a defining feature of urban sprawl and spatial inefficiency. As illustrated in Figure 13, the trajectories of built-up area and road network length significantly outpaced population growth over the study period.

Figure 13. Relative change in built-up area, population, and road network compared to 1981 (Authors' analysis)

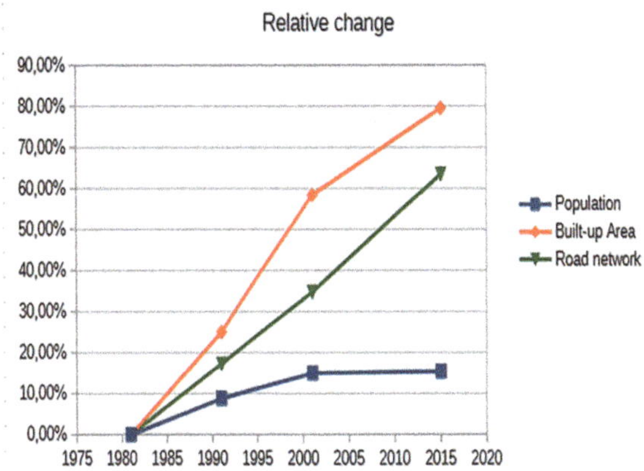

The quantitative trends are detailed in Table 4, which presents absolute values for built-up area, population, road network length, and derived urban density measures from 1981 to 2015.

Table 4. Urban indicators for the study area: 1981–2015

Year	1981	1991	2001	2015
Total area of the study area (ha)	4381.60			
Built-up area (ha)	449.79	562.66	712.63	807.78
Population (inhabitants)	151053	164369	173633	174224
Road network (km)	466	523	601	729
Average population density per built-up area (inhab/ha)	335.82	292.12	243.64	215.68
Constructed road network per hectare in the study area (km/ha)	0.10	0.12	0.13	0.16

Source: Authors' analysis.

Despite the enactment of two significant General Urban Plans—one in 1986 (Government Gazette 1061/D/7.11.1986), which aimed to establish a polycentric urban model, and another in 2011 (Government Gazette 358/AAP/30.12.2011), which incorporated sustainability principles such as urban compactness, containment of ex-urban development, and agricultural land protection—implementation was slow and inconsistent. Consequently, these plans failed in reversing prevailing sprawl patterns or in guiding cohesive spatial development.

The LULC categories defined in Section 3.1.1 formed the analytical basis for tracking transformations over time. Using this classification, land use/land cover maps were generated for four reference years—1981, 1996, 2008, and 2015. These maps illustrate the spatial footprint, intensity, and direction of urban expansion and land conversion in the Patras Urban Complex. By visualizing change across discrete time points, they clearly depict urban sprawl dynamics, especially the transition from agricultural to urban land uses. Figure 14 presents the map series showing the spatial distribution of land use categories for each reference year.

Figure 14. Land use/land cover maps of the Patras Urban Complex for 1981, 1996, 2008, and 2015 (Authors' digitization and classification of aerial/orthophoto data)

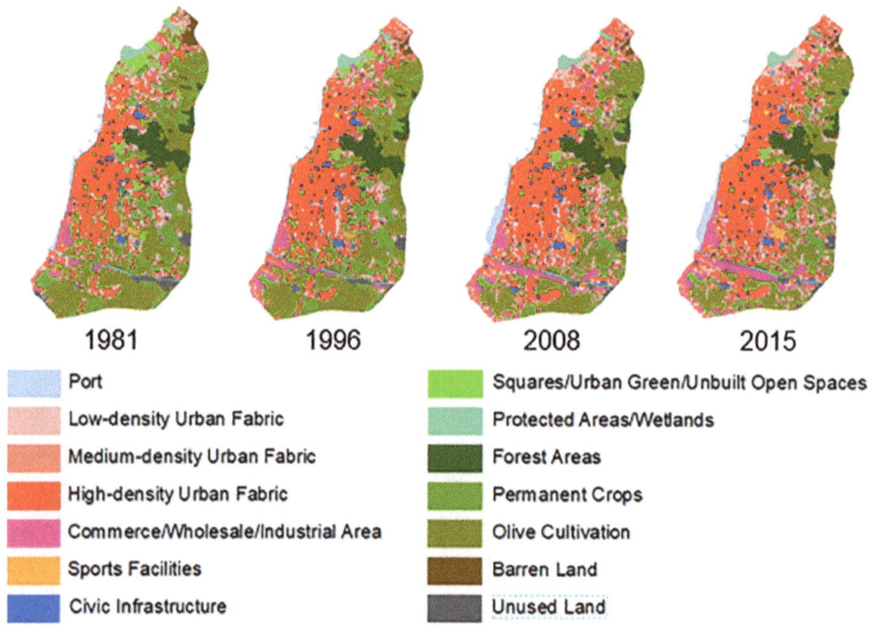

Classification reliability was confirmed through an accuracy assessment (see Section 3.1.1), which reported overall accuracies above 93% and Kappa values exceeding 0.93. This provides confidence in the robustness of the spatial dataset and the subsequent spatiotemporal analysis.

The classified maps reveal a heterogeneous and increasingly urbanized landscape between 1981 and 2015. A clear spatial trend emerges: agricultural land steadily declined, while urban fabric expanded, particularly in the north, south, and east of the study area. This transformation was most pronounced in lowland and peri-urban zones, where fragmentation and discontinuity of the urban form increased. While some eastern agricultural zones remained relatively stable, the overall pattern indicates the conversion of arable and olive-growing land into low- and medium-density residential areas.

Commercial and industrial development also intensified during this period, concentrated along the southern and eastern transport corridors and former industrial zones. Civic infrastructure, including schools, sports facilities, and civic institutions, increased in line with General Urban Plan directives. Meanwhile, urban green spaces—including parks and undeveloped open plots—declined due to infill development, especially within previously designated expansion zones.

In contrast, forested areas and protected ecological zones (e.g., Agyia marsh, riparian corridors) remained largely unchanged, demonstrating the relative effectiveness of environmental protections. However, the surrounding landscapes were increasingly subjected to encroaching development, contributing to broader ecological fragmentation.

Overall, the spatial footprint of Patras evolved toward a more dispersed, low-density urban structure, characteristic of Mediterranean sprawl. These spatial changes set the foundation for the quantitative metrics and temporal analysis that follow.

Quantitative Metrics of Land Use/Land Cover Change

Spatial metrics are crucial for identifying key spatial and temporal changes, providing a bridge between landscape complexity and decision-making. Metrics were calculated using FRAGSTATS 4.2 (McGarigal & Marks, 1995; McGarigal & Cushman, 2002). Four landscape-level and four class-level metrics were used to characterize spatial structure and dynamics.

The selected landscape-level metrics—TCA, CONTAG, SHDI, and PAFRAC—provide insights into scale, spatial aggregation, diversity, and shape complexity. CONTAG (Contagion Index) quantifies spatial aggregation or dispersion; lower values indicate fragmentation. TCA (Total Core Area) measures the sum of non-edge (core) areas across all land use patches and reflects the intact area of land uses. PAFRAC (Perimeter-Area Fractal Dimension) assesses patch shape complexity using a regression of perimeter and area; higher values suggest more irregular, fragmented patches. SHDI (Shannon's Diversity Index) captures land use diversity across the landscape. Together, these metrics indicate growing fragmentation, spatial complexity, and diversification.

The results of landscape-level spatial metrics across the study period are summarized in Table 5 highlighting key trends in fragmentation, diversity, and shape complexity.

Table 5. Landscape-level spatial metrics for the Patras Urban Complex (1981–2015)

	CONTAG	TCA	PAFRAC	SHDI
1981	47.2392	810.0900	1.3734	2.0248
1996	47.2193	804.0600	1.3789	2.0357
2008	44.7942	612.4500	1.4105	2.0862
2015	43.8654	514.3500	1.4152	2.1000

Source: Authors' analysis using FRAGSTATS 4.2 (McGarigal & Cushman, 2002).

The temporal evolution of these landscape-level metrics is visualized in Figures 15 through 18 highlighting trends in spatial aggregation, core area reduction, shape complexity, and land use diversity between 1981 and 2015.

Figure 15. CONTAG (Contagion Index) over time (Authors' analysis using FRAGSTATS 4.2 (McGarigal & Cushman, 2002))

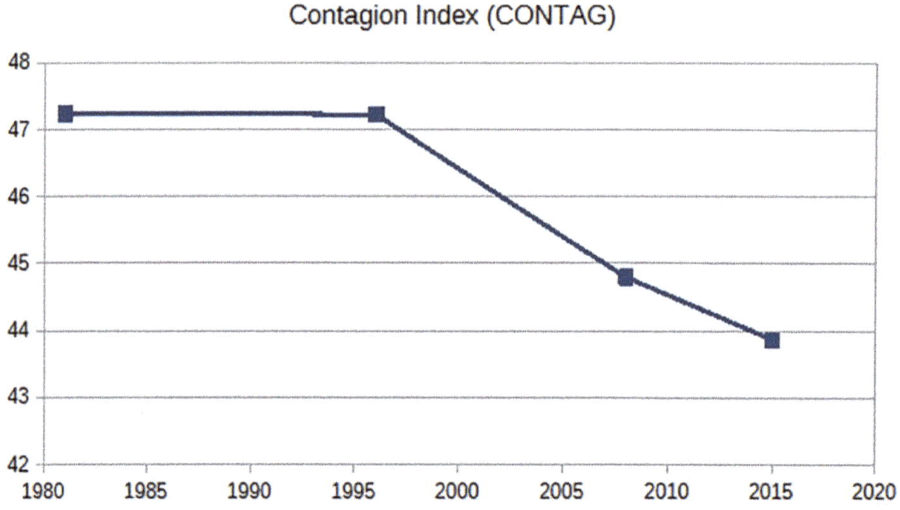

Figure 16. TCA (Total Core Area) over time (Authors' analysis using FRAGSTATS 4.2 (McGarigal & Cushman, 2002))

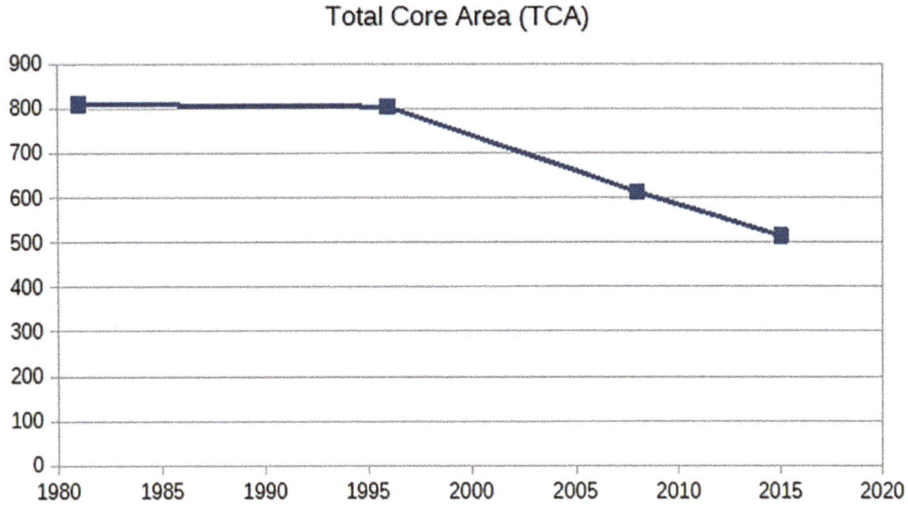

Figure 17. PAFRAC (Fractal Dimension) over time (Authors' analysis using FRAGSTATS 4.2 (McGarigal & Cushman, 2002))

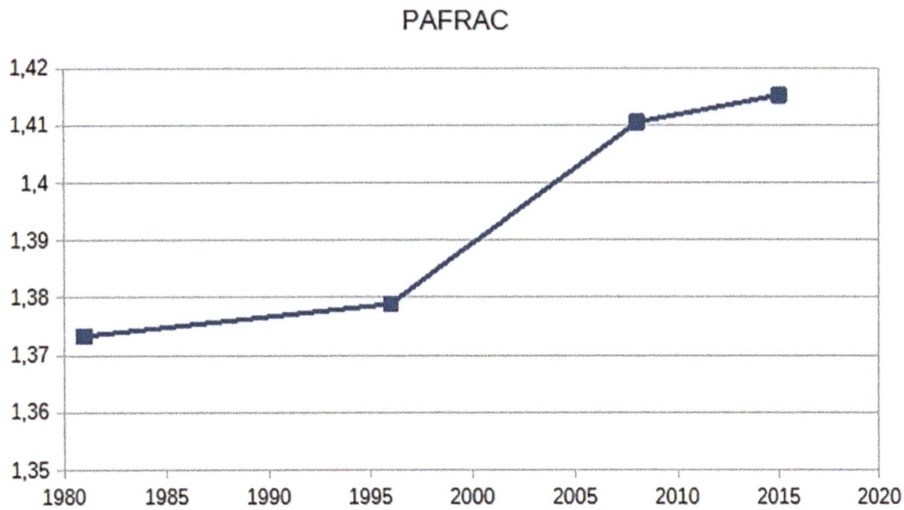

Figure 18. SHDI (Diversity Index) over time (Authors' analysis using FRAGSTATS 4.2 (McGarigal & Cushman, 2002))

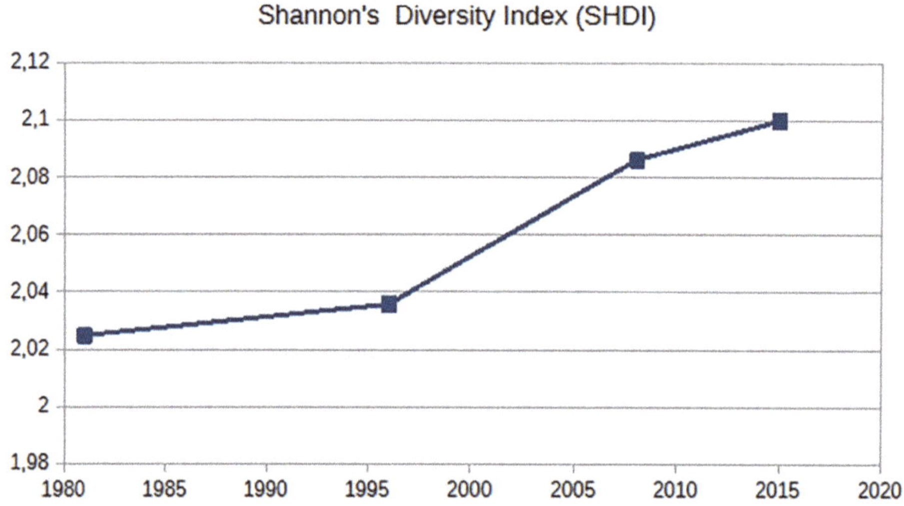

At the class level, four spatial metrics were used to analyze the internal dynamics of each land use/land cover category. PD (Patch Density) indicates the degree of fragmentation. LPI (Largest Patch Index) represents the percentage of the landscape occupied by the largest patch. High values in urban fabric classes show dominance or compactness. COHESION quantifies the physical connectedness of patches. Stable values suggest continuity of land use. AI (Aggregation Index) reflects the degree of clustering among similar land use/land cover types. Declining AI in green/recreational areas may reflect dispersion or loss of spatial integrity. Class-level metrics are detailed in Appendix Table A.2, with definitions provided in Table A12.

Together with the landscape-level indices, these metrics enrich the spatial analysis and support interpretation of urban growth patterns and land transformation across the 1981–2015 period. The temporal trends of these class-level metrics are illustrated in Figures 19 through 22 providing insight into changes in spatial cohesion, fragmentation, and clustering across LULC categories from 1981 to 2015.

Figure 19. PD (Patch Density) over time (Authors' analysis using FRAGSTATS 4.2 (McGarigal & Cushman, 2002))

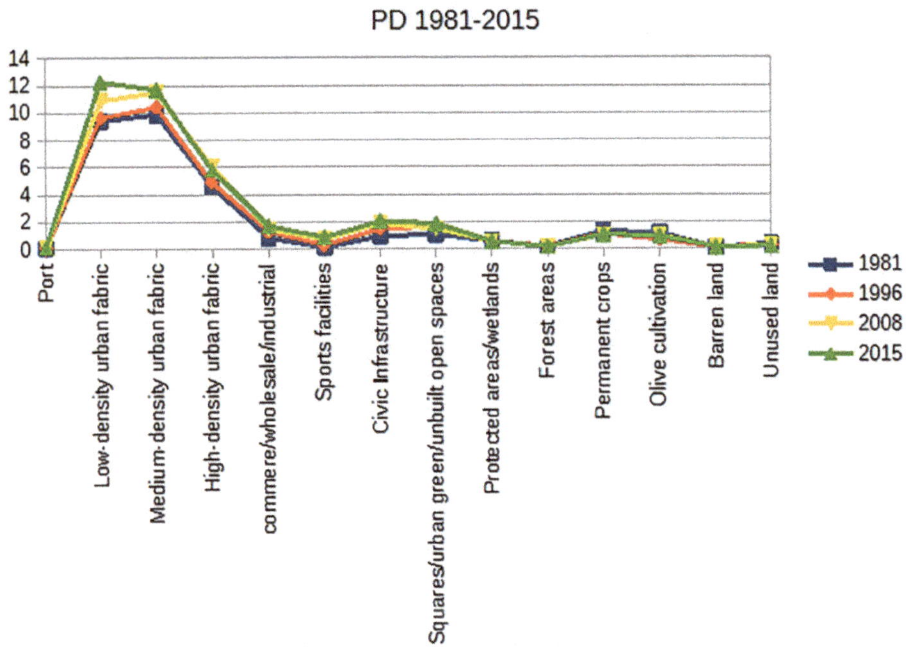

Figure 20. LPI (Largest Patch Index) over time (Authors' analysis using FRAGSTATS 4.2 (McGarigal & Cushman, 2002))

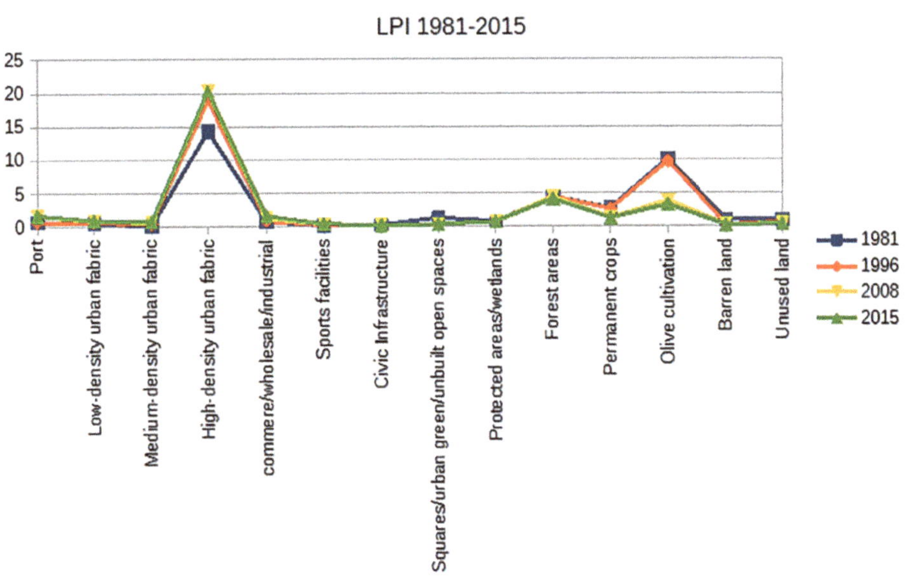

Figure 21. COHESION over time (Authors' analysis using FRAGSTATS 4.2 (Mc-Garigal & Cushman, 2002))

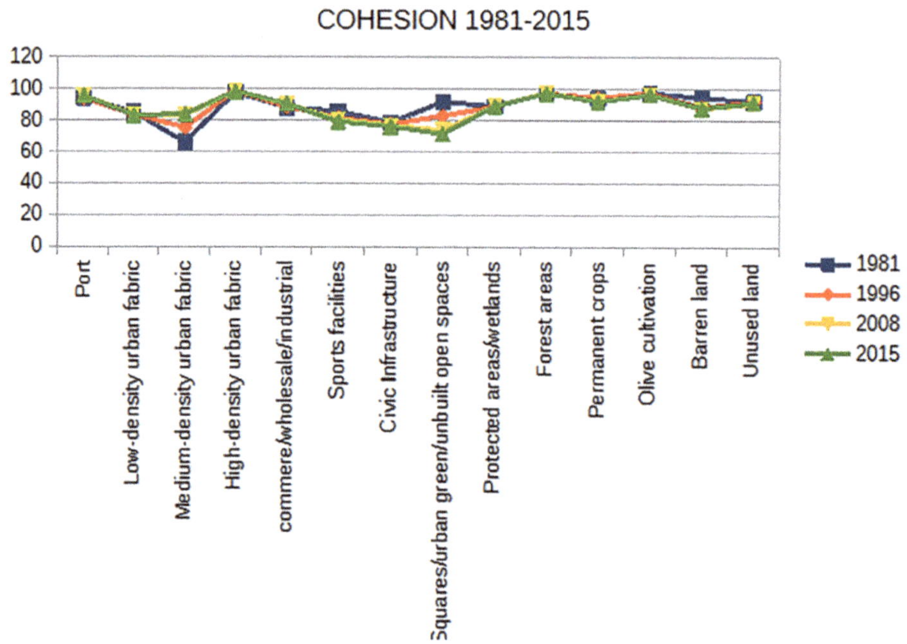

Figure 22. AI (Aggregation Index) over time (Authors' analysis using FRAGSTATS 4.2 (McGarigal & Cushman, 2002))

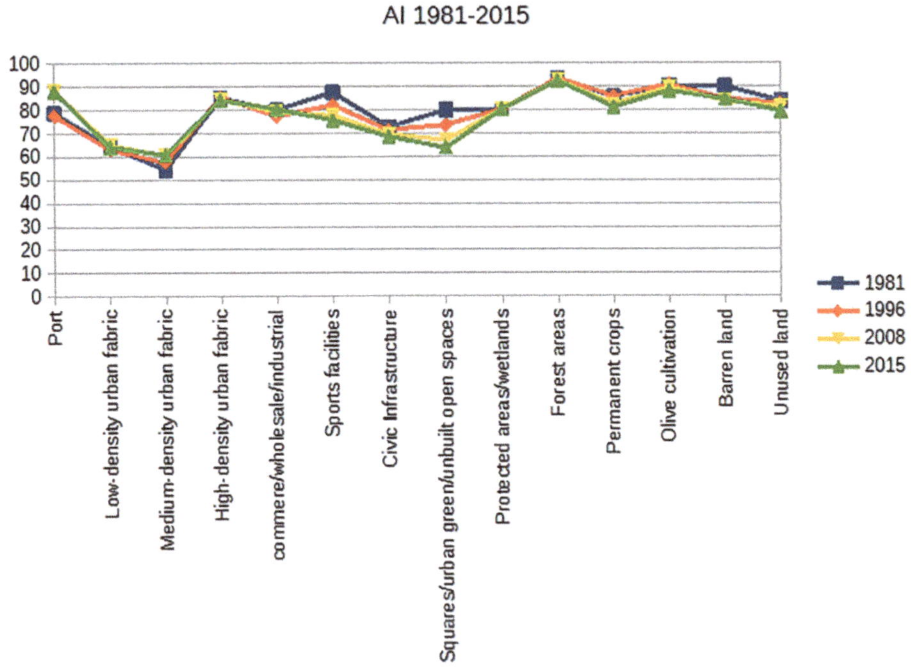

Increasing PD values indicate fragmentation in low and medium density urban fabrics. LPI confirms the dominance of high density urban fabric, while declines in agricultural classes suggest land loss. COHESION remains stable overall, while AI reflects increasing dispersion in open green and recreational areas.

Spatiotemporal Dynamics in Land Use/Land Cover

Land use/land cover change (LULCC) detection identifies and quantifies changes over time and space (Green et al., 1994). In this study, a post-classification comparison was conducted in ArcGIS to analyze changes across three key intervals—1981–1996, 1996–2008, and 2008–2015—as well as the entire 1981–2015 period. This method is widely used in urban land studies due to its effectiveness in detecting spatial transformation across discrete time steps (Weng, 2007).

The method allowed for a detailed assessment of urban dynamics, including the scale, direction, and intensity of land conversion. Table 6 summarizes the results by showing area change, percentage change, and annual growth rates across all land use/land cover categories.

Table 6. Spatiotemporal land use/land cover changes in the study area (1981–2015)

LULC Category	Area Change 1981–1996 (%)	Area Change 1996–2008 (%)	Area Change 2008–2015 (%)	Total Change 1981–2015 (%)	ACR (%)
Port	0.00	0.00	0.00	0.00	0.00
Low Density Urban Fabric	0.26	36.97	4.48	43.48	1.07
Medium Density Urban Fabric	8.80	51.22	6.36	74.98	1.66
High Density Urban Fabric	17.36	11.94	0.22	31.67	0.81
Commercial/Wholesale/ Industrial Area	61.55	35.94	15.00	152.57	2.76
Sports Facilities	15.70	101.13	7.96	151.23	2.75
Civic Infrastructure	35.25	18.07	4.53	66.92	1.52
Squares/Urban Green/Unbuilt Open Spaces	-42.99	-54.08	12.07	-70.66	-3.54
Protected Areas/Wetlands	0.00	0.00	0.00	0.00	0.00
Forest Areas	0.00	-0.38	-3.74	-4.11	-0.12
Permanent Crops	-11.18	-34.06	-6.65	-45.33	-1.76
Olive Cultivation	-7.24	-26.12	-7.89	-36.88	-1.34
Barren Land	-50.06	9.50	1.37	-44.57	-1.72
Unused Land	-27.09	1.49	-3.55	-28.63	-0.99

Note: ACR =Average Annual Change Rate, calculated as the compound annual growth rate (CAGR) over the entire study period (1981–2015).

Source: Authors' calculations based on classified LULC datasets (1981–2015).

Spatial Patterns and Land Use/Land Cover Dynamics

1981–1996. This period marked the early stages of outward expansion. High-density urban fabric increased by 17.4%, while commercial and industrial zones grew by 61.6%, concentrated in the southern sector. Civic infrastructure expanded (+35.3%), while urban green spaces declined by 43%, largely due to infill. Agricultural land began to show moderate losses: permanent crops decreased by 11.2%, and olive cultivation by 7.2%, though overall change remained spatially dispersed. Forests, wetlands, and port areas remained stable.

1996–2008. This was the most transformative phase. Low-density urban fabric increased by 37% and medium density by 51.2%, particularly in the north and east. Commercial and industrial zones grew by 35.9%, reflecting intensified development along new transport corridors. The decline of agricultural land accelerated: permanent crops dropped by 34% and olive cultivation by 26%. Urban green spaces shrank further (–54%), and civic infrastructure saw substantial expansion—sports facilities grew by 101%, while civic buildings rose by 18%. Forested areas experienced a slight decrease near new infrastructure.

2008–2015. Urban growth slowed but continued. Low- and medium-density urban fabrics grew by 4.5% and 6.4%, respectively. Commercial/industrial expansion continued at 15%, though at a slower pace. Agricultural decline persisted (permanent crops –6.6%, olive cultivation –7.9%), while urban green/open space showed a slight increase (+12%), potentially reflecting land reclassification, or small-scale planning initiatives. Marginal losses occurred in forest areas near development fronts.

Overall Trends 1981–2015. Over the full 34-year period, the study area experienced significant and sustained urban sprawl. The conversion of agricultural land was the most notable transformation: permanent crops declined by 45.3% and olive cultivation by 36.9%. Low-density urban fabric expanded by 43.5%, medium- density by 75%, and commercial/industrial areas more than doubled (+152.6%). At the same time, urban green/open space declined by 70.7%, highlighting the spatial inefficiency and fragmentation typical of Mediterranean urbanization. Figure 23 presents in a map series, only those areas where land use transitions occurred, providing visual emphasis on active landscape transformation.

Figure 23. Land use/land cover changes by reference period (Authors' analysis based on classified LULC datasets (1981–2015))

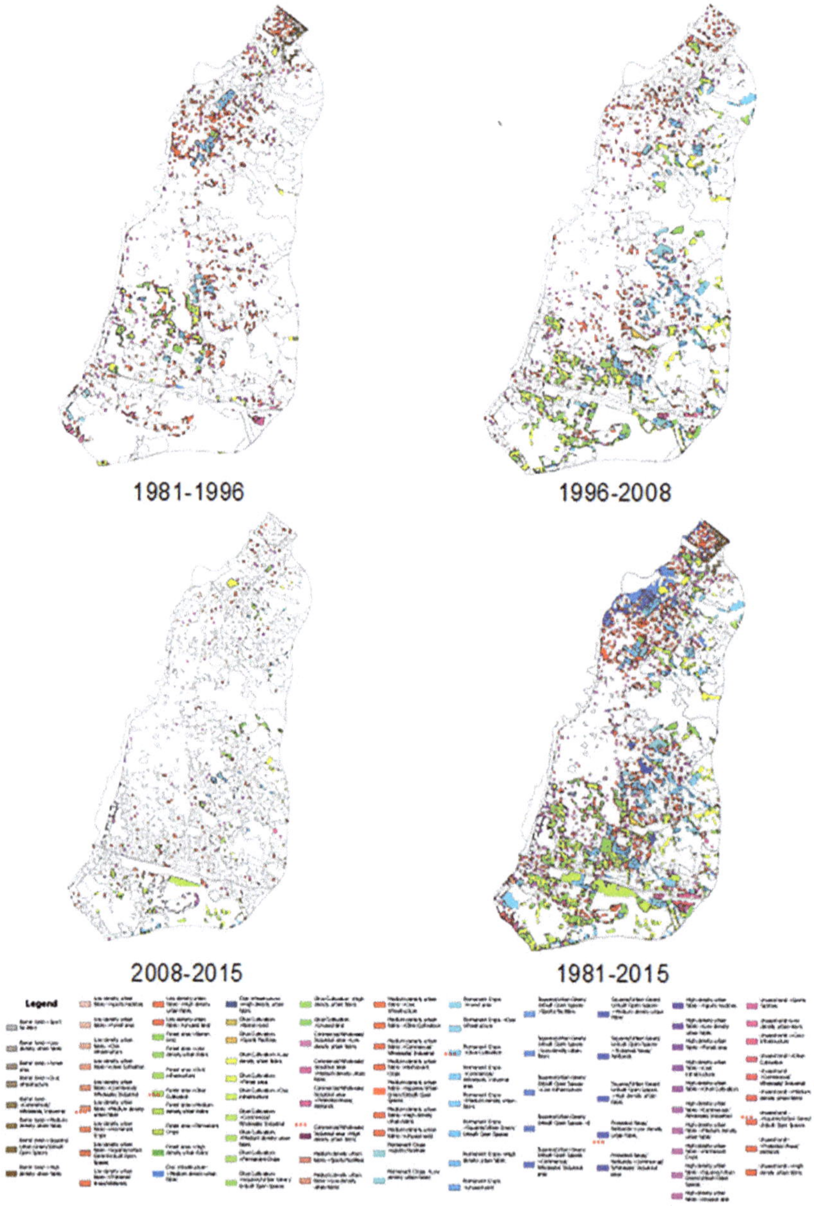

This map series visualizes spatiotemporal LULCC across four reference intervals. Each map highlights only those areas where land use transitions occurred, omitting unchanged categories to emphasize the dynamics of transformation. This visual filtering enhances interpretability by drawing attention to the spatial extent, intensity, and distribution of change throughout the study area.

From a spatial perspective, the maps reveal apparent trends of change with time. During the 1981–1996 period, changes are more dispersed, reflecting the early phases of urban development. The 1996–2008 period exhibited a high level of land conversion intensification, particularly in peripheral regions, which indicates rapid urban growth and infrastructure development. During 2008-2015, transformation is less extensive and subdued, consistent with a period of economic constraint. The cumulative map 1981–2015 clearly outlines the long-term urbanization footprint, with clusters of transformation concentrated along the northern, eastern, and southern development corridors.

The accompanying legend classifies a wide range of LULC transitions using a detailed color-coded scheme. Each color corresponds to a specific type of land use/land cover change (e.g., agricultural to residential, green space to infrastructure), allowing for fine-grained interpretation of spatial processes.

This series supports a deeper understanding of the land transformation process, providing visual evidence of the directional spread of urbanization, the decline of agricultural land, and the gradual replacement of open space with built-up surfaces. By visualizing only the areas of change, the maps function as analytical tools that isolate active landscape dynamics, facilitating the detection of vulnerable zones and the timing of major land use shifts.

Profile of Elite Interviewees

Nine experts participated in the elite interviews, representing a broad mix of professional backgrounds and institutional affiliations. This multidisciplinary composition ensured that both technical expertise and institutional experience were reflected in the findings. Several participants had been directly involved in the preparation or review of General Urban Plans (1986 and 2011), while others contributed to infrastructure policy, environmental regulation, or provided academic perspectives on urban development. Table 7 summarizes the roles and sectors of the interviewees.

Table 7. Profiles of expert interviewees and professional sectors

Experts ID	Role / Sector
E1	Real estate agent (private sector)
E2	Urban planner, Municipality of Patras – Urban and Traffic Planning Dept
E3	Urban planner, supervisor of local master plans
E4	Urban planner, Municipality of Patras

continued on following page

Table 7. Continued

Experts ID	Role / Sector
E5	Architect/Urban planner (private sector)
E6	Transport and GIS consultant (private sector)
E7	Agronomist, Ministry of Rural Development – Spatial Planning Unit
E8	Professor, University of Patras
E9	Professor, University of Patras

Source: Authors' fieldwork (expert interviews).

This diverse composition of expertise ensured a balanced representation of institutional, technical, and academic perspectives. Several participants had direct involvement in the preparation or review of General Urban Plans (1986 and 2011), while others contributed to environmental and infrastructure policies or advised local government. Academic participants provided a theoretical and critical perspective on the evolution of urban form and planning governance.

Key Drivers from Expert Interviews

Thematic analysis of the interview data revealed several key drivers behind urban sprawl in the Patras area. These were organized according to the five driving force categories proposed by Bürgi et al. (2004):

Political and Institutional Drivers

Weak enforcement of planning regulations, permissive zoning frameworks, and the legal tolerance of "out-of-plan" construction were consistently identified as primary drivers of sprawl. Several interviewees emphasized how delays in formal plan implementation created a regulatory vacuum that enabled unplanned development to proceed unchecked. Additionally, transport policy especially state-led initiatives to expand highways and interchanges, was highlighted as a structural driver that facilitated decentralization and reinforced low-density expansion.

Cultural Drivers

Population density and preferences for single-family housing, combined with a historical tradition of private landownership and inheritance-based land fragmentation, supported suburban dispersion. The desire for privacy, access to outdoor space, and avoidance of dense urban environments were cited as enduring cultural patterns.

Technological Drivers

Improved accessibility, particularly through expanded road networks and arterial infrastructure, encouraged residential development in peri-urban areas. The city's strong dependence on private car use further reinforced low-density, dispersed urban patterns, exacerbating sprawl.

Economic Drivers

Land price differentials, deindustrialization, and speculative development were mentioned as factors influencing the outward shift of urban growth. Former industrial zones and lower land values in agricultural areas created favorable conditions for development. Moreover, large-scale road infrastructure played an economic role by enhancing accessibility, raising land values, and enabling new real estate markets to emerge beyond the urban core.

Natural and Spatial Drivers

Although less influential, topographic features (e.g., elevation), proximity to coastal amenities, and the availability of flat agricultural land were seen as influencing development patterns.

Elite interviewees consistently noted that sprawl patterns in Patras were not purely market-driven but deeply rooted in institutional practices and planning cultures. Out-of-plan development was highlighted as both a symptom and a cause of fragmented urban form—reinforcing a spatial logic of low-density, infrastructure-deficient expansion. As previously shown in Figure 1, the conceptual framework of land use change drivers (Hersperger & Bürgi, 2007) provides a structured lens for analyzing sprawl dynamics. Figure 24 illustrates how this framework was adapted to the case of Patras, based on insights derived from thematic analysis.

Figure 24. Case-specific adaptation of the Hersperger & Bürgi framework to sprawl dynamics in Patras (Adapted from Hersperger & Bürgi (2007), elaborated by the authors)

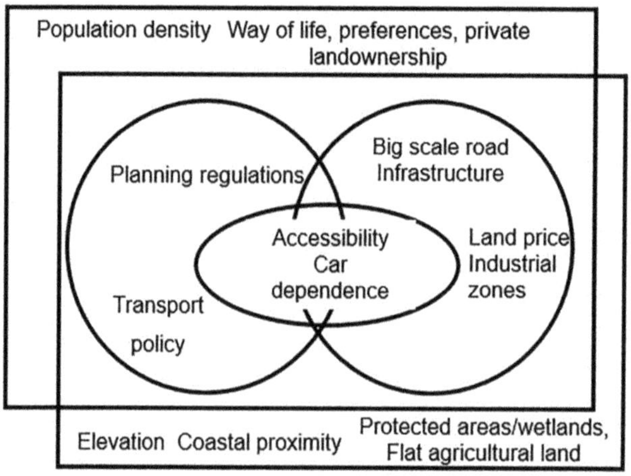

ANALYSIS AND DISCUSSION

This section presents the results of a multi-year multiple linear regression (MLR) analysis investigating the primary drivers of urban sprawl in the study area from 1981 to 2015. Urban sprawl is quantified using built-up surface density derived from created spatial datasets, modeled against a range of explanatory variables reflecting socio-economic, infrastructural, environmental, and policy factors. The MLR framework enables year-specific models that capture temporal dynamics and spatial heterogeneity and provides actionable insights into the evolving structure of urban growth.

Variable Framework and Modeling Approach

The selection of independent variables was informed by expert interviews, existing literature, and the conceptual framework previously discussed. Variables were grouped into five broad categories:

- **Cultural Drivers**: Population density, distance to schools
- **Economic Drivers**: Distance from city center, industrial zones
- **Technological Drivers**: Road network density, distance from ring road interchanges

· **Environmental Drivers**: Elevation, distance from coastlines and protected areas
· **Institutional/Policy Drivers**: Date of inclusion in the urban plan

These variables were operationalized using Geographic Information Systems (GIS) and remote sensing sources and further validated by local expert consultation. Land use/land cover categories were treated as dummy variables, with 'urban fabric' serving as the reference category. All continuous variables were standardized to ensure comparability and reduce scale-related bias. Multicollinearity was evaluated using the Variance Inflation Factor (VIF). No variable exceeded the critical VIF threshold of 5, indicating acceptable collinearity. To ensure the validity of the regression results, Moran's I was calculated on the residuals of each MLR model to test for spatial autocorrelation. Results indicated no significant clustering ($p > 0.05$), supporting the independence assumption of OLS regression and eliminating the need for spatial error correction models.

Separate regression models were constructed for the years 1981, 1991, 2001, and 2015, each incorporating time-appropriate datasets to reflect the socio-spatial conditions of those periods. Robust standard errors were used to correct for heteroskedasticity and potential outlier effects. Model specification was guided by both theoretical expectations and empirical diagnostics, including adjusted R^2, root mean square error (RMSE), Akaike/Bayesian information criteria (AIC/BIC), and significance tests. Stepwise selection procedures were applied to refine models for interpretability and parsimony, ensuring that only the most impactful variables were retained.

Model Performance Overview

The models demonstrated consistently high explanatory power across all four years, confirming the robustness of the MLR framework in modeling urban sprawl patterns. The adjusted R^2 values—ranging from 83.5% in 1981 to 86.7% in 2015—indicate that a substantial proportion of variation in built-up surface density was captured by the predictors. This strong performance suggests that the chosen variables successfully reflect both structural and contextual dimensions of urban expansion.

1981: Adjusted R^2 = 83.5%, RMSE = 0.06195
1991: Adjusted R^2 = 84.3%, RMSE = 0.05987
2001: Adjusted R^2 = 85.3%, RMSE = 0.05764
2015: Adjusted R^2 = 86.7%, RMSE = 0.05382

In addition to statistical fit, model parsimony and generalizability were ensured through stepwise regression optimization, which yielded stable results across time periods. The increasing R^2 values over time may reflect the improved explanatory value of spatial infrastructure variables as urbanization patterns matured and became more structured. These improvements also align with policy changes, such as the implementation of the General Urban Plan in 1986 and the expansion of the road network in the early 2000s. The evolution of these models also underscores the value of temporal disaggregation, which allows the capture of lagged or evolving effects—especially crucial in urban development processes that unfold over decades.

Key Predictors and Temporal Shifts

The analysis of regression coefficients across all models revealed several important patterns. Road network density was consistently the strongest positive predictor, significant in all models. Its coefficient peaked in 1991 ($\beta = 2.55$), reflecting the influence of transportation accessibility on peripheral urban growth. By 2015, the effect declined to $\beta = 1.02$, suggesting a shift in the role of roads from initiating growth to accommodating existing development. This trend may also reflect the saturation of key infrastructural corridors and the transition from expansion to densification. Road accessibility also supported leapfrog development, allowing built-up zones to form disconnected but strategically linked nodes. Population density exerted strong positive effects in the earlier models ($\beta = 2.15$ in 1981) but its influence steadily declined, becoming insignificant by 2001 and excluded altogether in 2015. This evolution reflects a transformation from population-constrained development within the urban core to infrastructure-led spatial dispersion. As the city acquired a more polycentric structure, raw population density lost predictive power, implying that housing preferences were increasingly shaped by lifestyle factors and infrastructure provision rather than demographic concentration alone.

The role of planning variables proved remarkably stable. Pre-existing urban plan status exhibited a consistently strong and positive relationship with built-up density across all years, confirming the impact of early planning frameworks on urban form. Zones incorporated after 1980 (1980–1989, 1990–1999) lagged significantly in development, indicating a delay in infrastructure rollout and regulatory implementation. These findings highlight the inertia embedded in institutional systems and the long gestation periods often required for planning interventions to yield physical outcomes. Furthermore, the dominance of older urban plan areas underscores the cumulative advantage they enjoy through sustained investment, proximity to services, and institutional familiarity.

Other variables gained importance later in the period. Distance from schools, for instance, emerged as a significant factor from 2001 onward. The negative coefficient in 2015 ($\beta = -0.0000234$) suggests that proximity to educational infrastructure increasingly shaped residential choices as the city matured. This points to the rising importance of service-based amenities and social drivers in the evolution of urban sprawl.

Finally, environmental and agricultural constraints—including natural and semi-natural areas, olive cultivation, and permanent crops—retained consistently negative associations with urban expansion. While their coefficients diminished slightly over time, they nevertheless highlight ongoing development pressure on ecologically or agriculturally valuable land. These findings reflect broader socio-political shifts, such as changes in land tenure, economic incentives for conversion, and weakening enforcement of protective regulations. Although vulnerable to circumvention, these ecological contexts remain critical constraints on urbanization, even as their effectiveness is gradually eroded.

Taken together, these patterns illustrate a transition from a monocentric, demand-driven urban expansion model to one increasingly mediated by infrastructure and institutional frameworks. Accessibility and regulatory regimes have emerged as the principal determinants of spatial growth, while population pressure and environmental constraints, though still relevant, have receded in explanatory power.

Spatial Projections and Predictive Mapping

To move from abstract statistical relationships to a tangible representation of sprawl dynamics, predictive spatial mapping was employed using kriging interpolation. This geostatistical technique allowed for the generation of continuous surfaces depicting built-up surface density predictions for each of the four reference years. The use of kriging was justified by its ability to model spatial autocorrelation and produce reliable estimations at unsampled locations based on the regression-predicted values of the 300 spatial sample points.

However, it is important to emphasize that while the maps are grounded in a stable predictive model—derived from multiple linear regression—the kriging method inherently involves a degree of uncertainty. This stems from underlying assumptions about spatial continuity and the distribution of values across the surface. As such, the predictive maps should be interpreted not as precise representations of sprawl, but rather as generalized spatial trends, capturing the probable intensity and directional spread of urban expansion in the study area. The resulting spatial patterns of predicted built-up density are presented in Figure 25. These maps highlight the estimated spatial intensity and directional spread of urban sprawl, based on the regression-derived values and kriging interpolation.

Figure 25. Kriging-interpolated predictive maps of built-up density (1981, 1991, 2001, 2015) (Authors' analysis using GIS-based regression modeling and kriging interpolation)

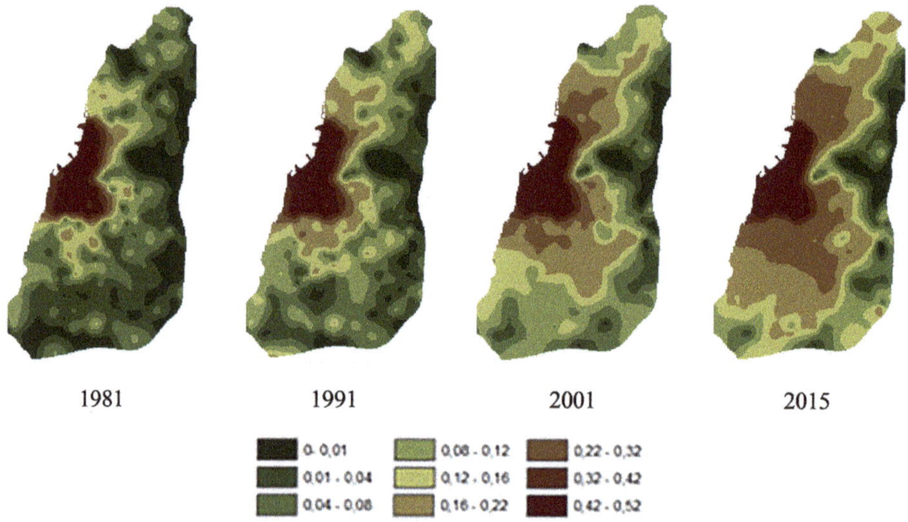

In **1981**, urban development was tightly concentrated within the historical core and adjacent zones of Patras, with limited diffusion beyond the pre-existing urban plan boundaries. The peripheral areas remained predominantly agricultural or underutilized, and sprawl was mostly absent outside of institutional boundaries.

By **1991**, the pattern had shifted markedly. High built-up density values began to radiate outward from the center along established transportation corridors, particularly toward the north, south, and east. The influence of the expanding road network became spatially evident, with leapfrogging development evident in otherwise non-contiguous patches.

The **2001** map shows increased consolidation of suburban development, particularly in areas that had been integrated into the urban plan during the 1980s and early 1990s. A belt of medium-to-high density development emerged, encompassing both older urban cores and newly developed suburban enclaves. The proximity to educational infrastructure, especially in new residential zones contributed to this expansion.

In **2015**, spatial fragmentation increased further. Built-up densities appeared higher along ring road interchanges, newly upgraded transportation links, and around urban service nodes. Predictive mapping showed densification around areas previously on the urban fringe, confirming the continued diffusion and decentralization of urban development.

These maps serve as visual extensions of the regression results and highlight the critical role of spatial infrastructure, institutional frameworks, and ecological boundaries in shaping the urban footprint. They also provide a framework for identifying development hotspots and areas of potential conflict with agricultural preservation or environmental protection goals.

Discussion

To assess how each driving force evolved across the study period, a comparison of standardized regression coefficients (β) was conducted. Table 8 summarizes the magnitude and direction of influence for key variables in each regression model. The models demonstrated consistently strong fit, with Adjusted R^2 values ranging from 0.83 to 0.87, confirming their robustness.

Table 8. Comparison of standardized regression coefficients (β) for key sprawl drivers across reference years (1981–2015)

Variable	1981	1991	2001	2015
Population Density	+2.15	+0.88	+0.51	
Road Network Density	+2.31	+2.55	+1.77	+1.02
Pre-existing Urban Plan	+0.23	+0.25	+0.24	+0.22
Distance from Schools			−0.0000163	−0.0000234
Natural/Semi-Natural Areas	−0.07	−0.09	−0.11	−0.11
Olive cultivation	−0.05	−0.06	−0.06	−0.08
Permanent Crops	−0.05	−0.06	−0.10	−0.11
Out-of-Plan Areas	−0.02	−0.05	−0.09	−0.11
Distance from Ring Road Interchanges				+5.98e-06

Note: Values represent standardized β coefficients from multiple linear regression models (MLR). Model fit statistics (Adjusted R^2) are reported in the text.

blank = not statistically significant ($p>0.05$) or excluded by stepwise selection.

Source: Authors' analysis based on multiple linear regression models.

The analysis reveals several clear shifts in the sprawl regime. Population density, which was a strong predictor in the early years (β = 2.15 in 1981), steadily declined in influence and had become insignificant by 2015. This trend suggests that once core areas reached saturation, further growth depended less on demographic concentration and more on infrastructural availability and land-use opportunities.

Road network density consistently emerged as the most powerful driver across all models. Road Network Density peaked in 1991, during a period of significant transport investment, and remained a critical but slightly diminishing factor thereafter. This supports the idea that infrastructure initially facilitates sprawl, but its marginal returns may decline as the network matures.

Planning variables, particularly the pre-existing urban plan, remained strong and positive throughout, underscoring the institutional anchoring of development. The consistently negative coefficients for later or out-of-plan areas reflect long-term underdevelopment and delayed absorption into the urban system.

Environmental and Agricultural Constraints -including natural/semi-natural areas, permanent crops, and olive cultivation- maintained their negative influence, though diminishing slightly. Their continued significance emphasizes the friction between ecological preservation and expansion pressures.

New social drivers became evident later in the period. Distance from schools, for example, emerged as a significant factor from 2001 onward, with the 2015 model confirming the growing weight of educational infrastructure in residential location choices. This shift reflects the increasing importance of service-based amenities in shaping patterns of suburban development. Similarly, distance from ring road interchanges, although only marginally significant in 2015, suggests a lagged but conceptually important role for major transportation infrastructure in influencing sprawl dynamics.

These findings collectively reinforce the view that urban sprawl in Patras is driven less by population pressure and more by infrastructural expansion, land use regulation, and social amenity access. The 2015 model confirms the continued relevance of spatial planning frameworks and suggests that the effects of large-scale infrastructure projects, such as the Patras Ring Road, may only fully emerge over extended periods.

The strong influence of road network density in Patras confirms findings from other Mediterranean cities. In Athens, Chorianopoulos et al. (2010) showed how large-scale transport investments before the 2004 Olympic Games—new highways, the airport, and the suburban rail—intensified peripheral growth in the absence of an integrated planning framework. Similarly, Salvati and Carlucci (2014) documented how infrastructure-led expansion shaped fragmented urban development in Rome. Our results extend this evidence by demonstrating that even in a medium-sized city like Patras, road accessibility remains the most consistent predictor of sprawl across three decades. By contrast, socio-cultural factors such as lifestyle preferences and housing aspirations, which Muñoz (2003) and García-Palomares (2010) identified as central drivers in Spanish metropolitan regions, appear less decisive in the Patras case. Here, institutional practices and planning delays exerted stronger effects than purely cultural dynamics, underscoring the distinctive influence of regulatory frameworks in semi-peripheral Mediterranean contexts. Comparable trends have been reported in other coastal Mediterranean contexts, where the combination of transport infrastructure, flexible planning frameworks, and tourism development has fueled sprawl. García-Ayllón (2023) illustrates how mass tourism, demand for second homes, and weak governance have gradually converted agricultural and coastal landscapes into fragmented urban areas. In Patras, tourism pressures are less pronounced, yet the evidence points to a similar underlying mechanism: dispersed growth is primarily driven by infrastructure expansion and institutional shortcomings. This positions Patras as a semi-peripheral variant of the broader Mediterranean sprawl model, where regulatory gaps and transport corridors weigh more heavily than lifestyle or tourism-related factors.

Policy and Planning Implications

These comparative insights provide not only a clearer understanding of the Patras case, but also a basis for deriving policy and planning recommendations. Building on these findings, the next section outlines critical implications for urban policy, land use planning, and sustainable development in Mediterranean and semi-peripheral urban regions like Patras.

1. **Strategic Infrastructure Planning:** Road network density remains the most reliable predictor of sprawl, but its diminishing marginal effect suggests a tipping point beyond which new infrastructure may contribute more to fragmentation than cohesion. Policymakers must now weigh the benefits of expansion against the risks of uncoordinated growth.
2. **Reinforcement of Institutional Boundaries:** The enduring influence of early urban plan areas reveals the critical importance of timely, well-integrated planning frameworks. Delays in infrastructure rollout or plan ratification can leave newly integrated zones lagging for decades. Improving institutional responsiveness and ensuring enforcement mechanisms for development zones are essential.
3. **Incorporation of Social Infrastructure into Planning:** The rising influence of educational proximity points to the need for proactive placement of schools and services in expansion zones. Strategic integration of such infrastructure could guide growth more sustainably and equitably.
4. **Protection of Agricultural and Natural Areas:** Despite consistent resistance to development, agricultural zones and natural land are being incrementally converted. Strengthening regulatory buffers, enforcing zoning restrictions, and incentivizing agricultural continuity (e.g., subsidies, urban growth boundaries) could mitigate irreversible land loss.
5. **Use of Predictive Spatial Modeling in Policy Design:** The utility of regression-based and geostatistical mapping in revealing latent growth patterns supports its use in future scenario planning. Municipal authorities should consider institutionalizing such tools into regular land use assessment workflows.

CONCLUSION

The study examined the spatial dynamics and underlying drivers of urban sprawl in the Urban Complex of Patras and its surroundings between 1981 and 2015, applying a mixed-methods approach that integrated land use/land cover classification, spatial metrics, regression analysis, and expert interviews. The findings show that sprawl in Patras was not driven by demographic pressure alone, but emerged from a complex interplay of infrastructural development, weak regulatory enforcement, and socio-cultural preferences.

The consistent influence of road network density and institutional planning variables underscores the importance of early zoning frameworks and accessibility in shaping urban form. At the same time, the increasing role of social infrastructure—particularly proximity to schools—highlights evolving residential preferences in medium-sized Mediterranean cities. The study also demonstrated the utility of predictive spatial mapping and regression modeling as tools for anticipating urban expansion trends.

Key limitations include the reliance on historical aerial imagery (with a one-year proxy for 1981) and the inherent assumptions of kriging interpolation; however, cross-validation and accuracy assessment mitigate these concerns.

While specific to Patras, the methodology and insights are transferable to other semi-peripheral urban contexts experiencing uncoordinated growth. Future research should expand this approach to other medium-sized cities, assess post-2015 developments (e.g., green recovery plans, smart city initiatives), and explore participatory planning frameworks particularly in cities facing similar institutional and territorial pressures.

REFERENCES

Angel, S., Parent, J., Civco, D. L., Blei, A., & Potere, D. (2011). The dimensions of global urban expansion: Estimates and projections for all countries, 2000–2050. *Progress in Planning*, *75*(2), 53–107. DOI: 10.1016/j.progress.2011.04.001

Asprogerakas, E. (2003). *Characteristics and development potential of Greek medium-sized cities* [Unpublished doctoral dissertation]. University of Thessaly. [in Greek].

Bhatta, B., Saraswati, S., & Bandyopadhyay, D. (2010). Urban sprawl measurement from remote sensing data. *Applied Geography (Sevenoaks, England)*, *30*(4), 731–740. DOI: 10.1016/j.apgeog.2010.02.002

Boyatzis, R. E. (1998). Transforming qualitative information: Thematic analysis and code development. *Sage (Atlanta, Ga.)*.

Braun, V., & Clarke, V. (2006). Using thematic analysis in psychology. *Qualitative Research in Psychology*, *3*(2), 77–101. DOI: 10.1191/1478088706qp063oa

Brueckner, J. K. (2000). Urban sprawl: Diagnosis and remedies. *International Regional Science Review*, *23*(2), 160–171. DOI: 10.1177/016001700761012710

Burchell, R. W., Shad, N. A., Listokin, D., Phillips, H., Downs, A., Seskin, S., . . . Gall, M. (1998). *The costs of sprawl—Revisited* (TCRP Report 39). Transportation Research Board. https://digitalcommons.usf.edu/cgi/viewcontent.cgi?article=1189&context=cutr_reports

Bürgi, M., Hersperger, A. M., & Schneeberger, N. (2005). Driving forces of landscape change—Current and new directions. *Landscape Ecology*, *19*(8), 857–868. DOI: 10.1007/s10980-005-0245-3

Calthorpe, P., & Fulton, W. (2001). *The regional city: Planning for the end of sprawl*. Island Press.

Cheng, J., & Masser, I. (2003). Urban growth pattern modeling: A case study of Wuhan city, PR China. *Landscape and Urban Planning*, *62*(4), 199–217. DOI: 10.1016/S0169-2046(02)00150-0

Chin, N. (2002). *Unearthing the roots of urban sprawl: A critical analysis of form, function and methodology* (CASA Working Paper No. 47). University College London. https://discovery.ucl.ac.uk/id/eprint/249/1/Paper47.pdf

Chorianopoulos, I., Pagonis, T., Koukoulas, S., & Drymoniti, S. (2010). Planning, competitiveness and sprawl in the Mediterranean city: The case of Athens. *Cities (London, England)*, *27*(4), 249–259. DOI: 10.1016/j.cities.2009.12.011

Couch, C., & Karecha, J. (2006). Controlling urban sprawl: Some experiences from Liverpool. *Cities (London, England), 23*(5), 353–363. DOI: 10.1016/j.cities.2006.05.003

Couch, C., Leontidou, L., & Arnstberg, K. O. (2008). Introduction: Definitions, theories and methods of comparative analysis. In Couch, C., Petschel-Held, G., & Leontidou, L. (Eds.), *Urban sprawl in Europe: Landscape, land-use change and policy* (pp. 3–34). Blackwell.

Devi, N. N., Sridharan, B., & Kuiry, S. N. (2019). Impact of urban sprawl on future flooding in Chennai city, India. *Journal of Hydrology (Amsterdam), 574*, 486–496. DOI: 10.1016/j.jhydrol.2019.04.041

Díaz, S., Settele, J., Brondízio, E., Ngo, H., Guèze, M., Agard, J., & Zayas, C. (2019). *Summary for policymakers of the global assessment report on biodiversity and ecosystem services*. IPBES. https://doi.org/DOI: 10.5281/zenodo.3553579

Downs, A. (1999). Some realities about sprawl and urban decline. *Housing Policy Debate, 10*(4), 955–974. DOI: 10.1080/10511482.1999.9521356

Eurofound. (2012). *Medium-sized cities in Europe (summary)*. https://www.eurofound.europa.eu/en/publications/all/medium-sized-cities-europe-summary

European Commission, Joint Research Centre. (2016). *Atlas of the human planet 2016: Mapping human presence on Earth with the Global Human Settlement Layer* (EUR 28116 EN). Publications Office of the European Union. https://doi.org/DOI: 10.2788/889483

European Environment Agency, & FOEN. (2016). *Urban sprawl in Europe: Joint EEA–FOEN report* (EEA Report No. 11/2016). European Environment Agency & Swiss Federal Office for the Environment. https://www.eea.europa.eu/en/analysis/publications/urban-sprawl-in-europe

Ewing, R. (1997). Is Los Angeles-style sprawl desirable? *Journal of the American Planning Association, 63*(1), 107–126. DOI: 10.1080/01944369708975728

Ewing, R., Pendall, R., & Chen, D. (2002). *Measuring sprawl and its impact*. Smart Growth America. https://jacksontetonplan.com/DocumentCenter/View/767/Ewing-R-R-Pendall-and-D-Chen-2002-PDF

Foley, J. A., DeFries, R., Asner, G. P., Barford, C., Bonan, G., Carpenter, S. R., Chapin, F. S., Coe, M. T., Daily, G. C., Gibbs, H. K., Helkowski, J. H., Holloway, T., Howard, E. A., Kucharik, C. J., Monfreda, C., Patz, J. A., Prentice, I. C., Ramankutty, N., & Snyder, P. K. (2005). Global consequences of land use. *Science, 309*(5734), 570–574. DOI: 10.1126/science.1111772

Frenkel, A. (2004). The potential effect of national growth-management policy on urban sprawl and the depletion of open spaces and farmland. *Land Use Policy, 21*(4), 357–369. DOI: 10.1016/j.landusepol.2003.12.001

Galster, G., Hanson, R., Ratcliffe, M. R., Wolman, H., Coleman, S., & Freihage, J. (2001). Wrestling sprawl to the ground: Defining and measuring an elusive concept. *Housing Policy Debate, 12*(4), 681–717. DOI: 10.1080/10511482.2001.9521426

Government Gazette. (1986). [Approval of the General Urban Plan of Patras (Municipality of Patras)] (ΦΕΚ 1061/Δ/07.11.1986). [in Greek].

Government Gazette. (2011). [Approval of the General Urban Plan of the Municipal Unit of Patras, Municipality of Patras, Achaia Prefecture] (ΦΕΚ 358/ΑΑΠ/30.12.2011). [in Greek].

Government Gazette. (2012). [Amendment of Decision No. 11247/28.12.2012 regarding the 2011 Population and Housing Census results for the permanent population of the country] (ΦΕΚ 3465/Β΄/28.12.2012). [in Greek].

Green, K., Kempka, D., & Lackey, L. (1994). Using remote sensing to detect and monitor land-cover and land-use change. *Photogrammetric Engineering and Remote Sensing, 60*(3), 331–337. https://www.asprs.org/wp-content/uploads/pers/1994journal/mar/1994_mar_331-337.pdf

Harvey, W. S. (2011). Strategies for conducting elite interviews. *Qualitative Research, 11*(4), 431–441. DOI: 10.1177/1468794111404329

Hastaoglou-Martinidis, V. (1995). City form and national identity: Urban designs in nineteenth-century Greece. *Journal of Modern Greek Studies, 13*(1), 99–123. DOI: 10.1353/mgs.2010.0397

Hersperger, A. M., & Bürgi, M. (2007). Driving forces of landscape change in the urbanizing Limmat valley, Switzerland. In Koomen, E., Stillwell, J., Bakema, A., & Scholten, H. (Eds.), *Modelling land-use change: Progress and applications* (pp. 45–60). Springer., DOI: 10.1007/978-1-4020-5648-2_3

Hersperger, A. M., & Bürgi, M. (2009). Going beyond landscape change description: Quantifying the importance of driving forces of landscape change in a Central Europe case study. *Land Use Policy, 26*(3), 640–648. DOI: 10.1016/j.landusepol.2008.08.015

Joffe, H., & Yardley, L. (2003). Content and thematic analysis. In D. F. Marks & L. Yardley (Eds.), *Research methods for clinical and health psychology* (pp. 56–68). Sage.

Laboratory of Architectural Technology and Spatial Design. (1986). *Methodological approach to the analysis of challenges, prospects, and planning process for the Greek urban center. Special application: Urban regeneration of Patras* [Research report, in Greek]. University of Patras.

Lambin, E. F., Geist, H. J., & Lepers, E. (2003). Dynamics of land-use and land-cover change in tropical regions. *Annual Review of Environment and Resources, 28*(1), 205–241. DOI: 10.1146/annurev.energy.28.050302.105459

Leontidou, L. (1990). *The Mediterranean city in transition: Social change and urban development.* Cambridge University Press. DOI: 10.1017/CBO9780511522208

Leontidou, L., Afouxenidis, A., Kourliouros, E., & Marmaras, E. (2007). Infrastructure-related urban sprawl: Mega-events and hybrid peri-urban landscapes in Southern Europe. In Couch, C., Petschel-Held, G., & Leontidou, L. (Eds.), *Urban sprawl in Europe: Landscape, land-use change and policy* (pp. 71–101). Blackwell. DOI: 10.1002/9780470692066.ch3

Li, G., & Li, F. (2019). Urban sprawl in China: Differences and socioeconomic drivers. *The Science of the Total Environment, 673*, 367–377. DOI: 10.1016/j.scitotenv.2019.04.080

Lu, X., Chen, D., & Wang, Y. (2020). Is urban sprawl decoupled from the quality of economic growth? Evidence from Chinese cities. *Sustainability (Basel), 12*(1), 218. DOI: 10.3390/su12010218

McGarigal, K., & Cushman, S. A. (2002). Comparative evaluation of experimental approaches to the study of habitat fragmentation effects. *Ecological Applications, 12*(2), 335–345. DOI: 10.1890/1051-0761(2002)012[0335:CEOEAT]2.0.CO;2

McGarigal, K., & Marks, B. J. (1995). *FRAGSTATS: Spatial pattern analysis program for quantifying landscape structure* (Gen. Tech. Rep. PNW-GTR-351). U.S. Department of Agriculture, Forest Service, Pacific Northwest Research Station. https://doi.org/DOI: 10.2737/PNW-GTR-351

Michalak, W. Z. (1993). GIS in land use change analysis: Integration of remotely sensed data into GIS. *Applied Geography (Sevenoaks, England), 13*(1), 28–44. DOI: 10.1016/0143-6228(93)90078-F

Mikecz, R. (2012). Interviewing elites: Addressing methodological issues. *Qualitative Inquiry, 18*(6), 482–493. DOI: 10.1177/1077800412442818

Papadatou-Giannopoulou, Ch. (1997). [Investigation of transfer zone designation for building coefficient allocation in the Urban Complex of Patras]. Municipality of Patras, Directorate of Urban Planning and Implementation. [in Greek].

Pappas, V. (2006). *Small metropolitan areas in rapid transition: The case of Patras*. University of Patras. [in Greek]

Pielke, R. A.Sr, Pitman, A., Niyogi, D., Mahmood, R., McAlpine, C., Hossain, F., Goldewijk, K. K., Nair, U., Betts, R., Fall, S., Reichstein, M., Kabat, P., & de Noblet, N. (2011). Land use/land cover changes and climate: Modeling analysis and observational evidence. *Wiley Interdisciplinary Reviews: Climate Change*, *2*(6), 828–850. DOI: 10.1002/wcc.144

Power, A. (2001). Social exclusion and urban sprawl: Is the rescue of cities possible? *Regional Studies*, *35*(8), 731–742. DOI: 10.1080/00343400120084713

Salvati, L., & Carlucci, M. (2014). Urban growth and land-use structure in two Mediterranean regions: An exploratory spatial data analysis. *SAGE Open*, *4*(4), 2158244014561199. DOI: 10.1177/2158244014561199

Schneeberger, N., Bürgi, M., Hersperger, A. M., & Ewald, K. C. (2007). Driving forces and rates of landscape change as a promising combination for landscape change research—An application on the northern fringe of the Swiss Alps. *Land Use Policy*, *24*(2), 349–361. DOI: 10.1016/j.landusepol.2006.04.003

Seto, K. C., Güneralp, B., & Hutyra, L. R. (2012). Global forecasts of urban expansion to 2030 and direct impacts on biodiversity and carbon pools. *Proceedings of the National Academy of Sciences of the United States of America*, *109*(40), 16083–16088. DOI: 10.1073/pnas.1211658109

Skånes, H. M., & Bunce, R. G. H. (1997). Directions of landscape change (1741–1993) in Virestad, Sweden—Characterised by multivariate analysis. *Landscape and Urban Planning*, *38*(1–2), 61–75. DOI: 10.1016/S0169-2046(97)00019-4

Sohn, J., Choi, S., Lewis, R., & Knaap, G. (2012). Characterising urban sprawl on a local scale with accessibility measures. *The Geographical Journal*, *178*(3), 230–241. DOI: 10.1111/j.1475-4959.2012.00468.x

Soule, D. C., & Peirce, N. R. (2006). *Urban sprawl: A comprehensive reference guide*. Greenwood Press.

Suditu, B. (2012). Urban sprawl—the legal context and territorial practices in Romania. *Human Geographies – Journal of Studies and Research in Human Geography, 6*(1), 73–77.

United Nations, Department of Economic and Social Affairs, Population Division. (2019). *World urbanization prospects: The 2018 revision* (ST/ESA/SER.A/420). United Nations. https://doi.org/DOI: 10.18356/b9e995fe-en

Van Eetvelde, V., & Antrop, M. (2004). Analyzing structural and functional changes of traditional landscapes—Two examples from southern France. *Landscape and Urban Planning*, *67*(1–4), 79–95. DOI: 10.1016/S0169-2046(03)00030-6

APPENDIX

Supplementary Tables

Table 9. Glossary of Class-Level Spatial Metrics

Metric	Full Name	Description
PD	Patch Density	Number of patches per 100 hectares — indicates fragmentation
LPI	Largest Patch Index	Percentage of total area occupied by the largest patch — indicates dominance
COHESION	Patch Cohesion Index	Measures physical connectedness of patches in the landscape
AI	Aggregation Index	Degree to which patches of the same class are clustered or dispersed

Source: Based on FRAGSTATS 4.2 documentation (McGarigal & Marks, 1995; McGarigal & Cushman, 2002).

Table 10. Class-Level Spatial Metrics for Each Reference Year (1981–2015)

Class	PD	LPI	COHESION	AI
Class-Level Metrics for 1981				
Port	0.1153	0.7012	94.0464	78.6925
Sparse Urban Fabric	9.6354	0.6016	84.9299	64.6674
Moderate Urban Fabric	9.8889	0.0996	66.0845	54.3193
Dense Urban Fabric	4.5872	14.3459	97.6488	85.0867
Commerce/Wholesale/Industrial	0.8298	0.7095	88.2562	79.9807
Sports Facilities	0.1153	0.2137	85.8168	87.5445
Public Building Infrastructure	0.9451	0.2033	78.6952	72.6839
Squares/Urban Green/Open Spaces	1.0604	1.3111	91.8685	79.9941
Protected Areas/Wetlands	0.5763	0.7801	89.0196	80.3876
Forest Areas	0.1844	4.4065	97.2896	93.4239
Permanent Crops	1.2909	2.7115	94.2795	85.7088
Olive Cultivation	1.1526	10.0846	97.3514	90.3840
Barren Land	0.0692	0.9771	94.6660	90.0858
No Use	0.3919	0.9626	92.3040	83.6808
Class-Level Metrics for 1996				
Port	0.1153	0.6497	93.9086	77.7311
Sparse Urban Fabric	9.6354	0,6016	83.3222	63,2505
Moderate Urban Fabric	10.4245	0.4151	75.4492	57.3359

continued on following page

Table 10. Continued

Class	PD	LPI	COHESION	AI
Dense Urban Fabric	4.8663	19,0547	98.3752	85,5732
Commerce/Wholesale/Industrial	1.2915	0.8178	88.6234	77.3019
Sports Facilities	0.2768	0,2138	81.865	82,0669
Public Building Infrastructure	1.4530	0.2242	77.7506	71.4385
Squares/Urban Green/Open Spaces	1.6836	0.3321	82.8120	73.6264
Protected Areas/Wetlands	0.5535	0.7805	89.1802	80.418
Forest Areas	0.1845	4.4295	97.2961	93.4493
Permanent Crops	1.1070	2.6278	94.3916	85.9210
Olive Cultivation	0.6919	9.7806	97.5752	90.9307
Barren Land	0.1153	0.2179	88.5795	84.6304
No Use	0.2768	0.6642	92.1660	82.7473
Class-Level Metrics for 2008				
Port	0.1142	1.6541	95.5727	88.2244
Sparse Urban Fabric	10.8901	0.8568	83.5473	65.0511
Moderate Urban Fabric	11.5294	0.6719	83.2353	60.4792
Dense Urban Fabric	6.0729	20.3234	98.3098	84.4483
Commerce/Wholesale/Industrial	1.4611	1.2308	90.3840	79.2531
Sports Facilities	0.7077	0.3575	80.6235	77.5660
Public Building Infrastructure	1.8721	0.2281	76.7987	69.6136
Squares/Urban Green/Open Spaces	1.4611	0.3288	73.8541	67.0127
Protected Areas/Wetlands	0.5479	0.7726	89.1781	80.4180
Forest Areas	0.1598	4.4300	97.5191	92.9164
Permanent Crops	1.0274	1.2616	92.4091	82.3678
Olive Cultivation	0.9589	3.8526	96.4784	89.6520
Barren Land	0.1370	0.1993	87.7978	83.7743
No Use	0.2283	0.4130	90.9909	81.3366
Class-Level Metrics for 2015				
Port	0.1598	1.6705	95.4880	88.0365
Sparse Urban Fabric	12.2825	0.9452	83.1658	64.1813
Moderate Urban Fabric	11.6889	0.7356	83.9066	60.7315
Dense Urban Fabric	5.8445	20.3682	98.3448	84.3236
Commerce/Wholesale/Industrial	1.6894	1.6191	90.5689	80.4383
Sports Facilities	0.8675	0.3575	79.2088	75.6198
Public Building Infrastructure	2.0319	0.2281	76.1694	68.5714
Squares/Urban Green/Open Spaces	1.8721	0.3329	71.4154	63.8697
Protected Areas/Wetlands	0.5479	0.7726	89.1672	80.2795

continued on following page

Table 10. Continued

Class	PD	LPI	COHESION	AI
Forest Areas	0.1598	4.0806	97.3108	92.4455
Permanent Crops	1.0502	1.2513	92.0030	81.1657
Olive Cultivation	0.9132	3.2115	96.4297	88.0975
Barren Land	0.1370	0.1993	87.9852	84.6021
No Use	0.2055	0.3966	91.3892	79.3570

Source: Authors' calculations based on FRAGSTATS analysis

Table 11. Glossary of Class-Level Spatial Metrics

Metric	Full Name	Description
PD	Patch Density	Number of patches per 100 hectares — indicates fragmentation
LPI	Largest Patch Index	Percentage of total area occupied by the largest patch — indicates dominance
COHESION	Patch Cohesion Index	Measures physical connectedness of patches in the landscape
AI	Aggregation Index	Degree to which patches of the same class are clustered or dispersed

Source: Based on FRAGSTATS 4.2 documentation (McGarigal & Marks. 1995; McGarigal et al.. 2002).

Chapter 4
Smart Farms in Agriculture 4.0:
Cybersecurity Challenges and Solutions

Georgios Kavallieratos
https://orcid.org/0000-0003-1278-1943
University of Oslo, Norway

Sokratis Katsikas
https://orcid.org/0000-0003-2966-9683
Norwegian University of Science and Technology, Norway

ABSTRACT

The rapid advancement of digital technologies has made digital transformation increasingly critical across various sectors, including agriculture. This sector encompasses a wide range of activities, such as farming, food manufacturing, processing, storage, and is interconnected with essential cyber-physical systems like water and wastewater management, and chemical processes. The growing adoption of Information Technology (IT) and Operational Technology (OT) in agriculture has given rise to the concept of "Smart Farming." These integrated technologies enhance functionality by optimizing seed traits, improving soil treatment, and streamlining farm management and production quality. However, the integration of IT and OT inevitably broadens and diversifies cyber risk exposure, exacerbating existing vulnerabilities and introducing new threats. This chapter provides an exploration of Cyber-Physical Systems (CPS) in smart farms, delves into the cybersecurity challenges and threats they encounter, and outlines strategies for mitigating these risks effectively.

DOI: 10.4018/979-8-3693-8237-0.ch004

INTRODUCTION

Stouffer et al. in (Stouffer, et al., 2023) define Operational Technology (OT) as programmable systems and devices designed to interact with the physical environment. When integrated with Information Technology (IT), OT gives rise to Cyber-Physical Systems (CPSs). These smart systems consist of interconnected networks of physical and computational components, combining computation, communication, sensing, and actuation to perform time-sensitive functions. CPSs can operate with varying levels of interaction with their environment, including human engagement. They are widely employed across diverse applications, such as monitoring, surveillance, tracking, event detection, and process management, in sectors like smart cities, smart power grids, autonomous vehicles, and intelligent transportation systems (Cyber-Physical Systems Public Working Group, 2016).

The Cyber-Physical Systems (CPS) paradigm is set to revolutionize agriculture, driving improvements in crop yield, food quality, operational efficiency, profitability, sustainability, and safety. A smart farm exemplifies a CPS, where functions and processes are monitored and controlled via a computer and communication network using interconnected agents (Barreto, Amaral, & Pereira, 2017), (Navarro, Costa, & Pereira, 2020). Smart farming technologies address the growing global demand for food while ensuring food security (Gupta, Abdelsalam, S., & Mittal, 2020). Notable examples include the use of remote soil sensors in Chilean blueberry farming (Alvino & Marino, 2017) and the integration of AI applications in sustainable precision agriculture (Linaza, et al., 2021). Additionally, smart farming significantly enhances the food supply chain by leveraging big data analytics to generate valuable insights throughout the agricultural process. These insights enable real-time decision-making, transforming traditional agricultural business models (Walter, Finger, Huber, & Buchmann, 2017).

The integration of Information Technology (IT) and Operational Technology (OT), a cornerstone of the digital transformation in smart farms, inevitably brings an expanded and more diverse set of cyber risks. These risks not only amplify existing vulnerabilities but also introduce new threats. This shift stems from the contrast between traditional operations, which were designed without cybersecurity considerations, and modern IT-driven processes that are accessible and controlled via external-facing information systems. These interfaces often lack robust security measures. Consequently, the food and agriculture sector has become a significant target for cyberattacks, ranking as the seventh most targeted industry in the United States, following closely behind manufacturing and financial services (Ewing-Chow, 2024).

The narrative of the chapter in capturing the state of the art concerning cybersecurity in smart farming is based on discussing CPSs in smart farms as the factors of change, recent cybersecurity incidents and attacks against smart farms; analyzing the cybersecurity threats against the CPSs of a typical smart farm; and reviewing mitigation strategies for the accordant risks. By leveraging the aforementioned results, future research paths are discussed to deal with the analyzed factors of change in smart farms and to combine the existing approaches. The sources used are from the scientific literature, technical reports, and media reports. The chapter aims to provide a comprehensive understanding of the landscape of cybersecurity issues in smart farms; to raise cybersecurity awareness in agriculture; and to provide future directions towards protecting smart farming infrastructure.

Concretely, this chapter provides:

- A comprehensive review and discussion of the main CPSs in smart farms.
- An overview of recent cybersecurity incidents and cyber threats in smart farms.
- Risk mitigation strategies, as well as future directions and trends in smart farming cybersecurity that aim to protect smart farming infrastructures.

The remainder of the chapter is structured as follows: Section 2 discusses architectural models of smart farms, with a classification of the main CPSs that these architectures comprise. Section 3 discusses recent cybersecurity incidents and attacks against smart farms and provides a detailed analysis of the cybersecurity threats against the CPSs of a typical smart farm. Section 4 presents cybersecurity risk mitigation strategies. Finally, Section 5 summarizes the findings and concludes this chapter.

SMART FARMS CPSS

Architectural Overview

Numerous studies have explored and analyzed the architectures of Cyber-Physical Systems (CPSs) in smart farms, categorizing them based on technologies, functionalities, and operations (Mhapsekar, Umrani, Faizan, Ali, & Abraham, 2024), (An, et al., 2017), (Gaire, et al., 2013), (Sarkar, et al., 2024), (Rajasekaran & Anandamurugan, 2019), (Gupta, Abdelsalam, S., & Mittal, 2020), (Friha, Ferrag, Maglaras, & Shu, 2022). For instance, Mohamed et al. (Mohamed A, et al., 2022) propose a four-layer IoT-based architecture for smart farming, comprising the farm perception layer, sensors and actuators layer, communication network layer, and application layer. Similarly, Alahe et al. (Alahe, et al., 2024) suggest a four-layer architecture encompassing perception, network, edge, and application layers. Another notable approach is a three-layer architecture proposed by Ashibani and Mahmoud (Ashibani & Mahmoud, 2017), which consists of the perception layer, transmission layer, and application layer. This classification is based on the distinct functions performed by each layer (Figure 1).

Despite the diversity of perspectives regarding the number of layers, the three-layer architecture—perception, transmission, and application—emerges as the most suitable framework. It strikes an optimal balance between simplicity, functionality, and efficiency, making it well-aligned with the demands of most smart farming applications while avoiding unnecessary complexity. For this reason, the three-layer model is adopted as the foundation for this work.

The *application layer* is the most interactive component, tasked with processing data received from the transmission layer and generating commands for physical components such as sensors and actuators. By employing advanced decision-making algorithms, this layer analyzes information from the perception layer and issues precise controls. It relies on cloud computing and sophisticated data mining techniques to optimize the management of interconnected systems.

The *transmission layer* acts as the intermediary, enabling seamless interaction between the perception and application layers. Utilizing various communication technologies, including local area networks, internet protocols, and sensor-related systems (e.g., Bluetooth, 4G, 5G, UMTS, ZigBee), this layer ensures efficient data exchange and processing.

The *perception layer* focuses on collecting real-time data, interpreting the physical environment, and executing commands from the application layer. It incorporates a wide range of equipment, such as sensors, actuators, cameras, GPS devices, laser scanners, intelligent systems, and RFID tags with 2-D barcode labels and readers. This layer gathers diverse data types, including sound, light, mechanical, chemical, thermal, electrical, biological, and locational information. Sensors monitor variables such as temperature, acceleration, humidity, vibration, location, and air chemical composition, providing the foundation for informed decision-making in the system.

Several CPSs are employed in the different layers of the smart farm architecture. By leveraging the three-layer architecture, the following CPSs in smart farms have been identified and categorized as depicted in Figure 1.

Figure 1. CPSs in smart farms

Application Layer

The primary role of the application layer is to process data received from the transmission layer and generate commands for execution by physical devices, including sensors and actuators (Peng, et al., 2013). By employing sophisticated decision-making algorithms, this layer analyzes aggregated data to derive accurate decisions and control instructions, facilitating necessary corrective actions. Key processes and application domains supported by Cyber-Physical Systems (CPSs) within this layer include:

- **Crop Management (CM)** is revolutionized with AI recommendations tailored to crop choices and resource allocation.
- **Precision Farming (PF)** is another application area where CPS and AI can be leveraged to improve the results. It integrates computational intelligence into complex systems, such as precision agriculture, using cooperative sensor swarms and layered architectures.
- **Farm Management System (FMS)** is one of the applications that serves as the key element of farm optimization, integrating vital functions such as, in the case of dairy farms, herd management, milk production monitoring, and financial administration into a centralized software solution.
- **User** (farmer, scientist)

- **Environmental Information Gathering (EIG)**: the system measures the state of a complex, dynamic system comprising climate, soil, pasture, and animals. These data support prediction of the land's future state and improved management outcomes through closed loop control. Wireless sensor networks are applied to monitor the farmland environment for irrigation and fertigation scheduling. Additionally, they monitor climate conditions and key environmental parameters, such as the temperature, humidity and illumination to acquire real time data or information and make decisions.

Transmission Layer

It facilitates seamless data exchange and processing between the perception and application layers. This interaction and transmission of data are enabled through various networks, including local area networks, communication networks, and the Internet, leveraging a range of technologies such as Bluetooth, 4G, 5G, UMTS, Wi-Fi, Infrared, and ZigBee, depending on the requirements of the sensor devices. Key processes and application areas supported by CPSs in this layer include:

- A **Gateway** collects the signals received from all the sensors and the signal receiver. It is located on the farm and sends all signals to smart farm servers through a high-speed broadband network.
- A **Smart Farm Server (SFS)** is designed to manage, process, and analyze data collected from various devices and sensors deployed on a smart farm and received from the gateway.
- A **Signal receiver** is responsible for receiving signals from various sensors, devices, and controllers deployed in the field, enabling data collection, transmission through the gateway, and subsequent processing by the SFS.

Perception Layer

Devices at this layer have the ability to collect real-time data that is needed for different purposes (e.g. monitoring and tracking), interpret what they receive from the physical world and perform commands from the application layer. Some of the processes and/or application areas that are supported by CPSs in this layer are:

- **Irrigation Management (IM)** is the systematic planning, scheduling, and control of water application to crops to optimize plant growth, maximize yields, conserve water, and reduce environmental impacts. It is enhanced through AI-managed irrigation systems, integrating real-time data for efficient resource utilization.

- **Soil Management (SM)** utilizes techniques like management-oriented modeling and artificial neural networks for predictive performance on soil parameters.
- **Disease Management (DM)** benefits from AI by enabling early detection through image analysis of crops and animals.
- **Crop Yield Prediction (CYP)** refers to the quantity of crops harvested per unit area of land. It is a critical indicator of agricultural productivity and a key factor in assessing the success of farming practices, crop variety performance, and the effectiveness of resource inputs like water, fertilizers, and pesticides.
- **Crop Health Monitoring (CHM)** involving the continuous tracking and analysis of plant health to optimize crop yield, reduce costs, and minimize environmental impacts.
- **Pesticide Management (PM)** is the strategic use, handling, and application of chemical and biological substances designed to control pests, diseases, and weeds in agricultural settings. AI aids in the accurate monitoring and identification of infected plants through techniques such as digital image analysis and sensor ML.
- **Fertilizer Management (FM)** is the strategic approach to applying nutrients in a way that optimizes crop growth, improves soil health, reduces environmental impact, and maximizes agricultural productivity.
- **Soil Nutrient Measurement (SNM)** is the process of analyzing soil to determine its essential nutrient content, which is vital for effective crop growth and optimal agricultural practices. By understanding soil nutrient levels, farmers can make informed decisions about fertilizer application, crop selection, and soil management to enhance crop yield, reduce environmental impact, and maintain soil health over time.
- **Livestock Monitoring (LM)** is the continuous tracking and analysis of animal health, behavior, and productivity to optimize livestock management, improve animal welfare, and enhance farm profitability. This process employs a combination of wearable devices, sensors, data analytics, and remote monitoring systems to provide real-time insights into livestock conditions, enabling proactive management decisions.
- **Sensor systems**
- Soil sensor: soil sensor node contains sensors measuring ground temperature, soil temperature, volumetric water content (VWC) and electric conductivity (EC).
- Weather sensor: contains sensors measuring air temperature, photo-synthetically active radiation (PAR), pressure, wind, rain and hail measurements.

- **Cattle Tags (CT)** send radio signals to base stations. Based on the time lapsed to receive the signal at three base stations, the locations of cattle are determined.
- **UAV-based CPSs**: remote sensing systems based on an easily transportable helicopter platform that is capable of acquiring multispectral images at desired locations and times. The ground and aerial measurements are used for estimating nitrogen levels on-demand across a farm and guide fertilizer application with these estimates in turn.
- **GIS-based CPSs**: geographic information systems (GISs) are mainly used to store, retrieve and transform spatial information related to productivity and agronomy. This information can be derived from a number of data sources including digital maps, digitized maps and photographs, soil and crop surveys, sensor data with positioning information, point analytical data and/or yield maps. GIS has also been adopted to estimate average soil loss that would generally result from splash, sheet, and rill erosion from agricultural plots.

An Example: Dairy Farm CPSs

The architectural overview presented in Section 2.1 applies to a wide variety of smart farms. However, not all smart farms will follow the full architectural model. For example, an analysis of the CPSs employed in dairy farms is presented in (Vatn, Kavallieratos, & Katsikas, 2024). The functions, data flows, and interdependencies between CPSs were analyzed and a model diagram was proposed. The dependencies on other internal systems were also provided. The CPSs of a typical Dairy farm are:

- **Farm Management System (FMS)**: The FMS in a dairy farm facilitates the management and monitoring of functions and operations such as herd management, milk production, animal health, and nutrition. The FMS is highly interconnected with other CPSs on a dairy farm.
- **Automatic Milking System (AMS)**: The AMS is a robotic system designed to milk cows efficiently, optimizing the milking process, reducing labor, and enhancing productivity and animal welfare. It integrates with other systems for smooth operation.
- **Herd Management System (HMS)**: The HMS collects data from sensors attached to cows, such as eating habits, lying time, stand-up counts, step counts, and temperature. These sensors, placed on the cow's neck or leg, also serve as electronic ID tags, aiding in interactions with systems like segregation gates or the AMS. The data supports health monitoring, including detecting abnormal walking patterns or changes in rumination.

- **Automatic Cleaning System (ACS)**: The ACS is a self-driving robot that cleans cow living spaces by removing manure.
- **Automatic Feed-Pushing System (AFPS)**: The AFPS is a robot that pushes feed closer to cows in the feed alley for easier access.
- **Automatic Feeding System (AFS)**: The AFS delivers feed efficiently and precisely, enhancing productivity and resource use.
- **Automatic Lighting System (ALS):** The ALS adjusts barn lighting based on a schedule or input from light sensors.
- **Environmental Ventilation System (EVS):** The EVS regulates barn temperature through ventilation.
- **Segregation System (SeSy):** The SeSy controls cow movement through farm gates.
- **Video Surveillance System (VSS):** The VSS monitors the dairy farm.

Figure 2 depicts the dairy farm's CPSs mapped onto the three-layer architecture.

Figure 2. CPSs in dairy farms

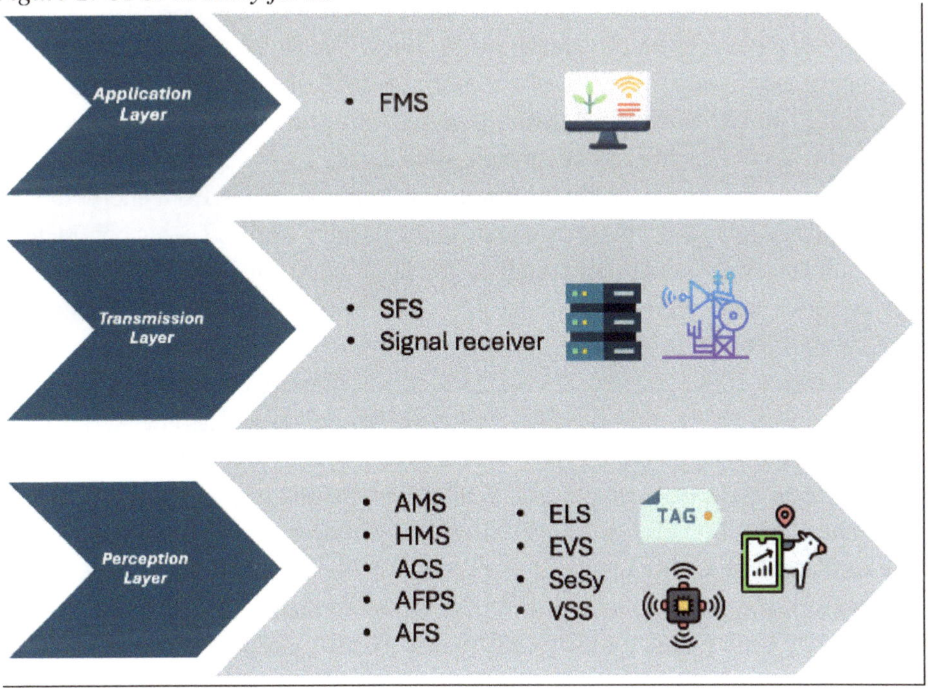

CYBERSECURITY OF SMART FARMS

Cybersecurity Incidents

Cybersecurity incidents in the agriculture sector were reviewed and analyzed in (Kulkarni, et al., 2024). This survey analyzed thirty cybersecurity incidents and categorized them according to the year, type of incident, country of the victim, attack cause, financial loss, data loss, and service interruption. The majority of the incidents analyzed were related to data breach, malware, DDoS, ransomware, and software bugs. In the sequel we discuss incidents not included in (Kulkarni, et al., 2024). Note that not all of these incidents are specific to smart farms; they rather pertain to the agriculture sector at large. Nevertheless, they are useful in identifying trends and methods of attack that are relevant to the special case of smart farms.

In 2017, an agricultural equipment manufacturer suffered a data breach that compromised the sensitive information of more than 1,000 customers, including their names, addresses, phone numbers, and license plate numbers (Snode Technologies, 2023). In 2020 malware attack occurred in wool sales in Australia and New Zealand. The industry used Talman Software, a widely used tool across Australia and New Zealand, was hit by a ransomware attack. The attack impact was to disrupt the buying and trading system (Becker, 2020). In 2020, the non-profit food provider Loaves & Fishes revealed that sensitive customer information had been exfiltrated as part of the larger Blackbaud data breach. Blackbaud, a software and cloud hosting solutions provider, successfully thwarted a ransomware attack from encrypting its files but ultimately paid a ransom to prevent the hackers from exposing confidential client information (Snode Technologies, 2023). In 2020, the Home Chef, a startup that provides food ingredients, meal kits, and recipes to customers, announced the leak of information about an undisclosed number of its customers (Whittaker, 2020). In May 2020, Harvest Sherwood food distributors company was targeted by ransomware attack, and the attackers stole critical data. The attackers gained access to 2,600 files including information such as cash-flow analysis, distributor data, business insurance content, and vendor information (Cyware Labs, 2020). In November 2020, Fonterra, a New Zealand dairy cooperative, fell victim to a ransomware attack that compromised its personal files, though no ransom was paid. In July 2021, the New Zealand government disclosed that the country's agriculture industry had been targeted by cybercriminals (Herald, 2020). In December 2020, several Australian winemakers reported being targeted by a ransomware attack that disrupted their operations and resulted in data loss (Snode Technologies, 2023).

A ransomware attack on an Iowa farming co-op was launched in 2021. The attack targeted coorporative data and disrupted the computer networks used to keep food supply chains and feeding schedules on track for millions of chickens, hogs and cattle (Bogage & Reiley, 2021). A cyber-attack targeted the Minnesota-based farm supply and grain marketing cooperative Crystal Valley in 2021. The ransomware attack targeted computer systems and disrupted the daily operations (Greig, 2021). In 2021, cybersecurity researchers discovered a sprawling cyber-attack against six U.S. states. The attackers initially gained access to state networks through USA Herds, an app that helps state governments track livestock diseases (Greenberg, 2022). In March 2021, JFC International, a leading distributor and wholesaler of Asian food products, fell victim to a ransomware attack that disrupted several of its systems. The company promptly informed law enforcement, employees, and business partners about the incident and was able to restore normal operations shortly thereafter (Paganini, 2021).

In 2022, HP Hood Dairy, was hit by a cyber-attack disrupting all the plants of the company. Specifically, 13 dairy plants around the country went offline, impacting the delivery time and the production process (Gardizy, 2022). A ransomware attack hit U.S. agricultural equipment maker AGCO in 2022. The attack disrupted the operations and delayed tractor sales at a critical point in the planting season (Warminsky, 2022). Adversaries lay the groundwork to attack an agricultural producer in Ukraine. In April 2022, Microsoft reported that an advanced persistent threat reportedly placed a file encryptor on the network of a Ukrainian agricultural producer during the early stages of the war. Though not yet activated, this file encryptor means that this agricultural company has been targeted (United States Agency for International Development, 2023). In February 2022, a feed milling company reported two incidents involving unauthorized access to its systems by a malicious actor, who may have attempted to initiate a ransomware attack. The attempts were detected and thwarted before any encryption could take place (Snode Technologies, 2023). In March 2022, a multi-state grain company faced a Lockbit 2.0 ransomware attack. The company, which plays a vital role in the spring planting season by offering grain processing, seed, fertilizer, and logistics services, was targeted during a critical time (Snode Technologies, 2023).

Dole systems were hit by a ransomware attack in 2023. The attack had a limited overall impact on its operations, with the main disruption occurring on its fresh vegetables and Chilean business (Jones, 2023). Mondelez suffered a data breach after an attack at third-party law firm in 2023. Personal information from past and present employees were affected (Cluley, 2023).

Figure 3 presents a timeline of cyber-attacks targeting CPSs in agriculture, as listed above.

Figure 3. Cyber Attacks against CPSs in agriculture

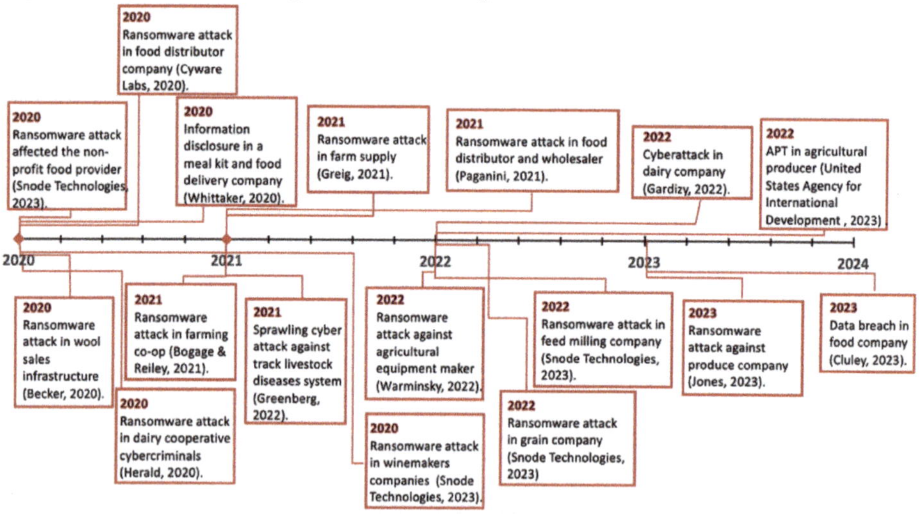

Cybersecurity Threats in Smart Farms

A systematic literature review for cybersecurity aspects in agriculture is presented in (Campoverde-Molina & Luján-Mora, 2024) where general threat descriptions for agriculture are also discussed. Cyber threat intelligence approaches are analyzed in (Bui, et al., 2024) and several cyber threats against smart farm infrastructure are discussed based on four-layers architecture. The major security threats are analyzed in (Alahe, et al., 2024) following a 4 layers smart farm architecture, including perception layer, network layer, edge layer, and application layer and the confidentiality, integrity, and availability cybersecurity objectives. The cybersecurity threats in agriculture 4.0 and 5.0 are discussed in (Maraveas, Rajarajan, Arvanitis, & Vatsanidou, 2024) where attacks are classified based on the attack's target. Several categories of cyber threats against smart farms are discussed in (Barreto & Amaral, 2018) focusing on privacy, social engineering, ransomware, denial of service, Cyber-Espionage, Agroterrorism, Confidential Information and Intellectual Property, and supply chain.

As smart farms increasingly adopt CPSs to enhance productivity and sustainability, they also become more susceptible to several cybersecurity threats. The STRIDE methodology—encompassing Spoofing, Tampering, Repudiation, Information Disclosure, Denial of Service, and Elevation of Privilege—provides a systematic framework for analyzing these threats. By leveraging the STRIDE classification, this section delves into the threats against a typical smart farm's CPS – as these have been described in Section 3 –, examining how attackers might exploit them and the potential consequences. This section aims to provide a comprehensive mapping between the existing threats, the security objectives that these threats violate, and the CPSs that are affected by these threats. This analysis underscores the critical need for comprehensive threat mitigation strategies to protect the integrity, confidentiality, availability, authenticity, possession and control, utility, and non-repudiation of smart farming CPSs.

The STRIDE threats are described in detail by Adam Shostack in (Shostack, 2014). *Spoofing* refers to an adversary's ability to impersonate someone or something else. *Tampering* involves altering or disrupting a system's disk, network, or memory. *Repudiation* is a threat where an individual denies performing an action that impacts the system's operation or claims they are not responsible for the outcomes of their actions. *Information Disclosure* occurs when confidential information is exposed to unauthorized individuals. *Denial of Service (DoS)* disrupts system availability by exhausting the resources needed for proper operation. Finally, *Elevation of Privilege* enables an adversary to perform unauthorized actions, gaining access to privileges beyond their intended level. Each threat category in the STRIDE model corresponds to a compromise of one of the security objectives in the CIAAAN (Confidentially, Integrity, Availability, Authentication, Authorization, Non-Repudiation) model of security objectives. However, when considering security objectives for CPSs, those known as the *Parkerian Hexad* (Parker, 2002) have been argued to provide better insight, particularly when it comes to analyzing safety along with security (Kavallieratos, Katsikas, & Gkioulos, 2020). These objectives are: confidentiality, integrity, availability, authenticity, possession and control, utility, and non-repudiation. *Confidentiality* ensures that information exchanged and communication links between CPSs and their services are protected from unauthorized access. *ntegrity* safeguards information, services, CPSs, and communication links from unauthorized modifications or tampering. *Availability* guarantees that authorized entities can access information, services, CPSs, and communication links when needed. *Authenticity* ensures that only authorized entities manage, configure, and operate smart farms CPSs and their services. *Possession* and *Control* protect against unauthorized possession or control of confidential data and communication links. *Utility* emphasizes the practical usability of information and communication links, while *Non-Repudiation* ensures that CPSs cannot deny their actions or responsibilities. Together, these objectives aim to ensure the security of the CPSs in smart farms. Table 1 provides an overview of cybersecurity threats in smart farms by classifying these into different STRIDE threat categories, the corresponding security objectives, and the CPSs that are affected. Note that, even though the list of threats is quite extensive, it has been compiled based on the available literature. This means that not all threats in Table 1 will be relevant to all smart farm situations and it is possible that threats not included in Table 1 will be relevant to specific smart farm situations. The standard approach to analyze threats in any specific implementation scenario is through a thorough, systematic risk assessment process, that includes a sub-process for identifying threats.

Table 1. Threats in CPSs of smart farm

Threat	Description	Security objective	CPS	Layer
Spoofing	Data interception and manipulation (Kim, Lim, & Cho, 2018), (Alwateer, Loke, & Zuchowicz, 2019).	Integrity, Availability, Possession and Control	UAV	Perception Layer
	Masquerade (Koduru & Koduru, 2022).	Authenticity, Availability, Confidentiality, Utility	CM, FMS	Application Layer
	Unauthorized access to the FMS. (Vatn, Kavallieratos, & Katsikas, 2024).	Authenticity, Confidentiality, Possession and Control	FMS	Application Layer
	ID tags spoofing (Vatn, Kavallieratos, & Katsikas, 2024).	Availability, Possession and Control, Integrity	AMS	Perception Layer
	Spoof cattle identity (Vatn, Kavallieratos, & Katsikas, 2024).	Integrity, Possession and Control	HMS	Perception Layer
	Spoof the identity of the ACS (Vatn, Kavallieratos, & Katsikas, 2024).	Integrity, Possession and Control	ACS	Perception Layer
	Spoof a legitimate connection from the FMS and change the settings or routes of the AFPS (Vatn, Kavallieratos, & Katsikas, 2024).	Integrity, Possession and Control	AFPS	Perception Layer
	Impersonate the FMS (Vatn, Kavallieratos, & Katsikas, 2024).	Authenticity, Integrity, Possession and Control	AFS	Perception Layer
	Impersonate an authorized user to manipulate the automatic lighting system (Vatn, Kavallieratos, & Katsikas, 2024).	Authenticity, Integrity, Possession and Control	ALS	Perception Layer
	Spoof the Bluetooth connection to the system, impersonating a legitimate user (Vatn, Kavallieratos, & Katsikas, 2024).	Authenticity, Integrity	EVS	Perception Layer
	Clone an RFID (Vatn, Kavallieratos, & Katsikas, 2024).	Utility, Possession and Control	SeSy	Perception Layer

continued on following page

Table 1. Continued

Threat	Description	Security objective	CPS	Layer
	Replace the live video feeds with recorded or altered footage (Vatn, Kavallieratos, & Katsikas, 2024).	Integrity, Possession and Control	VSS	Perception Layer
Tampering	Inject malware payloads into the UAVs memory or the ground station itself (Kim, Wampler, Goppert, Hwang, & Aldridge, 2012).	Integrity, Availability, Possession and Control	UAV	Perception Layer
	Attack on the integrity of the data (Rajasekaran & Anandamurugan, 2019).	Integrity	All	All
	Alter the data of the IoT-based platforms (Ferrag, Shu, Yang, Derhab, & Maglaras, 2020), (Gupta, Abdelsalam, S., & Mittal, 2020), (Demestichas, Peppes, & Alexakis, 2020).	Integrity, Availability, Utility	Sensors systems, Signal receivers	Perception Layer
	Attack on the integrity of the Agri-bio data (Friha, Ferrag, Shu, Maglaras, & Wang, 2021)	Integrity, Confidentiality	IM, SM, DM, CYP, CHM, PM, FM, SNM	Perception Layer
	Tampering the data to harm agriculture organisms (Drape, et al., 2021).	Integrity, Availability, Utility	PM, FM, SNM	
	Introducing false information about soil moisture levels (Gupta, Abdelsalam, S., & Mittal, 2020).	Integrity, Possession and Control, Availability, Utility	SNM	Perception Layer
	Disseminate false information about a smart farm, falsely reporting a disease outbreak in crops or livestock (Gupta, Abdelsalam, S., & Mittal, 2020).	Integrity, Availability, Utility	LM, DM	Perception Layer
	Inject false data, potentially compromising various compliance certification processes (Gupta, Abdelsalam, S., & Mittal, 2020).	Integrity, Utility	FMS	Application Layer
	Alter data stored in the FMS (Vatn, Kavallieratos, & Katsikas, 2024).	Integrity, Availability, Utility	FMS	Application Layer
	Tampering with the milk quality control systems (Vatn, Kavallieratos, & Katsikas, 2024).	Integrity, Availability	AMS	Perception Layer

continued on following page

Table 1. Continued

Threat	Description	Security objective	CPS	Layer
	Alter data such as eating habits, temperature readings, and rumination activity (Vatn, Kavallieratos, & Katsikas, 2024).	Integrity, Possession and Control, Availability, Utility	HMS	Perception Layer
	Intercept or alter the data exchanged between the ACS, FMS, and HMS (Vatn, Kavallieratos, & Katsikas, 2024).	Integrity, Availability	ACS	Perception Layer
	Inject malicious data into the sensor network or location module of the AFPS (Vatn, Kavallieratos, & Katsikas, 2024).	Integrity, Availability, Possession and Control	AFPS	Perception Layer
	Manipulate the feeding intervals set in the AFS (Vatn, Kavallieratos, & Katsikas, 2024).	Integrity, Availability, Utility	AFS	Perception Layer
	Tamper - physically or remotely -with the light sensor inputs (Vatn, Kavallieratos, & Katsikas, 2024).	Integrity, Utility	ALS	Perception Layer
	Altering the data sent to the EVS from the app (Vatn, Kavallieratos, & Katsikas, 2024).	Integrity	EVS	Perception Layer
	Manipulate the gate control system to disable it or cause it to open/close unexpectedly (Vatn, Kavallieratos, & Katsikas, 2024).	Integrity, Possession and Control, Availability, Utility	SeSy	Perception Layer
	Manipulate the video recordings stored on the recorder or PC (Vatn, Kavallieratos, & Katsikas, 2024).	Integrity, Utility	VSS	Perception Layer
Repudiation	Alter data in the FMS, such as milk production figures or animal health records, and deny their actions (Vatn, Kavallieratos, & Katsikas, 2024).	Non-repudiation	FMS	Application Layer
	Deny making harmful changes in the AMS, such as altering milk schedules or medication doses (Vatn, Kavallieratos, & Katsikas, 2024).	Non-repudiation	AMS	Perception Layer
	Deny making harmful changes in the HMS (Vatn, Kavallieratos, & Katsikas, 2024).	Non-repudiation	HMS	Perception Layer

continued on following page

Table 1. Continued

Threat	Description	Security objective	CPS	Layer
	Deny making unauthorized changes to sensor data, such as feeding or rumination patterns (Vatn, Kavallieratos, & Katsikas, 2024).	Non-repudiation	ACS	Perception Layer
	Unauthorized commands to the AFPS without leaving any trace (Vatn, Kavallieratos, & Katsikas, 2024).	Non-repudiation	AFS	Perception Layer
	Inability to trace actions in the ALS to a specific user (Vatn, Kavallieratos, & Katsikas, 2024).	Non-repudiation	ALS	Perception Layer
	Deny making specific EVS adjustments via the mobile app (Vatn, Kavallieratos, & Katsikas, 2024).	Non-repudiation	EVS, VSS	Perception Layer
Information leakage	Leak/sell smart farm's data (Gupta, Abdelsalam, S., & Mittal, 2020).	Confidentiality, Availability	FMS	Application Layer
	Cloud data leakage in smart farming data (Gupta, Abdelsalam, S., & Mittal, 2020).	Confidentiality, Availability	FMS	Application Layer
	Leak of sensitive and confidential data (Gupta, Abdelsalam, S., & Mittal, 2020).	Confidentiality, Availability	All	All
	Leak confidential information transmitted between smart farm's CPSs (Yazdinejad, et al., 2021).	Confidentiality, Availability, Utility	SFS, Gateway	Transmission Layer
	Ransomware attack (Friha, Ferrag, Maglaras, & Shu, 2022).	Confidentiality, Availability	FMS	Perception Layer
	Disclosure of soil, crop, and agricultural purchase data (Gupta, Abdelsalam, S., & Mittal, 2020).	Confidentiality, Availability, Utility	SM, CYP, PM, FM, SNM	Perception Layer
	Disclosure of information about agriculture anti-jamming devices (Gupta, Abdelsalam, S., & Mittal, 2020).	Confidentiality, Availability, Possession and Control, Utility	FMS	Application Layer
	Compromise third party systems and redirect data to external servers (Gupta, Abdelsalam, S., & Mittal, 2020).	Confidentiality, Availability	FMS/Supply chain	Application Layer

continued on following page

Table 1. Continued

Threat	Description	Security objective	CPS	Layer
	The leakage of confidential data and exploit against them in the commodity market (Gupta, Abdelsalam, S., & Mittal, 2020).	Confidentiality, Availability	FMS	Application Layer
	Leak cow health information (Vatn, Kavallieratos, & Katsikas, 2024).	Confidentiality	FMS	Application Layer
	Leak data on milk production, quality analysis, or health monitoring (Vatn, Kavallieratos, & Katsikas, 2024).	Confidentiality	AMS	Perception Layer
	Intercept data transmitted from sensors to receivers and from receivers to the FMS (Vatn, Kavallieratos, & Katsikas, 2024).	Confidentiality	HMS	Perception Layer
	Gain access to operational data, such as cow locations, cleaning schedules, or facility layout (Vatn, Kavallieratos, & Katsikas, 2024).	Confidentiality	ACS	Perception Layer
	Intercept sensors data collected by the AFPS (Vatn, Kavallieratos, & Katsikas, 2024).	Confidentiality, Utility	AFPS	Perception Layer
	Expose data such as feed types and feeding schedules (Vatn, Kavallieratos, & Katsikas, 2024).	Confidentiality, Availability	AFS	Perception Layer
	Gain unauthorized access to the ALS lighting schedules and configurations, revealing the farm's operating patterns (Vatn, Kavallieratos, & Katsikas, 2024).	Confidentiality, Availability	ALS	Perception Layer
	Intercepting communications and leak information about the farm's operating conditions (Vatn, Kavallieratos, & Katsikas, 2024).	Confidentiality	EVS	Perception Layer
	Gather gate usage statistics for malicious purposes (Vatn, Kavallieratos, & Katsikas, 2024).	Confidentiality	SeSy	Perception Layer
	Leak feeds or recordings, revealing sensitive information about the farm's operations or personnel (Vatn, Kavallieratos, & Katsikas, 2024).	Confidentiality	VSS	Perception Layer

continued on following page

Table 1. Continued

Threat	Description	Security objective	CPS	Layer
Denial of Service	De-authentication attack on a UAV (Kim, Wampler, Goppert, Hwang, & Aldridge, 2012).	Availability, Authenticity	UAV	Application Layer
	Sinkhole attack against of wireless sensor network of the smart farm (Rajasekaran & Anandamurugan, 2019).	Availability	Sensor systems	Perception Layer
	Jamming attack against wireless sensor network of the smart farm (Rajasekaran & Anandamurugan, 2019).	Availability	Sensors systems, Signal receivers	Perception Layer
	Brute-force, sinkhole, selective forwarding, and black hole attacks targeted the smart farms network (Friha, Ferrag, Maglaras, & Shu, 2022).	Availability	Sensor systems, Gateway, SFS	Perception Layer, Transmission Layer
	Flood agriculture apps with fake requests (Demestichas, Peppes, & Alexakis, 2020).	Availability	FMS	Application Layer
	Disrupt GNSS by deploying numerous distributed low-power jammers, interfering with GNSS signals over large areas (Gupta, Abdelsalam, S., & Mittal, 2020).	Availability	Sensor systems, GIS-based CPSs	Perception Layer
	Traffic overload by overwhelming the FMS (Vatn, Kavallieratos, & Katsikas, 2024).	Availability	FMS	Application Layer
	Disrupt communication between the AMS and the FMS (Vatn, Kavallieratos, & Katsikas, 2024).	Availability	AMS	Perception Layer
	Jam the wireless signals between the sensors and receivers, causing a denial of service (Vatn, Kavallieratos, & Katsikas, 2024).	Availability	HMS	Perception Layer
	Disable the robot by targeting key components, such as the power supply or communication systems, or by overwhelming the system with excessive or incorrect data (Vatn, Kavallieratos, & Katsikas, 2024).	Availability	ACS	Perception Layer
	Disable the AFPS by overwhelming it with requests or exploiting a system vulnerability (Vatn, Kavallieratos, & Katsikas, 2024).	Availability	AFPS	Perception Layer

continued on following page

Table 1. Continued

Threat	Description	Security objective	CPS	Layer
	Overloading the AFS with requests (Vatn, Kavallieratos, & Katsikas, 2024).	Availability	AFS	Perception Layer
	Overload the lighting system controllers with requests (Vatn, Kavallieratos, & Katsikas, 2024).	Availability	ALS	Perception Layer
	Block the weather data source (Vatn, Kavallieratos, & Katsikas, 2024).	Availability	EVS	Perception Layer
	Jam RFID signals near the gates (Vatn, Kavallieratos, & Katsikas, 2024).	Availability	SeSy	Perception Layer
	Overload the VSS (Vatn, Kavallieratos, & Katsikas, 2024).	Availability	VSS	Perception Layer
Elevation of Privileges	Masquerade attack and gain access to restricted information (Koduru & Koduru, 2022).	Authenticity, Possession and Control, Non-Repudiation	FMS	Application Layer
	Inject malware payloads into the UAVs memory or the ground station (Kim, Wampler, Goppert, Hwang, & Aldridge, 2012), (Koduru & Koduru, 2022).	Possession and Control	UAV	Application Layer
	Compromise edge gateways (Friha, Ferrag, Shu, Maglaras, & Wang, 2021).	Possession and Control	Gateway/SFS	Transmission Layer
	Access to private data by leveraging third party applications (Demestichas, Peppes, & Alexakis, 2020).	Possession and Control	FMS	Application Layer
	Privilege escalation within the FMS, and perform actions normally reserved for authorized users (Vatn, Kavallieratos, & Katsikas, 2024).	Possession and Control	FMS	Application Layer
	Unauthorized access and control over AMS functions (Vatn, Kavallieratos, & Katsikas, 2024).	Possession and Control	AMS	Perception Layer
	Elevate access privileges in HMS and gain access or modify the transmitted data (Vatn, Kavallieratos, & Katsikas, 2024).	Possession and Control	HMS	Perception Layer
	Unauthorized access in the ACS (Vatn, Kavallieratos, & Katsikas, 2024).	Possession and Control	ACS	Perception Layer

continued on following page

Table 1. Continued

Threat	Description	Security objective	CPS	Layer
	Gain control of the AFPS operation and alter the robot's behavior (Vatn, Kavallieratos, & Katsikas, 2024).	Possession and Control	AFPS	Perception Layer
	Control of the AFS robot (Vatn, Kavallieratos, & Katsikas, 2024).	Possession and Control	AFS	Perception Layer
	Gain administrative access to the ALS and disrupt the lighting schedules (Vatn, Kavallieratos, & Katsikas, 2024).	Possession and Control	ALS	Perception Layer
	Disable the ventilation (Vatn, Kavallieratos, & Katsikas, 2024).	Possession and Control	EVS	Perception Layer
	Override the gate system, allowing any cow to pass through (Vatn, Kavallieratos, & Katsikas, 2024).	Possession and Control	SeSy	Perception Layer
	Unauthorized access to VSS (Vatn, Kavallieratos, & Katsikas, 2024).	Possession and Control	VSS	Perception Layer

By examining the information in Table 1 from the perspective of threats, it becomes apparent that Information Disclosure and Tampering are the most prominent ones for smart farms. Denial of Service and Elevation of Privileges come next, followed by Spoofing and Non-Repudiation.

The threats that CPSs in smart farms face may violate several security objectives and influence different systems, as shown in Table 1. Considering the security objectives as described above, Availability and Confidentiality are among the most violated ones, followed by Integrity and Utility, Possession and Control. Last come Authenticity and Non-repudiation. Considering the identified CPSs in Section 2, the FMS is influenced is the most influenced system, followed by the UAV, PM, FM, SNM, Gateway, and SFS and then by Sensors, Signal, SM, DM, and CYP; last come CM, IM, CHM, LM, and GIS. In the dairy farm example, the FMS is considered the most critical CPS, as it is central to the model and directly connected to the internet. The AMS is essential in farm operations and its potential to harm the cows if compromised is high based on the attacks. The dependence between the AMS and FMS implies that a breach in the AMS could affect other interconnected systems. Four out of ten CPSs in a dairy farm are considered critical (Vatn, Kavallieratos, & Katsikas, 2024), while three CPSs are characterized as less critical.

MITIGATION TECHNIQUES

A number of sources propose sets of countermeasures (Friha, Ferrag, Maglaras, & Shu, 2022), (Koduru & Koduru, 2022), (Manninen, 2018), (Nikander, Manninen, & Laajalahti, 2020), (Maraveas, Rajarajan, Arvanitis, & Vatsanidou, 2024) (Alahe, et al., 2024) (Campoverde-Molina & Luján-Mora, 2024) to mitigate cybersecurity risks in the agriculture sector at large and in smart farms specifically. Such sources focus on general security controls and specific types of threats and systems. The NIST Guide to Industrial Control Systems (ICS) Security (Stouffer, et al., 2023) provides the ICS overlay, which is a partial tailoring of the controls and control baselines in (Joint Task Force, 2020) that adds supplementary guidance, specific to ICS. The list below presents classes of potential mitigation techniques for the analyzed threats that CPSs in the agriculture sector and in smart farms face (Friha, Ferrag, Maglaras, & Shu, 2022), (Koduru & Koduru, 2022), (Manninen, 2018), (Nikander, Manninen, & Laajalahti, 2020). Note that, as is the case with threats, this list presents only broad mitigation techniques; specific controls and/or configurations of controls should be determined following a risk assessment of the situation of the specific smart farm.

Firewalls

For smart farms, firewall rules can be configured to ensure that only specific network segments have access to the farm's systems. Additionally, firewalls can provide authorized users with a secure way to connect to the farm's network remotely and protect sensitive farm operations and data from unauthorized access and cyber threats. Moreover, Next-Generation Firewalls (NGFW) aim to enhance the protection of the networks by leveraging deep-packet inspection, application surveillance and control, built-in intrusion prevention, and cloud-based threat intelligence (Friha, Ferrag, Maglaras, & Shu, 2022).

Backup Planning

Maintaining regular backups of critical data and isolating these backups from the primary system is a vital mitigation strategy. This prevents ransomware from compromising both primary and backup systems. Additionally, the backup of the network connection is vital in case of network outages to ensure the integrity and authenticity of the data and systems' operations (Manninen, 2018), (Nikander, Manninen, & Laajalahti, 2020). For example, providing an automated 4G backup connection ensures the internet access in case of a connection fault, ensures keeping periodic offline backups of the important files, and provides a power backup (e.g., UPS) in case of a power outage (Manninen, 2018).

Secure Maintenance

A secure maintenance strategy for smart farms involves implementing regular, scheduled maintenance for computers, network hardware, and automation systems. Manufacturers often include maintenance services with automation systems as part of an annual service plan, ensuring the reliability of critical components. However, the remote maintenance of the systems should be also ensured and performed in a secure manner. In addition, farm operators should maintain their local network and cyber-physical environments to safeguard against vulnerabilities. Regular updates and patches for software and hardware are essential to mitigate risks from potential cyber threats. This proactive approach not only secures the farm's digital assets but also ensures uninterrupted farm operations (Manninen, 2018), (Nikander, Manninen, & Laajalahti, 2020).

Supply Chain Security

It is important to protect all the links in the chain (e.g., from seed procurement to final product delivery) against disruptions that can occur due to cyberattacks, fraud, and inefficiencies.

Network Segmentation

Separating operational and administrative networks can limit the spread of malware or intrusions, ensuring that a breach in one system does not cascade through the entire smart farm infrastructure (Manninen, 2018).

Network Monitoring

The network should be inspected (e.g., network packets, network traffic) to identify potential malicious activity (Friha, Ferrag, Maglaras, & Shu, 2022). Smart farms rely on a diverse range of interconnected devices, such as IoT sensors, automation systems, surveillance cameras, and data processing units, making effective network monitoring such as Real-time Traffic Analysis, Device Health Monitoring, and Incident Response critical to operational success.

Security Awareness and Education

An effective security awareness program for farmers, staff, and all agricultural roles facilitates the identification, analysis, and protection of the infrastructures from emerging cyber threats. By leveraging simulators and penetration testing tools, the cybersecurity training of agriculture personnel can be enhanced. Further, the analysis of existing standards and regulations, and the development of cybersecurity guidelines considering the agriculture needs could foster a cybersecurity training and awareness program (Friha, Ferrag, Maglaras, & Shu, 2022), (Yazdinejad, et al., 2021), (Manninen, 2018), (Nikander, Manninen, & Laajalahti, 2020). The availability of cybersecurity services is growing, as more companies enter the field. However, there is a pressing need to raise awareness of these services among primary producers. Educating farmers and fostering a stronger cybersecurity culture within the agriculture industry are essential steps to ensure the effective adoption of these solutions. By improving knowledge and understanding, the industry can better protect itself against evolving digital threats (Manninen, 2018).

Security by Design

The application and deployment of secure products within infrastructure that meet the cybersecurity requirements of the relevant standards (Manninen, 2018), (International Electrotechnical Commission, 2019), (International Organization for Standardization (ISO), 2016), (National Institute of Standards and Technology, 2024) and guidelines (Manninen, 2018), (Cooper, 2015) can significantly improve the cybersecurity posture of smart farms.

Periodic Evaluation

Regular and thorough security assessments are crucial for safeguarding smart agricultural enterprises. Engaging penetration testing experts or conducting red-black teaming exercises helps identify vulnerabilities and strengthen defenses. Additionally, the frequent updating of software, firmware, and hardware is vital to keeping pace with the rapid advancements and evolving threats in the IT landscape (Friha, Ferrag, Maglaras, & Shu, 2022).

Cybersecurity Management

A central element of any cybersecurity improvement program is the establishment of a cybersecurity management system, according to the specifications of standards such as e.g., the ISO 27001. For example, smart farms should maintain a list of computers and smart devices to identify potential vulnerabilities and ensure the adequate level of antivirus protection (Manninen, 2018). Another example is the adoption of cybersecurity policies to enhance the secure authentication of users (Friha, Ferrag, Maglaras, & Shu, 2022), (Demestichas, Peppes, & Alexakis, 2020). Further still, the design and establishment of disaster recovery plans is paramount in keeping the impact of a potentially successful attack within pre-specified acceptable limits infrastructure (Manninen, 2018), (Cooper, 2015).

CONCLUDING REMARKS

Through cybersecurity mitigation strategies, a robust cybersecurity culture within the agriculture sector should be fostered and a collaboration with other critical infrastructure sectors to share knowledge and strengthen collective defenses should be established. Given the findings of the analysis, future research should focus on addressing the key vulnerabilities identified in smart farm CPSs. Efforts should be directed towards developing robust mechanisms to mitigate Information Disclosure and Tampering, which account for the highest number of threat scenarios. Research into advanced encryption techniques, secure communication protocols, and tamper-resistant hardware designs will be critical. Similarly, enhancing defenses against Denial of Service (DoS) and Elevation of Privilege threats, which also give rise to significant risks, should be prioritized. Additionally, a deeper exploration of Repudiation threats, despite their absence in the identified scenarios, may uncover potential vulnerabilities that are not yet evident. Researchers should also examine the interdependencies among CPSs to develop holistic solutions that analyze the propagation of the impacts, particularly in systems like FMS, which is affected in the majority of the threat scenarios. Finally, research should explore methods to ensure compliance with the least-violated security objectives, such as Non-Repudiation and Authenticity, ensuring a balanced security posture across all objectives.

REFERENCES

Alahe, M., Wei, L., Chang, Y., Gummi, S., Kemeshi, J., Yang, X., Won, K., & Sher, M. (2024). Cyber security in smart agriculture: Threat types, current status, and future trends. *Computers and Electronics in Agriculture*, *226*, 109401. DOI: 10.1016/j.compag.2024.109401

Alvino, A., & Marino, S. (2017). Remote Sensing for Irrigation of Horticultural Crops. *Horticulturae*, *3*(2), 3020040. DOI: 10.3390/horticulturae3020040

Alwateer, M., Loke, S., & Zuchowicz, A. (2019). Drone services: Issues in drones for location-based services from human-drone interaction to information processing. *Journal of Location Based Services*, *13*(2), 94–127. DOI: 10.1080/17489725.2018.1564845

An, W., Wu, D., Ci, S., Luo, H., Adamchuk, V., & Xu, Z. (2017). Chapter 25 - Agriculture Cyber-Physical Systems. In H. Song, D. Rawat, S. Jeschke, & C. (. Brecher, *Cyber-Physical Systems: Foundations, Principles and Applications* (pp. 399-417). Academic Press.

Ashibani, Y., & Mahmoud, Q. (2017). Cyber physical systems security: Analysis, challenges and solutions. *Computers & Security*, *68*, 81–97. DOI: 10.1016/j.cose.2017.04.005

Barreto, L., & Amaral, A. (2018). Smart farming: Cyber security challenges. *2018 International Conference on Intelligent Systems (IS)* (pp. 870--876). IEEE. DOI: 10.1109/IS.2018.8710531

Barreto, L., Amaral, A., & Pereira, T. (2017). Industry 4.0 implications in logistics: An overview. *Procedia Manufacturing*, *13*, 1245–1252. DOI: 10.1016/j.promfg.2017.09.045

Becker, J. (2020, February 27). *ABC News*. Retrieved from Cyber attack forces cancellation of wool sales across Australia: https://www.abc.net.au/news/rural/2020-02-27/ransomware-cyber-attack-cripples-australian-wool-sales/12007912

Bogage, J., & Reiley, L. (2021, September 21). *The Washington Post*. Retrieved from Russian hackers target Iowa grain co-op in $5.9 million ransomware attack: https://www.washingtonpost.com/business/2021/09/21/new-cooperative-hack-ransomware/

Bui, H. T., Aboutorab, H., Mahboubi, A., Gao, Y., Sultan, N. H., Chauhan, A., Parvez, M. Z., Bewong, M., Islam, R., Islam, Z., Camtepe, S. A., Gauravaram, P., Singh, D., Ali Babar, M., & Yan, S. (2024). Agriculture 4.0 and beyond: Evaluating cyber threat intelligence sources and techniques in smart farming ecosystems. *Computers & Security*, *140*, 103754. DOI: 10.1016/j.cose.2024.103754

Campoverde-Molina, M., & Luján-Mora, S. (2024). Cybersecurity in smart agriculture: A systematic literature review. *Computers & Security*, ●●●, 104284.

Cluley, G. (2023, June 21). *Bitdefender*. Retrieved from Oreo maker Mondelez staff hit by data breach at third-party law firm: https://www.bitdefender.com/en-gb/blog/hotforsecurity/oreo-maker-mondelez-staff-hit-by-data-breach-at-third-party-law-firm

Cooper, C. (2015). Cybersecurity in food and agriculture. In Leclair, J. (Ed.), *Protecting our future* (Vol. 2). Hudson Whitman/ Excelsior College Press.

Cyber Physical Systems Public Working Group. (2016). *Framework for Cyber-Physical Systems Release 1.0*. National Institute of Standards and Technology.

Cyware Labs. (2020). *REvil Ransomware Targeted Sanitary Components Supplier | Cyware Hacker News*. Retrieved from Cyware Labs: https://social.cyware.com/news/revil-ransomware-targeted-sanitary-components-supplier-e2a41c63

Demestichas, K., Peppes, N., & Alexakis, T. (2020). Survey on Security Threats in Agricultural IoT and Smart Farming. *Sensors (Basel)*, *20*(22), 6458. DOI: 10.3390/s20226458 PMID: 33198160

Drape, T., Magerkorth, N., Sen, A., Simpson, J., Seibel, M., Murch, R., & Duncan, S. (2021). Assessing the Role of Cyber biosecurity in Agriculture: A Case study. *Frontiers in Bioengineering and Biotechnology*, ●●●, 742. PMID: 34490231

Ewing-Chow, D. (2024, September 20). *Forbes*. Retrieved from Agri-Food Sector Under Increasing Threat From Cyber Attacks: https://www.forbes.com/sites/daphneewingchow/2024/09/20/agri-food-sector-under-increasing-threat-from-cyber-attacks/

Ferrag, M., Shu, L., Yang, X., Derhab, A., & Maglaras, L. (2020). Security and Privacy for Green IoT-Based Agriculture: Review, Blockchain Solutions, and Challenges. *IEEE Access : Practical Innovations, Open Solutions*, *8*, 32031–32053. DOI: 10.1109/ACCESS.2020.2973178

Friha, O., Ferrag, M., Maglaras, L., & Shu, L. (2022). Digital Agriculture Security: Aspects, Threats, Mitigation Strategies, and Future Trends. *IEEE Internet of Things Magazine*, *5*(3), 82–90. DOI: 10.1109/IOTM.001.2100164

Friha, O., Ferrag, M., Shu, L., Maglaras, L., & Wang, X. (2021). Internet of Things for the Future of Smart Agriculture: A Comprehensive Survey of Emerging Technologies. *IEEE/CAA Journal of Automatica Sinica*, 718-752.

Gaire, R., Lefort, L., Compton, M., Falzon, G., Lamb, D., & Taylor, K. (2013). Demonstration: Semantic Web Enabled Smart Farm with GSN. *International Workshop on the Semantic Web*. Retrieved from Demonstration: Semantic Web Enabled Smart Farm with GSN.

Gardizy, A. (2022, March 18). *Boston Globe*. Retrieved from Hackers hit Hood. Dairy shut down milk production this week after the event: https://www.bostonglobe.com/2022/03/18/business/school-milk-could-be-short-supply-after-hood-plants-hit-by-cyber-event/

Greenberg, A. (2022, March 8). *WIRED*. Retrieved from Chinese Spies Hacked a Livestock App to Breach US State Networks: https://www.wired.com/story/china-apt41-hacking-usaherds-log4j/

Greig, J. (2021, September 22). *ZDNET*. Retrieved from Crystal Valley Cooperative becomes latest agriculture business hit with ransomware: https://www.zdnet.com/article/crystal-valley-cooperative-becomes-latest-agriculture-business-hit-with-ransomware/

Gupta, M., Abdelsalam, M. S. K., & Mittal, S. (2020). Security and Privacy in Smart Farming: Challenges and Opportunities. *IEEE Access : Practical Innovations, Open Solutions*, *8*, 34564–34584. DOI: 10.1109/ACCESS.2020.2975142

Herald, N. (2020). *Fonterra Targeted by Hackers*. Retrieved from NZ Herald: https://www.nzherald.co.nz/business/fonterra-targeted-by-hackers/6PEVEZEO5AI6O2W2RS7HJ6WMTM/

International Electrotechnical Commission. (2019). *Industrial-process measurement control and automation, Framework for functional safety and security IEC TR 63069:2019*. International Electrotechnical Commission.

International Organization for Standardization (ISO). (2016). *Information technology — Security techniques — Information security management systems, ISO 27000 series*. Geneva: International Organization for Standardization (ISO).

Joint Task Force. (2020). *Security and Privacy Controls for Information Systems and Organizations. NIST SP 800-53r5*. National Institute of Standards and Technology.

Jones, D. (2023, May 18). *CYBERSECURITY DIVE*. Retrieved from Dole incurs $10.5M in direct costs from February ransomware attack: https://www.cybersecuritydive.com/news/dole-10m-costs-ransomware/650711/

Kavallieratos, G., Katsikas, S., & Gkioulos, V. (2020). SafeSec Tropos: Joint security and safety requirements elicitation. *Computer Standards & Interfaces*, *70*, 103429. DOI: 10.1016/j.csi.2020.103429

Kim, A., Wampler, B., Goppert, J., Hwang, I., & Aldridge, H. (2012). *Cyber Attack Vulnerabilities Analysis for Unmanned Aerial Vehicles. Infotech@Aerospace*. American Institute of Aeronautics and Astronautics.

Kim, S., Lim, G., & Cho, J. (2018). Drone flight scheduling under uncertainty on battery duration and air temperature. *Computers & Industrial Engineering*, *117*, 291–302. DOI: 10.1016/j.cie.2018.02.005

Koduru, T., & Koduru, N. P. (2022). An Overview of Vulnerabilities in Smart Farming Systems. *Journal of Student Research*.

Kulkarni, A., Wang, Y., Gopinath, M., Sobien, D., Rahman, A., & Batarseh, F. (2024). A Review of Cybersecurity Incidents in the Food and Agriculture Sector. *arXiv:2403.08036*.

Linaza, M. T., Posada, J., Bund, J., Eisert, P., Quartulli, M., Döllner, J., . . . Lucat, L. .. (2021). Data-Driven Artificial Intelligence Applications for Sustainable Precision Agriculture. *Agronomy*, no. 1227.

Manninen, O. (2018). *Cybersecurity in Agricultural Communication Networks: Case Dairy Farms*. JAMK University of Applied Sciences.

Maraveas, C., Rajarajan, M., Arvanitis, K., & Vatsanidou, A. (2024). Cybersecurity threats and mitigation measures in agriculture 4.0 and 5.0. *Smart Agricultural Technology*, *9*, 100616. DOI: 10.1016/j.atech.2024.100616

Mhapsekar, R., Umrani, M., Faizan, M., Ali, O., & Abraham, L. (2024). Building Trust in AI-Driven Decision Making for Cyber-Physical Systems (CPS): A Comprehensive Review. *2024 IEEE 29th International Conference on Emerging Technologies and Factory Automation (ETFA)* (pp. 1-8). Padova, Italy: IEEE.

Mohamed, A., Jose, A., Luis, G., Zuniga, M., Pedraza, M., Carvajal, G., Jara, N., & Carvajal, R. (2022). LoRa based IoT platform for remote monitoring of large-scale agriculture farms in Chile. *Sensors (Basel)*, •••, 2824. PMID: 35458808

National Institute of Standards and Technology. (2024). *The NIST Cybersecurity Framework (CSF) 2.0*. National Institute of Standards and Technology.

Navarro, E., Costa, N., & Pereira, A. (2020). A Systematic Review of IoT Solutions for Smart Farming. *Sensors (Basel)*, *20*(15), 4231. DOI: 10.3390/s20154231 PMID: 32751366

Nikander, J., Manninen, O., & Laajalahti, M. (2020). Requirements for cybersecurity in agricultural communication networks. *Computers and Electronics in Agriculture, 179*, 105776. DOI: 10.1016/j.compag.2020.105776

Paganini, P. (2021). *Distributor of Asian Food JFC International Hit by Ransomware*. Retrieved from security affairs: https://securityaffairs.com/115150/malware/jfc-international-ransomware-attack.html

Parker, D. B. (2002). Toward a New Framework for Information Security. In Bosworth, S., & Kabay, M. E. (Eds.), *The Computer Security Handbook* (4th ed.). John Wiley & Sons.

Peng, Y., Lu, T., Liu, J., Gao, Y., Guo, X., & Xie, F. (2013). Cyber-physical System Risk Assessment. *2013 Ninth International Conference on Intelligent Information Hiding and Multimedia Signal Processing* (pp. 442-447). Beijing, China: IEEE.

Rajasekaran, T., & Anandamurugan, S. (2019). Challenges and Applications of Wireless Sensor Networks in Smart Farming—A Survey. *Advances in Big Data and Cloud Computing. Proceedings of ICBDCC18.* (pp. 353–361). Singapore: Springer.

Sarkar, S., Ganapathysubramanian, G., Singh, A., Fotouhi, F., Kar, S., Nagasubramanian, K., Chowdhary, G., Das, S. K., Kantor, G., Krishnamurthy, A., Merchant, N., & Singh, A. (2024). Cyber-agricultural systems for crop breeding and sustainable production. *Trends in Plant Science, 29*(2), 130–149. DOI: 10.1016/j.tplants.2023.08.001 PMID: 37648631

Shostack, A. (2014). *Threat Modeling: Designing for Security* (1st ed.). Wiley Publishing.

Snode Technologies. (2023). *Cyber Security in Smart Agriculture Technology*. Snode.

Stouffer, K., Pease, M., Tang, C., Zimmerman, T., Pillitteri, V., Lightman, S., & Thompson, M. (2023). *Guide to industrial control systems (ICS) security. NIST special publication 800.82r3*. National Institute of Standards and Technology.

United States Agency for International Development. (2023). *Cybersecurity AGRICULTURE AND FOOD SECURITY.* Retrieved from USAID: https://www.usaid.gov/sites/default/files/2023-10/Cybersecurity%20Briefer_Agriculture%20and%20Food%20Security.pdf

Vatn, K., Kavallieratos, G., & Katsikas, S. (2024). Lecture Notes in Computer Science: Vol. 14398. *Threat Analysis in Dairy Farming 4.0. Computer Security. ESORICS 2023 International Workshops. ESORICS 2023*. Springer.

Walter, A., Finger, R., Huber, R., & Buchmann, N. (2017). Smart farming is key to developing sustainable agriculture. *Proceedings of the National Academy of Sciences U.S.A.*, 6148-6150. DOI: 10.1073/pnas.1707462114

Warminsky, J. (2022, May 6). *The Record*. Retrieved from Agricultural equipment maker AGCO reports ransomware attack: https://therecord.media/agco-ransomware-attack-agricultural-sector

Whittaker, Z. (2020). *Home Chef Confirms 8 Million User Records Stolen in Breach*. Retrieved from techcrunch: https://techcrunch.com/2020/05/20/home-chef-data-breach/

Yazdinejad, A., Zolfaghari, B., Azmoodeh, A., Dehghantanha, A., Karimipour, H., Fraser, E., Green, A. G., Russell, C., & Duncan, E. (2021). A Review on Security of Smart Farming and Precision Agriculture: Security Aspects, Attacks, Threats and Countermeasures. *Applied Sciences (Basel, Switzerland)*, *11*(16), 7518. DOI: 10.3390/app11167518

Chapter 5
Investigation of Post-Flood Soil Pollution by Heavy Metals and Nitrogen Compounds Based on GIS in Palama of Thessaly, Greece

Stavroula Galanopoulou
https://orcid.org/0009-0005-6405-5953
Agricultural University of Athens, Greece

Chrysoula Georgouli
Agricultural University of Athens, Greece

Spyridon Kaloudis
https://orcid.org/0000-0002-5432-5236
Agricultural University of Athens, Greece

ABSTRACT

The study examines the pollution of surface soils in the Municipality of Palamas, Karditsa, Greece, following the severe flooding caused by storm "Daniel" in September 2023. The heavy rainfall inundated nearly 73,000 hectares, severely affecting agriculture and infrastructure. The research investigates soil contamination by heavy metals and nitrogenous compounds and examines the potential anthropogenic and natural (geogenic) sources. Soil samples were analyzed for various pollutants, including chromium (Cr), copper (Cu), manganese (Mn), nickel (Ni), lead (Pb) and

DOI: 10.4018/979-8-3693-8237-0.ch005

zinc (Zn). Geochemical pollution indices were applied to support the assessment of contamination levels and infer potential sources. Results showed that most heavy metals were within normal levels, except for Cr and Ni, which were elevated in some samples. Geographic Information Systems (GIS) were used to create pollution maps, highlighting areas with higher concentrations of pollutants. The study concludes that flooding and associated runoff combined with weathering of metal-rich rocks, enhanced the transport and deposition of heavy metals in the soil.

INTRODUCTION

Intensive agricultural practices significantly contribute to soil contamination, particularly with heavy metals, nitrates, and phosphates. The use of agrochemicals and contaminated irrigation water can introduce heavy metals like cadmium, lead, and zinc into the soil, thereby compromising soil health and food safety (Alloway, 2012; Liu et al., 2005). Excessive nitrogen fertilization leads to nitrate leaching, which pollutes both groundwater and surface water, adversely affecting water quality (Abbasi & Sepaskhah, 2023). Similarly, the overapplication of phosphate fertilizers results in phosphate accumulation in soils and contributes to eutrophication, ultimately degrading aquatic ecosystems (Liu et al., 2021; López-Arredondo et al., 2014).

Additionally, climate change, caused by increased greenhouse gas emissions and global warming, has led to more intense and prolonged rainfall events (Bellos & Sawidis, 2005). As a result, water overflow and flooding have become more frequent (Marrugo-Negrete et al., 2019; Weber et al., 2023). Flooding affects agricultural soil in multiple ways. It degrades soil structure by breaking down aggregates and dispersing clay particles (Suchara et al., 2021). Excess water reduces aeration as it fills soil pores, limiting oxygen availability for roots and microorganisms (Rangasami et al., 2022). Chemical properties also change, with pH often increasing in acidic soils and the redox potential decreasing, which disrupts nutrient availability and organic matter decomposition (Wang et al., 2022). Floodwaters transport contaminants such as metals and organic pollutants from polluted areas or sediments (Antić-Mladenović et al., 2019; Weber et al., 2023). Sediment transport plays a crucial role in this process. It spreads pollutants over large areas while depositing nutrient-rich or contaminated materials (Ponting et al., 2021). This can either enhance soil fertility or create risks for soil health and crop production, depending on the sediment's composition and source (Hilscherova et al., 2007; Pulkrabová et al., 2008).

In early September 2023, an extreme and unprecedented weather event struck the Thessaly region of Greece, bringing 700 mm of rainfall within 48 hours (Iatrou et al., 2024). This event, hereafter referred to as storm “Daniel”, led to widespread flooding: within 60 hours from the onset of rain, 72,951 ha of land were submerged

(Iatrou et al., 2024), (Figure 1). Given that the mean annual precipitation in the area is approximately 700 mm (Loukas, 2010), the event represents an unprecedented hydrological anomaly. The consequences were severe, including sediment deposition on cultivated lands, crop destruction, particularly in tomato, cotton, corn, and alfalfa fields, biodiversity disruption, pollution, soil compaction, and extensive infrastructure damage.

Figure 1. A satellite image shows the widespread flood damage caused by storm Daniel across Thessaly (Copernicus, 2023)

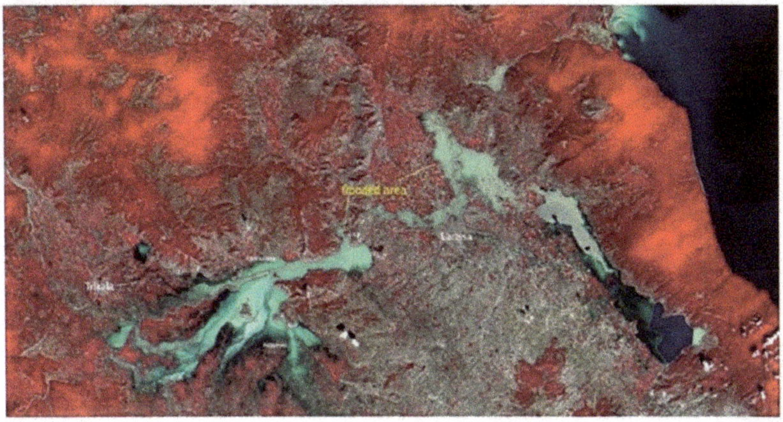

Note. Blue-green color: flooded areas, Red color: healthy vegetation, Brown-gray color: urban areas, non-vegetated land.

Environmental Issues in the Thessaly Plain

The Thessalian Plain faces severe soil degradation due to mechanized and intensive farming, rendering desertification a major concern (Kairis et al., 2022). However, land degradation is primarily driven by the limited availability of freshwater (Giakoumatos et al., 2024), while contamination from insecticides, pesticides, and over-application of fertilizers represents another significant environmental issue. Since plants can absorb only a limited amount of nitrogen, substantial quantities of nitrate ions leach into the environment. Carried by rainfall and irrigation, these pollutants often infiltrate groundwater and other surface water bodies, such as rivers and lakes, causing considerable water quality deterioration (Christakis, 2013). Consequently, groundwater degradation is an increasing concern, particularly in areas affected by intensive agriculture and high fertilizer and irrigation inputs. This pollution is more pronounced in vulnerable zones, where chemical compounds, notably nitrogen and phosphates from anthropogenic sources, pose a serious threat to water resources (Papaioannou et al., 2007).

The Thessalian Plain represents one of Greece's most significant agricultural regions, with substantial water requirements for both domestic use and irrigation. In this area, cotton, wheat, and maize dominate the lowlands, while the foothills of the eastern mountains are primarily occupied by fruit orchards, olive groves, and vineyards (Loukas, 2010). Nevertheless, studies on the Pinios River reveal concerning levels of water pollution (Giakoumatos et al., 2024; Sentas et al., 2018). Teou (2014) reports that this pollution largely originates from urban and industrial waste, as well as agricultural and livestock activities. Moreover, increasing irrigation demands, particularly in the summer, further reduce the river's flow intensifying pollution and water quality degradation (Sentas et al., 2018).

Cotton cultivation in Thessaly occupies approximately 30% of the region's arable land and is highly water-intensive, relying on frequent irrigation that places considerable demand on groundwater, primarily extracted from wells. This extensive water use contributes to the depletion of aquifers and alters the local hydrological balance. Additionally, the application of agrochemicals, including insecticides and herbicides poses a threat to biodiversity, affecting both flora and fauna (Konate et al., 2024). In contrast, wheat, which covers about 40% of the cultivated area, generally exerts a lower environmental impact, requiring less irrigation and fewer chemical inputs (Christakis, 2013). Other contributors to regional contamination include urban waste, livestock wastewater, and industrial discharges. The area hosts numerous industrial facilities, concentrated in zones, such as Larissa, Volos, and Karditsa, while smaller agro-processing units further contribute to the environmental load (Teou, 2014).

Environmental contamination in the Thessalian Plain is intensified by industries linked to agricultural production and by organic waste from dairy facilities near the Pinios River. Livestock farms and slaughterhouses-mainly pig and poultry, with some cattle-also contribute to the environmental load (Teou, 2014). Several wastewater treatment plants operate in the basin, many treating septic sludge and some industrial wastewater. Secondary treatment, including nitrogen removal, is applied, and the resulting sludge is either landfilled or reused as fertilizer in Karditsa and Larissa (Giakoumatos et al., 2024; Teou, 2014). Diffuse urban wastewater from settlements lacking infrastructure further contaminates water bodies, particularly in the Pinios and Almyros-Pelion basins (Teou, 2014). In addition, landfills in Larissa, Trikala, Volos, and Argalasti, as well as uncontrolled waste dumps, exacerbate environmental pressures (Teou, 2014).

This study aims to assess the impact of flooding from storm Daniel on surface soil pollution in areas experiencing the highest floodwater volumes. Agricultural soils from the municipality of Palamas, in the Thessalian Plain (Greece), were analyzed for their physicochemical properties and for concentrations of heavy metals and nitrogen compounds.

MATERIALS AND METHODS

Study Area

Thessaly, in central Greece, covers 14,036 km^2 (≈11% of the country) and borders Macedonia, Sterea Ellada, Epirus, and the Aegean Sea. The region combines mountainous and semi-mountainous areas with flat plains, including the fertile Thessalian Plain, crossed by the Pinios River, which has shaped the predominance of agricultural activities.

Administratively, Thessaly comprises four units: Larissa, Karditsa, Trikala, and Magnesia, the later including the Sporades islands (Figure 2), with Larissa as the regional capital. Key mountainous areas such as Mount Olympus, Mount Ossa, and Mount Pelion contribute to the region's ecological and cultural diversity, while Lake Tavropos represents a major infrastructural project.

The hydrographic network is dominated by the Pinios River fed by tributaries including the Enipeas, Sofaditis, Kalentzis, and Karabalis rivers, forming the ecologically valuable Pinios Delta.

The study area lies within the Municipality of Palamas, Karditsa (Figure 2) covering its northern and western parts, including the villages of Vlochos, Marathea, Metamorphosis, and Korda. The municipality spans 382.722 km^2, with 13,424 permanent residents (Hellenic Statistical Authority, 2023). The terrain is mainly flat, with scattered hills such as Asterio (313 m), Makryvouni (350 m), Lykorema (430 m), and Kastro (329 m), averaging 113 m in elevation.

Figure 2. The region of Thessaly and the sampling positions in the Palamas municipality area.

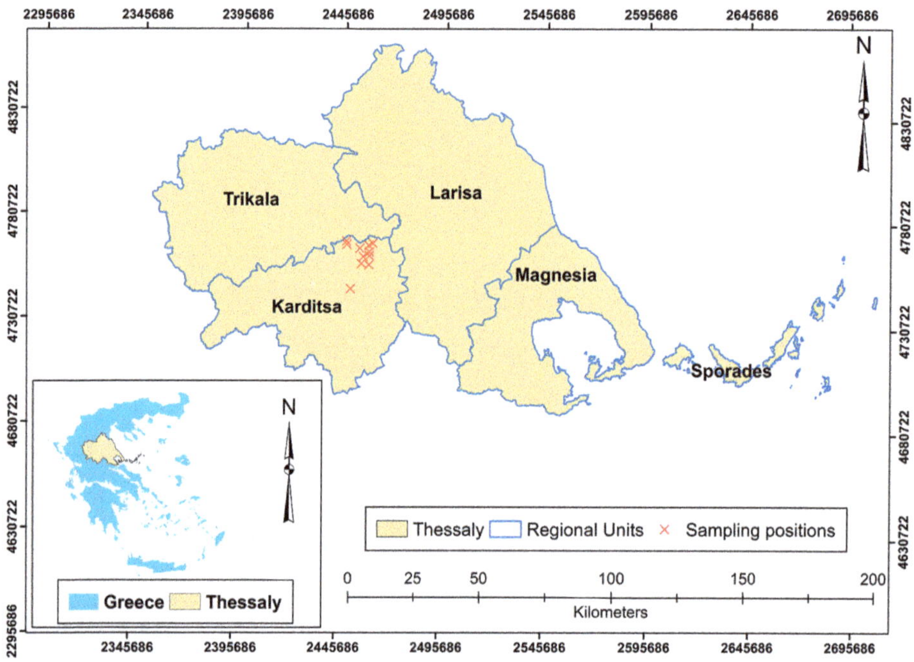

In the Karditsa regional unit, agricultural land accounts for 41.76% of the area, pastures cover 24.73%, and forests make up 23.79% (mostly found in the semi-mountainous areas). The agricultural land is generally highly productive, as it comes from alluvial deposits and drainage systems, and is also irrigated.

In the broader study area, there are small industrial and craft units, with the most notable being BIOPAL SA, a cotton ginning factory located in Palamas (Gkitsa, 2010).

GEOLOGICAL DATA

The broader study area is geologically divided into four tectonic zones: the Pelagonian, Sub-Pelagonian, Olonou-Pindos, and Gavrovo-Tripolis zones (Rozos et al., 1989). The Karditsa plain primarily consists of Quaternary alluvial deposits, bordered to the north by the Trikala plain, to the west by Mount Pindos, to the south by the massif of Leontario, which mainly consists of flysch, and to the east by the marble hills and the plain of Farsala (Tsinopoulos, 2007). More specifically, the region features the following formations (Figure 3) (Gkitsa, 2010):

Figure 3. Section of the geological map of Greece (Hellenic Survey for Geology and Mineral Exploration, 1983), corresponding to the study area. Colour processing was performed in a GIS environment by the authors.

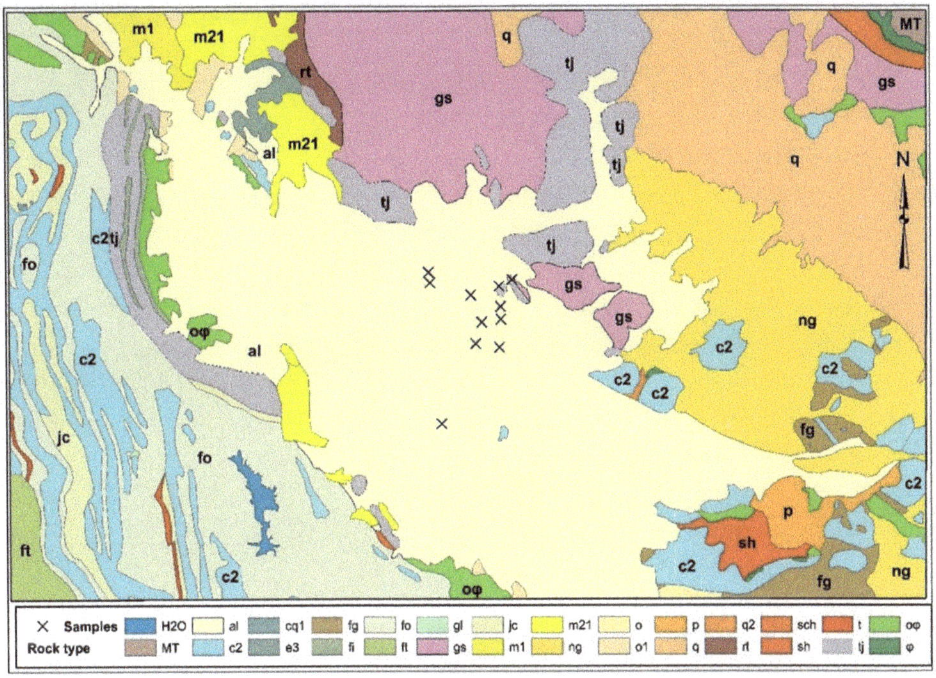

Note. **l:** *Alluvial deposits,* **c2:** *limestones,* **cq1:** *Lacustrine and continental deposits,* **e3:** *molasse (Crania series),* **fg:** *flysch,* **fi**: *flysch,* **fo:** *flysch,* **ft**: *flysch,* **gl:** *glacier drifts,* **gs:** *augen gneisses, gneisses, schists, amphibolites,* **H_2O:** *water,* **jc:** *first flysch,* **m1:** *molasse (Pentalofon series),* **m21:** *molasse (Tsotylion series),* **MT:** *marbles and dolomites,* **ng:** *lacustrine deposits (conglomerates, sandstones, marls, clays, sometimes with lignite beds),* **o:** *molassic formations,* **o1**: *molasse (Eptachorion series),* **p:** *lacustrine deposits (conglomerates, sand, clay, marls, marly limestones and often lignite beds),* **q:** *marine or continental formations undivided,* **q2:** *Scree,* **rt:** *limestones, greywackes, schists, prasinites, volcanites,* **sch:** *glaucophane schists,* **sh:** *same formation metamorphosed,* **t:** *siliceous schists and limstones with "Halobia",* **tj:** *limestones and dolomites,* **οφ:** *ophiolites,* **φ:** *flysch metamorphosed to phyllite.*

Ophiolitic Series: Ophiolites are extensively found, especially in the western part. These rocks, which are often associated with radiolarites, appear also south of Anavra. The dominant rocks are basic rocks such as peridotites, sometimes intersected with olivine-rich rocks. In the southern part of the complex, prevail diabases, dolerites, and locally, tuffs and breccias. These ophiolitic series, rich in chromium and nickel, are also known for containing chromite mineral deposits.

Radiolarites and Jurassic Limestones (Sedimentary Series): In the northwestern part of the region (e.g., Ellinokastro, Kryopigi, Amigdali, etc.) and the southwestern part, radiolarites alternate with Jurassic limestones (Koziax series).

Cretaceous Limestones (Upper Pindos Zone): Found from the northern part of Koziax to Smokovo, these limestones contain layers of quartz sandstones.

Cretaceous Limestones of the Cenomanian (Sub-Pelagonian Zone): These large thickness limestones appear in the area of Mataranga.

Flysch (Upper Cretaceous - Eocene): Located south and west of Karditsa, consisting of sandstones alternating with marly schists or pelites and less commonly with conglomerates.

Tertiary Formations: Found in the hills of Fanari, Kanali, Mitropolis, Anavra, and Leontario. These include marls and alternating marls and sandstones.

Alluvial Deposits (Pliocene - Quaternary Period): These widespread deposits in the Karditsa plain were formed in the Pliocene from the erosion of neighboring mountain masses. Near river mouths, coarser materials are found, while the rest of the plain is dominated by finer particles.

Metamorphic Rocks: At the northeastern edge of the Karditsa regional unit, metamorphic rocks appear, such as crystalline schists with limited surface exposure, and crystalline limestones – marbles.

These geological features reveal the complex nature of the region's terrain, which contributes to the area's agricultural productivity as well as its mineral wealth.

Soil Sampling

To assess soil pollution in Thessaly following the flooding caused by storm "Daniel", the municipality of Palamas in the Karditsa Regional Unit was selected for sampling, as it was among the most severely affected areas. Floodwaters persisted for an extended period, requiring substantial drainage time.

Sampling locations included the villages of Palamas, Vlochos, Metamorphosi, Marathea, and Korda within the Palamas municipality, as well as an area near Agioi Theodoroi outside the primary flood zone. Eleven (11) representative surface soil samples, from a depth of 0 to 5 cm, were collected, ten (10) from areas within the municipality of Palamas in the Karditsa Regional Unit, and one (1) from the area of Agioi Theodoroi near Karditsa, in early December 2023. Each sample was formed by collecting and combining five subsamples, after removing organic debris. After proper preparation of the samples, physicochemical analyses were performed, including granulometry, heavy metals, nitrogen compounds, organic matter, and pH. The exact sampling locations are shown in Figure 4.

Figure 4. Sampling points in the study area.

Note. Samples 1, 2, 5 and 6 were collected near Palamas; samples 3, 4 and 10 near Vlochos; samples 7 and 8 near Marathea and Korda; sample 9 near Metamorphosi; and sample 11 near Agioi Theodoroi.

Soil Sample Preparation

The soil samples were air-dried in a well-ventilated area, sieved through a 2 mm sieve, and part of the sample was powdered using a porcelain mortar.

Particle size analysis

The particle size distribution of the samples was determined using the hydrometer method (Bouyoucos, 1927, 1951, 1962).

Specifically, 52.5 g of each soil sample, which had been previously air-dried, were weighed, and a quantity of sodium hexametaphosphate and 150 ml of deionized water were added. After 24 hours, the samples were transferred to the cup of a mechanical stirrer and stirred for 10 minutes. The contents of the cup were then transferred to Bouyoucos glass cylinders. A hydrometer was placed inside the cylinders, and water was added up to the lower calibration mark of the cylinder. The suspension was stirred, and a hydrometer and thermometer were inserted into the cylinder. Two measurements (density and temperature) were taken: the first at 40 seconds after stirring the suspension and the second after 2 hours without further agitation. The density values were corrected based on temperature using specific correction tables, and the percentage composition of the soil in sand, silt, and clay was calculated.

pH determination

The pH of the soil samples was determined using an MW101 Pro pH meter (electrometric method) (Thomas et al., 1996). Twenty (20) grams of dried, sieved soil were weighed, and deionized water was added in a soil-to-water ratio of 1:2.5. The suspension was periodically stirred for 30 minutes and left undisturbed for another 30 minutes to allow partial sedimentation. After calibration of the pH meter using buffer solutions of known pH (4.01 and 7), the pH of the soil samples was measured.

Determination of organic matter

To determine the organic matter content, 1 g of powdered soil was accurately weighed and transferred into a pre-weighed crucible. The samples were heated at 105°C for several hours to remove moisture, weighed, and then incinerated at 450°C for 4 hours before being weighed again. The percentage of organic matter was then calculated (Papazotos et al., 2016).

Determination of major elements and heavy metals

The pseudo-total concentrations of heavy metals were determined using a Thermo Ice 3000 Series Atomic Absorption Spectrophotometer, following the US EPA 3050B method (U.S. Environmental Protection Agency, 1996).

The concentration of aluminium (Al) in surface soils of the study area was determined via X-ray fluorescence (XRF) spectrometry using a Bruker CTX 800 instrument.

Nitrogen content was measured using the Kjeldahl method, and nitrate nitrogen was determined using a cadmium reduction column (Bremner, 1996; Griffin et al., 1995).

RESULTS AND DISCUSSION

Particle Size Analysis

The particle size distribution of the surface soil samples collected from the Palamas Municipality was determined, specifically the percentages of sand, silt, and clay, as presented in Table 1. These percentages were plotted on a soil texture ternary diagram, enabling the classification of each sample's texture (Figure 5, Table 1).

According to the ternary diagram, the soils were classified as follows: four samples as "Loam", three as "Silty Loam", two as "Sandy Loam", and one each as "Clay Loam" and "Clay". Based on this, fine-grained materials, specifically silt and clay, dominate most of the area.

According to Gkitsa (2010), the broader study area mainly consists of flysch and limestone soils, while the riverine areas are dominated by alluvial deposits containing both coarse and fine materials. Therefore, both the flysch and river systems contribute clayey materials to the region.

Table 1. Granulometric composition of surface soils in the Palamas municipality area, Thessaly.

Sample	Sand%	Silt%	Clay%	Soil Texture Classification
1	31,12	54,29	14,59	Silty Loam
2	26,36	32,38	41,26	Clay
3	40,65	40,95	18,40	Loam
4	58,74	30,48	10,78	Sandy Loam

continued on following page

Table 1. Continued

Sample	Sand%	Silt%	Clay%	Soil Texture Classification
5	43,50	42,86	13,64	Loam
6	28,27	53,33	18,40	Silty Loam
7	23,50	44,76	31,73	Clay Loam
8	39,70	45,71	14,59	Loam
9	26,36	51,43	22,21	Silty Loam
10	46,36	38,10	15,54	Loam
11	53,03	40,00	6,97	Sandy Loam

Figure 5. Ternary diagram showing the texture classification of surface soil samples from the Thessaly region (Soil Survey Staff, 2024).

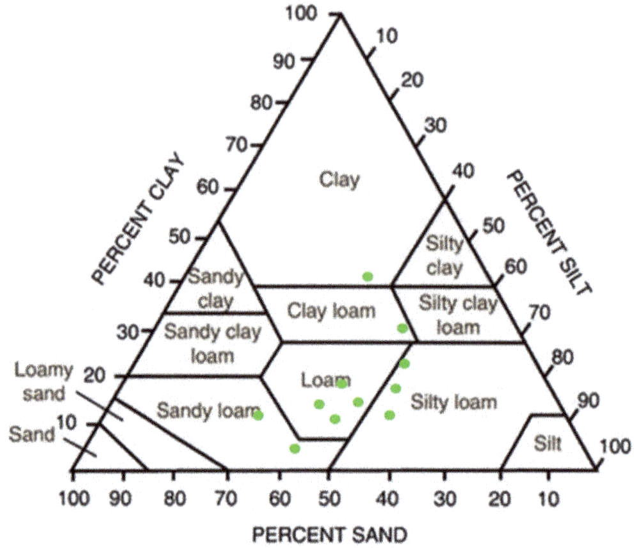

Determination of pH

Soils are classified as acidic, neutral, or alkaline based on their pH values. Soil pH levels around 7 are considered optimal for crop growth, nutrient availability, and microbial activity (McCauley et al., 2009). pH affects the mobility of heavy metals in the soil, regulating their solubility and uptake by plants (Naz et al., 2022). The more acidic the soil, the higher the solubility of metals, while in alkaline environments, metals become less mobile (Fijałkowski et al., 2012). In soils with low pH (acidic), metals can become toxic to plants due to their increased solubility. Soils with pH 6.6-7.3 are considered neutral, while those with pH 7.4-7.8 are slightly alkaline (Oshunsanya, 2018). In the study area, soil pH ranged from 6.78 to 7.63, with most samples exceeding pH 7 (Table 2).

Table 2. pH and organic matter values of the soil samples from the study area.

Sample	pH	Organic Matter %
1	7,45	1,65
2	7,44	3,44
3	6,88	7,98
4	7,00	3,20
5	7,53	2,45
6	7,42	5,00
7	7,51	5,89
8	7,53	6,67
9	7,42	4,04
10	7,63	2,45
11	6,78	7,35

According to the above classification, the soil samples are characterized as neutral to moderately alkaline in terms of pH. These values are expected for the study area, as the wider region features occurrences of limestone bedrock, which, through weathering and erosion, enriches the soil with bases. This aligns with Gkitsa (2010), who reported that the plains of Karditsa County are covered by a surface layer of soils ranging from neutral to slightly alkaline (pH 6.1–7.6).

Organic matter

Soil organic matter consists of plant and animal residues derived from various organisms inhabiting the soil. Through microbial activity, these residues undergo decomposition, ultimately forming humus – the stable end product of this process. Organic matter is an essential source of plant nutrients, a continuous energy source for microorganisms, and contributes to increasing soil moisture (Hussain et al., 2023). It also influences the physical and chemical properties of the soil. Furthermore, it impacts soil weathering, helps form a good soil structure, increases porosity, and regulates temperature. For these reasons, the organic matter content of the soil is an indicator of its fertility. In addition to fertilization, a common practice for increasing soil organic matter is the addition of animal manure, which provides the soil with nutrients such as nitrogen, phosphorus, and potassium. The organic matter content in the soils of the study area ranges from 1.65% to 7.98%, with an average value of 4.56% (Table 2). Most of the samples show high organic matter content and are classified as moderate, high, and rich in content. This is likely because the sampling points are located near the organic horizon, with samples collected from the top 5 cm of the soil, where plant roots are denser and crop residues trend to accumulate (Zhang et al., 2020).

Heavy metals

The total concentrations (mg/kg) of the heavy metals chromium (Cr), copper (Cu), manganese (Mn), nickel (Ni), lead (Pb), and zinc (Zn) were determined in the surface soil samples collected from the study area. Their results are presented in Table 3. Additionally, geochemical distribution maps of these heavy metals were created using ArcMap and Google Earth Pro. Figure 6 illustrates the spatial distribution of chromium, copper, manganese and nickel, while Figure 7 presents the surface distribution of zinc, lead, nitrogen, and nitrate nitrogen in the soils of Palamas municipality, Thessaly. The size of the symbols corresponds to the relative concentration of each pollutant.

Table 3. Concentrations of heavy metals in the surface soils of the Palamas municipality area, Thessaly in mg/kg.

Sample	Cr	Cu	Mn	Ni	Pb	Zn
1	80,92	28,63	573,08	145,42	-	60,09
2	169,07	46,85	869,10	232,11	-	85,73
3	68,51	48,51	868,52	96,71	8,97	104,86
4	28,10	36,32	997,47	37,73	4,01	43,17
5	74,93	22,67	502,36	150,82	-	47,86
6	79,80	31,22	645,90	134,34	-	67,55
7	140,78	43,99	1121,28	340,23	-	68,95
8	137,32	37,94	846,46	323,80	-	65,36
9	110,16	35,02	771,32	135,90	-	69,97
10	67,41	26,76	546,91	121,95	-	51,88
11	43,35	15,41	423,44	62,81	-	56,86

Figure 6. Geochemical maps showing the surface distribution of Cr, Cu, Mn and Ni in the soils of Palamas municipality, Thessaly. Symbol sizes correspond to the concentration levels of each pollutant.

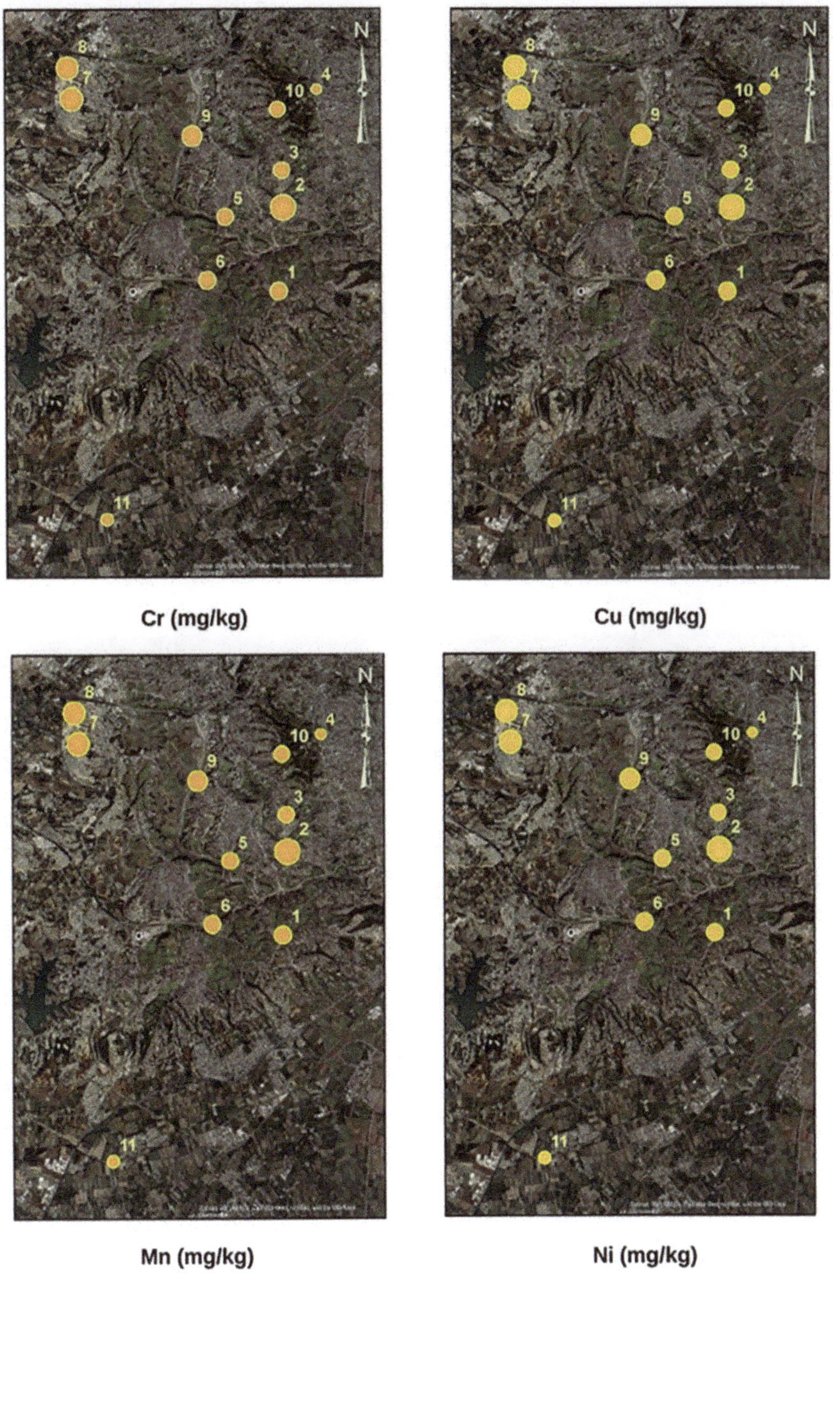

Figure 7. Geochemical maps showing the surface distribution of Zn, Pb, N and NO_3-N in the soils of Palamas municipality, Thessaly. Symbol sizes correspond to the concentration levels of each pollutant.

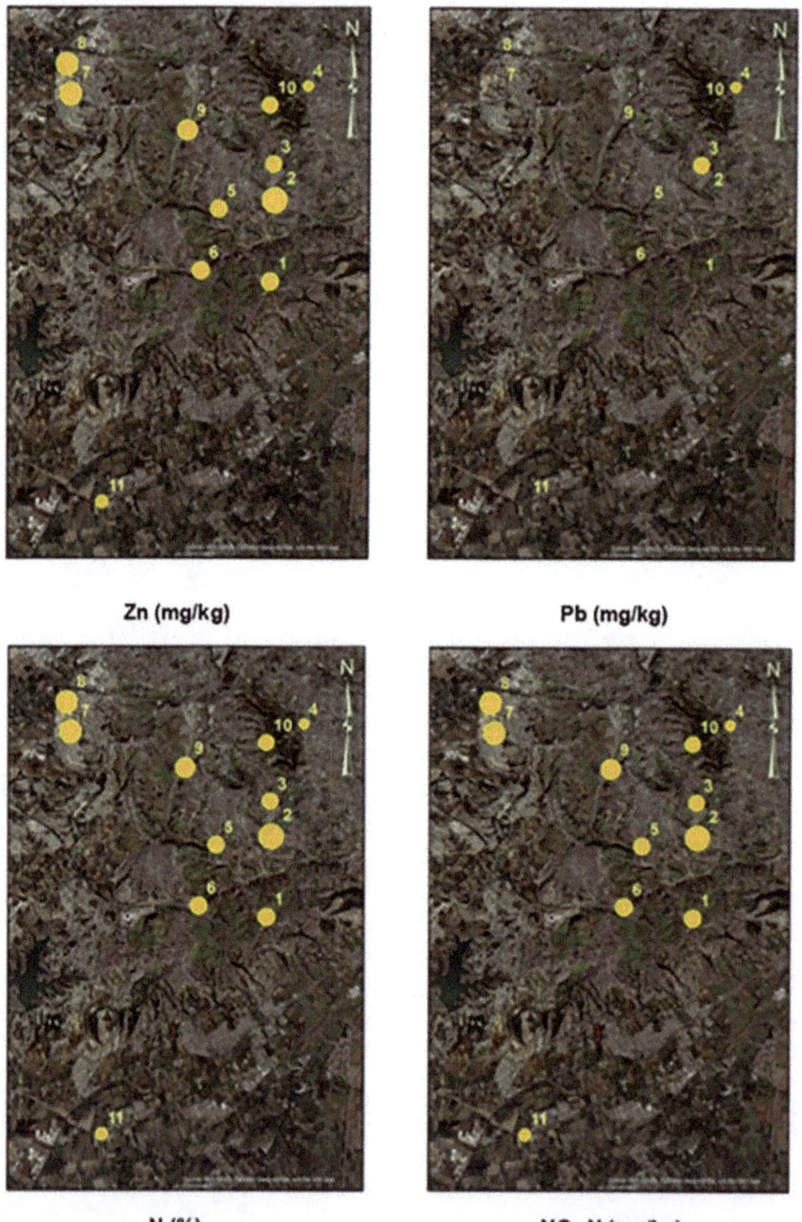

Chromium (Cr)

In the soil samples from the study area, chromium (Cr) concentrations ranged from 28.10 mg/kg to 169.07 mg/kg, with an average concentration of 90.94 mg/kg. The average Cr concentration aligns with the Earth's crust average (92 mg/kg) (Rudnick & Gao, 2003) and the European soil average (94.8 mg/kg) (Salminen et al., 2006; Yaroshevsky, 2006), but exceeds the global surface soils average (59.5 mg/kg) (Kabata-Pendias, 2010). It also surpasses previously reported values for the broader Karditsa region: 20.14 mg/kg (2010), 22.91 mg/kg (2011), and 29.88 mg/kg (2012) (Antoniadis & Golia, 2021) suggesting possible enrichment of surface soils with chromium due to flooding events.

According to Nacke et al. (2013) and Kabata-Pendias and Mukherjee (2007), one anthropogenic source of chromium is the widespread use of phosphate fertilizers. The most common phosphate fertilizer worldwide, diammonium phosphate (DAP) ($(NH_4)_2HPO_4$), may contain trace amounts of chromium and nickel (Nadarajan & Sukumaran, 2021). In the Marathea and Korda regions, sugar beet cultivation has historically dominated, but has recently been replaced by cotton farming, which requires nitrogenous fertilizers. Thus, the elevated chromium concentrations in the study area are likely partly attributable to anthropogenic factors such as fertilizer use.

Furthermore, chromium is abundant in ultramafic rocks (ophiolites), present in the broader study area with chromite mineralization also reported (Skordas, 2006). Therefore, part of the chromium in the soils may originate from flood-transported chromium-rich materials, as floodwaters can mobilize and deposit pollutants in different locations (Ciesielczuk et al., 2014).

The highest concentrations of chromium were found in the broader Palamas area and in the settlements of Marathea and Korda (Figure 6). These areas are characterized by a predominance of finer soil fractions, which possess higher cation exchange capacity (CEC) and grater specific surface area, allowing for more effective metal adsorption. Furthermore, the settlements of Marathea and Korda are located very close to the Pinios River. During the flooding event, river water from the Pinios inundated these areas, potentially playing a role in soil contamination through the deposition of sediments enriched with pollutants. This assumption is further supported by the fact that the least chromium contaminated samples (Samples 4 and 11) were located at the margin or outside the flooded zones, respectively.

Chromium concentrations in the study area exceed those reported in similar regions, such as northwestern Ohio (19 mg/kg) (Neupane & Roberts, 2009) and the Evros Basin, Spain (20.27 mg/kg) (Martín et al., 2006), but are lower than those in Huainan, China (117.95 mg/kg) (Liu et al., 2024) and the Ota agricultural region, southwestern Nigeria (156.20 mg/kg) (Kayode et al., 2022).

Copper (Cu)

In the surface soil samples from the study area, copper (Cu) concentrations ranged from 15.41 mg/kg to 48.51 mg/kg, with a mean of 33.94 mg/kg. This is comparable to global surface soils (38.90 mg/kg) (Kabata-Pendias, 2010) and the Earth's crust (28 mg/kg) (Rudnick & Gao, 2003). However, it exceeds the average concentrations reported for European soils (17.3 mg/kg) (Salminen et al., 2006; Yaroshevsky, 2006) and those reported by Golia and Diakoloukas (2022) for the same area, where surface soil samples (0-30 cm depth) from Alfisols, Inceptisols, Endisols, and Ertisols showed 21.40, 20.60, 11.40, and 2.10 mg/kg, respectively. It is also exceeds the concentrations reported by Bargotakis (2005) in the Karditsa region (3.46–29.2 mg/kg).

In comparison to the average copper concentration in agricultural surface soils in northwestern Ohio (23 mg/kg) (Neupane & Roberts, 2009), the copper levels in the Palamas municipality soils are also higher.

Thus, soils in the Palamas municipality are not considered polluted with copper. However, the elevated concentrations observed, compared to previous studies may be linked to the recent severe flooding. Copper levels are influenced by various environmental factors, including chemical, physical, and mineralogical soil properties, as well as natural processes such as parent material weathering, intense rainfall, continental dust, and forest fires (Moreira et al., 2022). In agricultural soils, anthropogenic inputs often dominate, with repeated use of copper-based fungicides being the primary source of Cu accumulation, while sewage sludge, animal manure, fertilizers, and industrial emissions also contribute (Alengebawy et al., 2021; Kabata-Pendias, 2010). The behaviour of copper in soil is strongly influenced by its interaction with organic matter. Divalent copper forms both soluble and insoluble complexes with organic compounds which affect its mobility and bioavailability in the soil (Mishra et al., 2019; Vlček & Pohanka, 2018).

Manganese (Mn)

In the surface soil samples from the study area, manganese (Mn) concentrations ranged from 423.44 mg/kg to 1,121.28 mg/kg, with an average concentration of 742.35 mg/kg. This average exceeds the global surface soil Mn concentration (488 mg/kg) (Kabata-Pendias, 2010) and the Earth's crust average (438.59 mg/kg) (Rudnick & Gao, 2003), but remains lower than the value reported by Turekian and Wedepohl (1961) for shales (850 mg/kg). Generally, manganese is present in the Earth's crust, with concentrations reaching up to 1000 mg/kg, particularly in basic igneous rocks. Manganese is found in minerals such as pyrolusite (MnO_2), rhodochrosite ($MnCO_3$), and rhodonite ($MnSiO_3$) (Kusumaningrum et al., 2018). In an alkaline environment, like that of the study area, manganese becomes less soluble, leading to higher concentrations in the soil.

Golia and Diakoloukas (2022) reported lower manganese concentrations in 0-30 cm surface soils from the same region with Alfisols, Inceptisols, Endisols, and Ertisols containing 19.60, 20.20, 15.60, and 11.70 mg/kg, respectively. In comparison, manganese concentrations in the surface soil samples from the study area are higher but remain within normal levels, except for Marathea, where a concentration of 1,121.28 mg/kg was recorded. Concentrations measured by Bargotakis (2005) in the broader area ranged from 114.3 to 1009.3 mg/kg, slightly lower than those observed in the present study. While manganese in the soils is primarily of geogenic origin, it may also derive from fertilizers, typically applied as $MnSO_4$, MnO (Golia, 2003).Overall, the soil is not considered significantly polluted with manganese.

Nickel (Ni)

In the surface soil samples from the study area, nickel (Ni) concentrations ranged from 62.81 mg/kg to 340.23 mg/kg, with an average concentration of 161.98 mg/kg. This average concentration is notably higher than the global average for Ni in surface soils (29 mg/kg) (Kabata-Pendias, 2010), the average concentration in the Earth's crust (47 mg/kg) (Rudnick & Gao, 2003), and the average concentration in European soils (37 mg/kg) (Salminen et al., 2006; Yaroshevsky, 2006). These concentrations are also higher than those reported in a previous study of surface soils (0-30 cm depth) in the broader study area, where Alfisols, Inceptisols, Endisols, and Ertisols contained 19.9, 12.1, 9.5, and 16.6 mg/kg, respectively (Golia & Diakoloukas, 2022).

Compared to similar regions worldwide, the concentrations observed in the present study exceed those found in the surface soils of northwestern Ohio (21 mg/kg) (Neupane & Roberts, 2009), flooded agricultural land in Opole, Poland (0.90–16.13 mg/kg) (Ciesielczuk et al., 2014), the Evros Basin in Spain (20.50 mg/kg) (Martín et al., 2006), and the Ota agricultural region in southwestern Nigeria (47.12 mg/kg) (Kayode et al., 2022).

These findings suggest that the surface soils of the study area are contaminated with nickel, likely due to flooding events and the transport of pollutants via water and sediments from neighbouring regions containing ultramafic rocks, rich in nickel (Skordas, 2006). The geological background significantly influences the mineralogical and chemical composition of the soils. Ophiolitic complexes, in the study area, including regions in Pindos, Orthys, and eastern Thessaly, consist of ultramafic and basic rocks such as dunites, peridotites, pyroxenites, and diabases which are rich in chromium and nickel (Figure 3) (Melfos & Voudouris, 2012; Pomonis et al., 2005; Rassios et al., 2019).

Water can also contribute to the transport and deposition of fine-grained material from agricultural areas where fertilizers containing nickel have been applied. Anthropogenic sources of nickel include phosphate fertilizers and pesticides (Kabata-Pendias & Mukherjee, 2007). The highest nickel concentrations are observed in areas with a higher percentage of clay and silt, particularly in the broader Palamas region and the villages of Marathea and Korda (Figure 6), since fine-grained material tends to adsorb and retain heavy metals and nutrients. These regions, historically covered by extensive agricultural lands and located near the Pinios River, may have been affected by deposition of polluted floodwaters and sediments. Although a direct connection cannot be confirmed, previous studies indicate that the Pinios River can carry significant pollutant loads (Bellos & Sawidis, 2005; Giakoumatos et al., 2024); for instance, Teou (2014) reported particulate nickel concentrations in river water ranging from 91 to 340 mg/kg, while Skordas (2006) found elevated levels of nickel, chromium, copper, and manganese in sediments near Larissa.

In conclusion, flooding event have likely contributed to nickel contamination in the soils of the study area, as the least contaminated samples were from peripheral or non-flooded zones. Nickel is essential for plant growth at low concentrations; however, at high concentrations, it can cause chlorosis, reducing chlorophyll production and inhibiting root development (Minari et al., 2020).

Zinc (Zn)

In the surface soil samples, zinc (Zn) concentrations ranged from 43.17 mg/kg to 104.86 mg/kg, with an average of 65.66 mg/kg. This is close to the average Zn levels in surface soils (70 mg/kg) (Kabata-Pendias, 2010), the Earth's crust (67 mg/kg) (Rudnick & Gao, 2003), and European soils (68.1 mg/kg) (Salminen et al., 2006; Yaroshevsky, 2006). It is also similar to the concentration in northwestern Ohio (64 mg/kg) (Neupane & Roberts, 2009). However, it is higher than values reported in previous studies in the broader area. Golia and Diakoloukas (2022) found lower concentrations in surface soils (0–30 cm depth) of Alfisols (19.9 mg/kg), Inceptisols (18.8 mg/kg), Endisols (8.9 mg/kg), and Ertisols (11.4 mg/kg). Bargotakis (2005) reported 0.22–66.6 mg/kg in the same region.

Zinc in surface soils can originate from both geogenic sources, such as sulphide ores, and anthropogenic activities, including fertilizers, pesticides, soil amendments, insecticides, and manure. Shurson et al. (2022), reported Zn concentrations 50-1450 mg/kg in phosphate fertilizers, 10-450 mg/kg in lime, and 15 to 250 mg/kg in manure. Higher Zn levels are also observed in limestone and organic soils (Grammenou, 2020).

Zinc concentrations in the soil samples from the study area do not exceed normal levels. However, the observed increasing trend may be related to flooding events and agricultural practices.

Lead (Pb)

The concentrations of lead (Pb) in the soil samples from the study area are generally low, with most of the samples below the detection limit of the instrument. The measured concentrations range from 4.01 to 8.97 mg/kg. The average lead content in soils globally is 27 mg/kg (Kabata-Pendias, 2010) and 17 mg/kg in the Earth's crust (Rudnick & Gao, 2003).

When compared with concentrations from a previous study (Golia & Diakoloukas, 2022) in the wider study area (Alfisols, Inceptisols, Endisols, and Ertisols), the values were 4.70 mg/kg, 3.10 mg/kg, 2.80 mg/kg, and 1.80 mg/kg, respectively. There is no significant change, indicating that the soil is not contaminated with lead.

Nitrogen (N) – Nitrate Nitrogen (NO_3-N)

Soil nitrogen enrichment beyond natural processes occurs through the addition of nitrogenous fertilizers. Nitrogen exists in soil in both organic and inorganic forms. Organic nitrogen, which is part of organic matter in various forms, is a component of complex organic carbon molecules and cannot be directly absorbed by plant roots.

The main inorganic forms of nitrogen are nitrate and ammonium. Through nitrification, ammonium nitrogen is converted to nitrate nitrogen by the action of specific bacteria. Both ammonium and nitrate are forms of nitrogen that plants can directly use. Plants absorb nitrate more readily than other nitrogen forms (Hofman & Van Cleemput, 2004). However, due to its negative charge and high mobility in the soil, nitrate is susceptible to leaching into groundwater, leading to contamination (Almasri, 2007). Therefore, its use should be avoided during periods of heavy rainfall. The ammonium form of nitrogen is more stable, adsorbs onto soil particles, and is less susceptible to leaching. However, when soils become saturated with water, as in flooded agricultural fields, the process of nitrification is significantly inhibited. Nitrification occurs when the soil pH ranges from 5.5 to 10, with the optimal pH being 8.5 (Babakos, 2005).

The concentrations of nitrogen and nitrate nitrogen found in the soils of the study area are shown in Table 4.

Table 4. Nitrogen and ammonium nitrogen concentrations in the soils of the Palamas municipality, Thessaly.

Sample	N %	NO_3-N (mg/kg)
1	0,11	6,27
2	0,10	5,88
3	0,38	12,00
4	0,10	7,25
5	0,13	7,01
6	0,24	10,07
7	0,27	14,50
8	0,31	11,82
9	0,19	8,51
10	0,10	11,05
11	0,31	9,87

These range from 0.10% to 0.38%, with an average of 0.20% for nitrogen, and from 5.88 to 14.50 mg/kg in dry weight, with an average of 9.48 mg/kg for nitrate nitrogen. According to Cameron et al. (2013), surface soils (the top 15 cm layer) can naturally contain nitrogen levels ranging from 0.1% to 0.6%, depending on their texture. Nitrate levels in agricultural soils are primarily elevated due to fertilizer use.

In the European Union, nitrate levels in agricultural soils are indirectly regulated by the Council of the European Union (1991), which aims to reduce nitrate leaching into groundwater and surface water. The directive sets a threshold for groundwater at 50 mg/L of nitrate, which is equivalent to 11.3 mg/L of nitrate nitrogen. Soil nitrate concentrations must remain below levels that could cause this limit to be exceeded.

According to the above, the soil samples do not show significant contamination from nitrogenous compounds.

ASSESSMENT OF POLLUTION LEVELS USING GEOCHEMICAL POLLUTION INDICES

Geoaccumulation Index (Igeo)

The Geoaccumulation Index (Igeo) is used to estimate the extent of pollution in an area and is calculated using the following formula (Muller, 1979):

$$I_{geo}=Iog_2\ [(C_n/1.5)*B_n]$$

Where:

C_n: Concentration of individual heavy metal
B_n: Value of geochemical background
1,5: It is applied to account for potential fluctuations in the natural background levels of a given metal in the environment, as well as minimal anthropogenic impacts.

The Igeo index takes values ranging from 0 to 6, depending on the pollution level (Table 5).

Table 5. Classification of the Igeo Index - pollution degree (Muller, 1979).

Igeo	Igeo Class	Pollution Degree
<0	0	Unpolluted
0-1	1	Unpolluted to moderately polluted
1-2	2	Moderately polluted
2-3	3	Moderately to strongly polluted
3-4	4	Strongly polluted
4-5	5	Strongly to extremely polluted
>5	6	Extremely high polluted

The geoaccumulation index (Igeo) was calculated for all heavy metals analyzed in each surface soil sample from the municipality of Palamas (Table 6). The geochemical background values used for the calculations were the concentrations of heavy metals in the Earth's crust (average shales) as reported by Turekian and Wedepohl (1961).

Based on the data presented in Tables 5 and 6, the surface soils exhibit no contamination with respect to copper, manganese, lead, and zinc, and are categorized as unpolluted for these elements (Igeo Class 0).

Regarding nickel, most of the samples show pollution. Specifically, samples 1, 5, 6, 9, and 10, collected from the areas of Palamas, Metamorfosi, and Vlochos, fall into Class 1 and are classified as unpolluted to moderately polluted with respect to nickel concentrations. Samples 2, 7, and 8 collected from the areas of Palamas, Maratheas, and Kordas fall into Class 2 and are classified as moderately polluted with nickel.

Table 6. Geoaccumulation index (Igeo) values in the surface soils of the study area.

Sample	Igeo Cr	Igeo Cu	Igeo Mn	Igeo Ni	Igeo Pb	Igeo Zn
1	-0,74	-1,24	-1,15	**0,51**	-	-1,25
2	**0,32**	-0,53	-0,55	**1,19**	-	-0,73
3	-0,98	-0,48	-0,55	-0,08	-1,74	-0,44
4	-2,26	-0,89	-0,35	-1,43	-2,90	-1,72
5	-0,85	-1,57	-1,34	**0,56**	-	-1,57
6	-0,76	-1,11	-0,98	**0,40**	-	-1,08
7	**0,06**	-0,62	-0,19	**1,74**	-	-1,05
8	**0,02**	-0,83	-0,59	**1,67**	-	-1,12
9	-0,29	-0,95	-0,73	**0,41**	-	-1,03
10	-1,00	-1,33	-1,22	**0,26**	-	-1,46
11	-1,64	-2,13	-1,59	-0,70	-	-1,33

As for chromium, only samples 2, 7, and 8 from the areas of Palamas, Maratheas, and Kordas exhibit slight pollution and are categorized in Class 1 (unpolluted to moderately polluted).

Single Pollution Index (PI)

The Single Pollution Index (PI) is used to assess the most environmentally harmful heavy metals and is calculated using the following formula:

$$PI=C_n/GB$$

Where:

PI: Pollution index for each element, Cn: Average concentration of the element in the soil (mg/kg), obtained from at least five sampling points.
GB: Geochemical background value for the same element (mg/kg) (Earth's crust (average shales) as reported by Turekian and Wedepohl (1961)).

The pollution levels of soils are categorized as follows according to Konstantinova et al. (2019) and Xia et al. (2021) (Table 7):

Table 7. Classification of pollution levels based on PI values

PI Value Range	Interpretation
PI < 1	Unpolluted
1 ≤ PI < 2	Low pollution level
2 ≤ PI < 3	Moderate pollution level
3 ≤ PI < 5	Strong pollution level
PI ≥ 5	Very strong pollution level

The Single Pollution Index (PI) for six heavy metals was calculated in the surface soils of the study area, yielding the values presented in Table 8.

Table 8. Pollution Index (PI) Values.

PI(Cr)	PI(Cu)	PI(Mn)	PI(Ni)	PI(Pb)	PI(Zn)
1,01	0,75	0,87	2,38	0,06	0,69

The Pollution Index (PI) values for Cu, Pb, Zn, and Mn were found to be lower than 1, indicating that the soils are classified as non-polluted with respect to these elements. For Cr, the PI value was calculated to be 1.01, indicating that the soils exhibit a low level of pollution with this element.

Finally, the PI value for Ni was calculated to be 2.38, indicating that the soils exhibit moderate pollution levels concerning this element.

Pollution Load Index (PLI)

The Pollution Load Index (PLI) is a tool used to assess the overall level of heavy metal pollution in soil or sediment. It provides a single value that reflects the cumulative contamination of multiple metals in an area and is given by the following equation (Varol, 2011):

$$PLI = \sqrt[n]{PI_1 \bullet PI_2 \bullet PI_3 \bullet \bullet\bullet\bullet PI_n}$$

Where:

n: The number of analyzed heavy metals
PI: The Single Pollution Index values calculated for each metal

The classification of pollution levels is presented in Table 9:

Table 9. Classification of pollution levels based on PLI values

PLI Value Range	Interpretation
PLI<1	Unpolluted
PLI=1	Baseline level of pollution; no significant anthropogenic impact is evident
PLI>1	Polluted

For the surface soils in the study area, the PLI value was calculated to be 0.63, indicating unpolluted by heavy metals soils.

Multi - Element Contamination (MEC)

The MEC index provides an assessment of the origin of pollution, indicating whether it is anthropogenic or lithogenic. If the index value exceeds 1, the pollution is considered anthropogenic, whereas values below 1 suggest a lithogenic origin. The calculation of the MEC index is based on the following formula (Kowalska et al., 2016).

$$MEC = \left[\left(\frac{C1}{T1} + \frac{C2}{T2} + \frac{C3}{T3} + \cdots + \frac{C4}{T4}\right)\right]/n$$

Where:

C: The content of heavy metals
T: The tolerable levels of heavy metals given by Kloke (1979) (Table 10)
n: The number of heavy metals

Table 10. Tolerable levels of heavy metals in soils (Kloke, 1979).

Elements	Tolerable levels
Mn	20000
Cd	3
Cu	100
Ni	100
Pb	100
Zn	300

In the soils of the study area, the MEC index was calculated as 0.63, indicating a lithogenic (natural) origin of the heavy metals.

Enrichment factor (EF)

The Enrichment Factor (EF) is used to assess the presence of anthropogenic pollution from heavy metals in surface soils. It is calculated by normalizing the concentration of a metal in the topsoil relative to the concentration of a reference element, such as Al, Fe, Mn, or Rb, which is not significantly affected by anthropogenic activities. This method helps to distinguish between natural and human-induced sources of contamination.

In this study, aluminium (Al) was chosen as the reference element, as it is a significant component of clay minerals and has been successfully used by many researchers (Abrahim & Parker, 2008; Barbieri, 2016). The geochemical background values were taken from the concentrations of heavy metals and Al in the solid crust (average shales), according to Turekian and Wedepohl (1961).

The enrichment factor (EF) is given by the following equation:

$$EF = (Metal/RE)_{soil}/(Metal/RE)_{background}$$

Where:

RE = Reference element in the sample and in the unpolluted area.

Values between $0.5 \leq EF \leq 1.5$ indicate that the concentration of the trace element is entirely due to weathering processes (natural processes). EF values >1.5 suggest that a significant proportion of the trace element originates from anthropogenic activities (Zhang & Liu, 2002). The concentrations of aluminium and the enrichment factor values for chromium and nickel in the soil samples of the study area are presented in Table 11.

Table 11. Aluminium concentrations and enrichment factor values for chromium and nickel in the soil samples of the study area.

Sample	Al (mg/kg)	EF (Cr)	EF (Ni)
1	67,600.00	1,06	**2,53**
2	69,000.00	**2,18**	**3,96**
3	64,600.00	0,94	**1,76**
4	69,400.00	0,36	0,64
5	67,200.00	0,99	**2,64**
6	65,500.00	1,08	**2,41**

continued on following page

Table 11. Continued

Sample	Al (mg/kg)	EF (Cr)	EF (Ni)
7	61,200.00	**2,04**	**6,54**
8	58,300.00	**2,09**	**6,53**
9	71,600.00	1,37	**2,23**
10	67,500.00	0,89	**2,13**
11	62,900.00	0,61	1,17

According to Sappa et al. (2020) and Abrahim and Parker (2008), the enrichment factor index is classified into the following categories (Table 12):

Table 12. Classification of pollution levels based on EF values

EF Value Range	Interpretation
$EF < 2$	Deficiency to minimal enrichment
$2 \leq EF < 5$	Moderate enrichment
$5 \leq EF < 20$	Significant enrichment
$20 \leq EF < 40$	Very high enrichment
$EF \geq 40$	Extremely high enrichment

Based on the Enrichment Factor values determined for the soil samples in the study area, the following results were observed:

For chromium, most samples show minimal enrichment, except for samples 2, 7, and 8 from the areas of Palamas, Marathea, and Korda, which exhibit moderate enrichment. At these locations, since the factor exceeds 1.5, anthropogenic sources of pollution are also indicated.

For nickel, samples 3, 4, and 11 from the areas south and northeast of Vlochos and Agioi Theodoroi in Karditsa show deficiency to minimal enrichment in nickel. Samples 1, 2, 5, 6, 9, and 10 exhibit moderate enrichment, while samples 7 and 8 from the Marathea and Korda areas show significant enrichment. Moreover, the Enrichment Factor indicates an anthropogenic origin for nickel in nine out of the eleven soil samples.

CONCLUSIONS

The surface soils of the municipality of Palamas in Thessaly, affected by the severe flood of September 2023, were analyzed for their physicochemical characteristics to assess pollution from heavy metals and nitrogen compounds.

The results showed that the soils are neutral to moderately alkaline, rich in organic matter, and composed mainly of fine particles. Nitrogen compound concentrations are within normal limits.

Heavy metals such as copper, manganese, zinc, and lead were detected at levels consistent with typical surface soils and crustal values reported in the literature. Geochemical pollution indices confirmed the absence of contamination by these metals. When compared with previous studies in the wider area, their concentrations were slightly higher, except for lead, which remained low.

Chromium and especially nickel were exceptions. Their concentrations in some areas exceeded permissible limits based on average normal values for surface soils and the Earth's crust. Five geochemical indices-Igeo, PI, PLI, MEC, and EF-were used to classify the soils and determine whether the metals originated from natural (geogenic) or anthropogenic sources.

Chromium contamination was mainly observed north of Palamas, in agricultural lands, and in the areas of Marathea and Korda. Elevated concentrations in these locations appear to result from both agricultural activities and natural inputs through erosion and weathering. Nickel contamination is more widespread than chromium, with soils ranging from unpolluted to moderately polluted. The highest nickel concentrations occur in the same areas -north of Palamas, Marathea, and Korda-reflecting both anthropogenic influence and geological sources. Geochemical indices, including the Enrichment Factor (EF) and the Multi - Element Contamination (MEC), provide insight into the origin and degree of metal contamination. The EF values indicate moderate enrichment and evidence of anthropogenic influence in most samples, while a few samples show significant enrichment. The MEC suggests that natural processes also play a significant role. The results point to a mixed origin of metals, driven by both anthropogenic activities and local geology. Areas with fine-grained soils and intensive agricultural activity show the highest enrichment and metal concentrations, reflecting the combined effects of flooding, runoff, and sediment deposition with levels exceeding those reported in earlier studies.

The overflow of the Pinios River may also have contributed to local pollution. The river flows through formations rich in nickel and chromium, and floodwaters likely carried and deposited metal-enriched sediments. Additional inputs may have originated from urban and industrial wastewater, as well as agricultural runoff containing fertilizers, pesticides, and organic residues. The highest heavy metal concentrations were determined in areas located near the Pinios River, while lower values were recorded in zones that experienced limited or no flooding and therefore showed only minor contamination. These processes explain the higher heavy metal levels observed in this study compared with previous reports and the lower values recorded in areas unaffected by the flood.

These findings highlight the environmental risks associated with extreme weather events and emphasize the need for enhanced natural resource management and the development of resilient infrastructure. Continued monitoring of the affected area is crucial to assess temporal variations and evaluate the long-term impacts of such catastrophic events under changing climatic conditions.

REFERENCES

Abbasi, M. R., & Sepaskhah, A. R. (2023). Nitrogen leaching and groundwater N contamination risk in saffron/wheat intercropping under different irrigation and soil fertilizers regimes. *Scientific Reports, 13*(1), 6587. https://doi.org/https://doi.org/10.1038/s41598-023-33817-5

Abrahim, G., & Parker, R. (2008). Assessment of heavy metal enrichment factors and the degree of contamination in marine sediments from Tamaki Estuary, Auckland, New Zealand. *Environmental Monitoring and Assessment, 136*(1), 227–238. https://doi.org/10.1007/s10661-007-9678-2

Alengebawy, A., Abdelkhalek, S. T., Qureshi, S. R., & Wang, M.-Q. (2021). Heavy metals and pesticides toxicity in agricultural soil and plants: Ecological risks and human health implications. *Toxics, 9*(3), 42. https://doi.org/10.3390/toxics9030042

Alloway, B. J. (2012). *Heavy metals in soils: trace metals and metalloids in soils and their bioavailability* (Vol. 22). Springer Science & Business Media.

Almasri, M. N. (2007). Nitrate contamination of groundwater: A conceptual management framework. *Environmental Impact Assessment Review, 27*(3), 220–242. https://doi.org/10.1016/j.eiar.2006.11.002

Antić-Mladenović, S., Kresović, M., Čakmak, D., Perović, V., Saljnikov, E., Ličina, V., & Rinklebe, J. (2019). Impact of a severe flood on large-scale contamination of arable soils by potentially toxic elements (Serbia). *Environmental Geochemistry and Health, 41*(1), 249–266. https://doi.org/10.1007/s10653-018-0138-4

Antoniadis, V., & Golia, E. E. (2021). Spatial and temporal assessment of cadmium and chromium contamination in soils in the Karditsa region (Central Greece). *Environmental Science and Pollution Research International, 28*(4), 3820–3827.

Babakos, G. (2005). *Determination of nitrite and nitrate ion levels in soils of the Almira region, Magnesia* [Bachelor's Thesis, University of Thessaly, School of Agricultural Sciences]. Volos, Greece.

Barbieri, M. (2016). The Importance of Enrichment Factor (EF) and Geoaccumulation Index (Igeo) to Evaluate the Soil Contamination. *Journal of Geology & Geophysics, 5*(1), 1–4. https://doi.org/10.4172/2381-8719.1000237

Bargotakis, P. (2005). *Measurement of heavy metals copper (Cu), zinc (Zn), iron (Fe) and manganese (Mn) in areas of karditsa*. University of Thessaly.

Bellos, D., & Sawidis, T. (2005). Chemical pollution monitoring of the river pinios (Thessalia—Greece). *Journal of Environmental Management*, *76*(4), 282–292. https://doi.org/10.1016/j.jenvman.2005.01.027

Bouyoucos, G. J. (1927). The hydrometer as a new method for the mechanical analysis of soils. *Soil Science*, *23*(5), 343–354.

Bouyoucos, G. J. (1951). A recalibration of the hydrometer method for making mechanical analysis of soils. *Agronomy Journal*, *43*(9), 434–438. https://doi.org/10.2134/agronj1951.00021962004300090005x

Bouyoucos, G. J. (1962). Hydrometer method improved for making particle size analyses of soils. *Agronomy Journal*, *54*(5), 464–465. https://doi.org/10.2134/agronj1962.00021962005400050028x

Bremner, J. M. (1996). *Nitrogen-total* (Sparks, A. L. P. D. L., Helmke, P. A., Loeppert, R. H., Soltanpour, P. N., Tabatabai, M. A., Johnston, C. T., & Sumner, M. E., Eds.). Vol. 5)., https://doi.org/https://doi.org/10.2136/sssabookser5.3.c37

Cameron, K., Di, H. J., & Moir, J. (2013). Nitrogen losses from the soil/plant system: A review. *Annals of Applied Biology*, *162*(2), 145–173. https://doi.org/10.1111/aab.12014

Christakis, C. (2013). *Multifunctional Landscape; The Case of The Plain of Thessaly* [Master Thesis, University of Thessaly].

Ciesielczuk, T., Kusza, G., Poluszyńska, J., & Kochanowska, K. (2014). Pollution of flooded arable soils with heavy metals and polycyclic aromatic hydrocarbons (PAHs). *Water, Air, and Soil Pollution*, *225*(10), 2145. https://doi.org/10.1007/s11270-014-2145-0

Copernicus, E. (2023). *Image of floods in Thessaly, Greece*. Retrieved Accessed 28 March 2023 from https://x.com/CopernicusEU/status/1701155548744339642

Council of the European Union. (1991). *Council Directive 91/676/EEC of 12 December 1991 concerning the protection of waters against pollution caused by nitrates from agricultural sources* (CELEX:31991L0676). https://eur-lex.europa.eu/legal-content/EN/TXT/?uri=CELEX%3A31991L0676

Fijałkowski, K., Kacprzak, M., Grobelak, A., & Placek, A. (2012). The influence of selected soil parameters on the mobility of heavy metals in soils. *Inżynieria i Ochrona środowiska, 15*, 81-92.

Giakoumatos, S. D., Skoulikidis, N. T., Karavoltsos, S., Sakellari, A., & Dimitriou, E. (2024). Spatiotemporal Variations in Water Physicochemical Status in Pinios River Catchment, at Eastern Mediterranean Region. *Land, 13*(11), 1959. https://doi.org/ https://doi.org/10.3390/land13111959

Gkitsa, A. (2010). *Water quality and water balance of the Kalentzi River* [Master Thesis, Uinversity of Thessaly].

Golia, E. (2003). *Soil pollution in Thessaly by heavy metals (Zn, Cu, Mn, Fe, Cd, and Pb) and their uptake by tobacco varieties: Spatial variability of heavy metals in selected areas of the study region* [Doctoral dissertation, University of Thessaly].

Golia, E. E., & Diakoloukas, V. (2022). Soil parameters affecting the levels of potentially harmful metals in Thessaly area, Greece: A robust quadratic regression approach of soil pollution prediction. *Environmental Science and Pollution Research International, 29*(20), 29544–29561. https://doi.org/10.1007/s11356-021-14673-0

Grammenou, A. (2020). *Heavy metal levels in soil and cereal plants in Metalleia.* Domokos. [Bachelor's Thesis, University of Thessaly]

Griffin, G., Jokela, W., Ross, D., Pettinelli, D., Morris, T., & Wilf, A. (1995). *Recommended soil nitrate-N tests* (Vol. 493). Northeastern Regional Publication.

Hellenic Statistical Authority. (2023). *Permanent population by region and regional unit (2021 census).* Retrieved 1/10/2024 from https://www.statistics.gr

Hellenic Survey for Geology and Mineral Exploration. (1983). *Geological map of Greece.* Retrieved 5/1/2025 from https://eagme.gr/site/services

Hilscherova, K., Dusek, L., Kubik, V., Cupr, P., Hofman, J., Klanova, J., & Holoubek, I. (2007). Redistribution of organic pollutants in river sediments and alluvial soils related to major floods. *Journal of Soils and Sediments, 7*(3), 167–177.

Hofman, G., & Van Cleemput, O. (2004). *Soil and Plant Nitrogen.* International Fertilizer Industry Association (IFA).

Hussain, A., Bashir, H., Zafar, S., Rehman, R., Khalid, M., Awais, M., Sadiq, M., & Amjad, I. (2023). The importance of soil organic matter (som) on soil productivity and plant growth. *Biological and Agricultural Sciences Research Journal, 2023*(1), 11-11. https://doi.org/ https://doi.org/10.54112/basrj.v2023i1.11

Iatrou, M., Tziouvalekas, M., Tsitouras, A., Evangelou, E., Noulas, C., Vlachostergios, D., Aschonitis, V., Arampatzis, G., Metaxa, I., & Karydas, C. (2024). Analyzing the Impact of Storm 'Daniel'and Subsequent Flooding on Thessaly's Soil Chemistry through Causal Inference. *Agriculture*, *14*(4), 549. https://doi.org/https://doi.org/10.3390/agriculture14040549

Kabata-Pendias, A. (2010). *Trace Elements in Soils and Plants* (4 ed.). CRC press. https://doi.org/10.1201/b10158

Kabata-Pendias, A., & Mukherjee, A. B. (2007). *Trace elements from soil to human*. Springer.

Kairis, O., Karamanos, A., Voloudakis, D., Kapsomenakis, J., Aratzioglou, C., Zerefos, C., & Kosmas, C. (2022). Identifying degraded and sensitive to desertification agricultural soils in Thessaly, Greece, under simulated future climate scenarios. *Land (Basel)*, *11*(3), 395. https://doi.org/https://doi.org/10.3390/land11030395

Kayode, O., Ogunyemi, E., Odukoya, A., & Aizebeokhai, A. (2022). Assessment of chromium and nickel in agricultural soil: Implications for sustainable agriculture. *IOP Conference Series. Earth and Environmental Science*.

Kloke, A. (1979). Content of Arsenic, Cadmium, Chromium, Fluorine, Lead, Mercury, and Nickel in Plants Grown on Contaminated Soils.

Konate, F.-O., Dembele, B. S., Dembele, M., & N'diaye, B. F. (2024). Evaluation of Effects of Cotton Growing on the Environment and the Population in the Koutiala Circle. *Advances in Social Sciences and Management*, *2*(10), 24–38. https://doi.org/https://doi.org/10.63002/assm.210.647

Konstantinova, E., Minkina, T., Sushkova, S., Konstantinov, A., Rajput, V. D., & Sherstnev, A. (2019). Urban soil geochemistry of an intensively developing Siberian city: A case study of Tyumen, Russia. *Journal of Environmental Management*, *239*, 366–375. https://doi.org/https://doi.org/10.1016/j.jenvman.2019.03.095

Kowalska, J., Mazurek, R., Gąsiorek, M., Setlak, M., Zaleski, T., & Waroszewski, J. (2016). Soil pollution indices conditioned by medieval metallurgical activity–A case study from Krakow (Poland). *Environmental Pollution*, *218*, 1023–1036. https://doi.org/https://doi.org/10.1016/j.envpol.2016.08.053

Kusumaningrum, R., Rahmani, S. A., Widayatno, W. B., Wismogroho, A. S., Nugroho, D. W., Maulana, S., Rochman, N. T., & Amal, M. I. (2018). Characterization of Sumbawa Manganese Ore and Recovery of Manganese Sulfate as Leaching Products. *AIP Conference Proceedings*.

Liu, L., Zheng, X., Wei, X., Kai, Z., & Xu, Y. (2021). Excessive application of chemical fertilizer and organophosphorus pesticides induced total phosphorus loss from planting causing surface water eutrophication. *Scientific Reports, 11*(1), 23015. https://doi.org/ https://doi.org/10.1038/s41598-021-02521-7

Liu, W.-h., Zhao, J.-z., Ouyang, Z.-y., Söderlund, L., & Liu, G.-h. (2005). Impacts of sewage irrigation on heavy metal distribution and contamination in Beijing, China. *Environment International, 31*(6), 805–812. https://doi.org/10.1016/j.envint.2005.05.042

Liu, Y., Shen, W., Fan, K., Pei, W., & Liu, S. (2024). Spatial distribution, source analysis, and health risk assessment of heavy metals in the farmland of Tangwang Village, Huainan City, China. *Agronomy (Basel), 14*(2), 394. https://doi.org/https://doi.org/10.3390/agronomy14020394

López-Arredondo, D. L., Leyva-González, M. A., González-Morales, S. I., López-Bucio, J., & Herrera-Estrella, L. (2014). Phosphate nutrition: Improving low-phosphate tolerance in crops. *Annual Review of Plant Biology, 65*(1), 95–123. https://doi.org/10.1146/annurev-arplant-050213-035949

Loukas, A. (2010). Surface water quantity and quality assessment in Pinios River, Thessaly, Greece. *Desalination, 250*(1), 266–273. https://doi.org/10.1016/j.desal.2009.09.043

Marrugo-Negrete, J., Pinedo-Hernández, J., Combatt, E. M., Bravo, A. G., & Díez, S. (2019). Flood-induced metal contamination in the topsoil of floodplain agricultural soils: A case-study in Colombia. *Land Degradation & Development, 30*(17), 2139–2149. https://doi.org/https://doi.org/10.1002/ldr.3398

Martín, J. A. R., Arias, M. L., & Corbí, J. M. G. (2006). Heavy metals contents in agricultural topsoils in the Ebro basin (Spain). Application of the multivariate geoestatistical methods to study spatial variations. *Environmental Pollution, 144*(3), 1001–1012. https://doi.org/https://doi.org/10.1016/j.envpol.2006.01.045

McCauley, A., Jones, C., & Jacobsen, J. (2009). Soil pH and organic matter. *Nutrient management module, 8*(2), 1-12.

Melfos, V., & Voudouris, P. C. (2012). Geological, mineralogical and geochemical aspects for critical and rare metals in Greece. *Minerals (Basel), 2*(4), 300–317. https://doi.org/https://doi.org/10.3390/min2040300

Minari, G. D., Saran, L. M., Constancio, M. T. L., da Silva, R. C., Rosalen, D. L., de Melo, W. J., & Alves, L. M. C. (2020). Bioremediation potential of new cadmium, chromium, and nickel-resistant bacteria isolated from tropical agricultural soil. *Ecotoxicology and Environmental Safety*, *204*, 111038. https://doi.org/https://doi.org/10.1016/j.ecoenv.2020.111038

Mishra, A., Shukla, D., Vaghela, K., & Saraf, M. (2019). Copper: Its biological role and toxicity. *Journal of the Indian Botanical Society*, *98*(1), 26–36.

Moreira, A., Moraes, L. A., de Melo, T. R., Heinrichs, R., & Moretti, L. G. (2022). Management of copper for crop production. *Advances in Agronomy*, *173*, 257–298. https://doi.org/https://doi.org/10.1016/bs.agron.2022.02.005

Muller, G. (1979). Schwermetalle in den Sedimenten des Rheins: Veranderungen seit 1971. *Umschau*, *24*, 778–783.

Nacke, H., Gonçalves, A.Jr, Schwantes, D., Nava, I., Strey, L., & Coelho, G. (2013). Availability of heavy metals (Cd, Pb, and Cr) in agriculture from commercial fertilizers. *Archives of Environmental Contamination and Toxicology*, *64*(4), 537–544. https://doi.org/10.1007/s00244-012-9867-z

Nadarajan, S., & Sukumaran, S. (2021). Chemistry and toxicology behind chemical fertilizers. In *Controlled Release fertilizers for sustainable agriculture* (pp. 195–229). Elsevier.

Naz, M., Dai, Z., Hussain, S., Tariq, M., Danish, S., Khan, I. U., Qi, S., & Du, D. (2022). The soil pH and heavy metals revealed their impact on soil microbial community. *Journal of Environmental Management*, *321*, 115770. https://doi.org/https://doi.org/10.1016/j.jenvman.2022.115770

Neupane, G., & Roberts, S. J. (2009). Quantitative comparison of heavy metals and As accumulation in agricultural and forest soils near Bowling Green, Ohio. *Water, Air, and Soil Pollution*, *197*(1), 289–301. https://doi.org/10.1007/s11270-008-9811-z

Oshunsanya, S. O. (2018). Introductory chapter: Relevance of soil pH to agriculture. *Soil PH for nutrient availability and crop performance*, 3-6. https://doi.org/http://dx.doi.org/10.5772/intechopen.82551

Papaioannou, A., Plageras, P., Dovriki, E., Minas, A., Krikelis, V., Nastos, P. T., Kakavas, K., & Paliatsos, A. (2007). Groundwater quality and location of productive activities in the region of Thessaly (Greece). *Desalination*, *213*(1-3), 209–217. https://doi.org/doi:10.1016/j.desal.2006.03.610

Papazotos, P., Chalkiadaki, O., Chatzistamatiou, E., Georgopoulos, G., Gkiouleka, I., Katsikis, I., Zygouri, E., Kelepertzis, E., & Argyraki, A. (2016). Heavy metals in urban park soils from Athens, Greece. *Deltio tis Ellinikis Geologikis Etaireias, 50*(4), 2251–2260. https://doi.org/https://doi.org/10.12681/bgsg.14281

Pomonis, P., Tsikouras, B., & Hatzipanagiotou, K. (2005). Geological evolution of the Koziakas ophiolitic complex (western Thessaly, Greece). *Ofioliti, 30*(2), 77–86.

Ponting, J., Kelly, T. J., Verhoef, A., Watts, M. J., & Sizmur, T. (2021). The impact of increased flooding occurrence on the mobility of potentially toxic elements in floodplain soil–A review. *The Science of the Total Environment, 754*, 142040. https://doi.org/https://doi.org/10.1016/j.scitotenv.2020.142040

Pulkrabová, J., Suchanová, M., Tomaniová, M., Kocourek, V., & Hajšlová, J. (2008). Organic pollutants in areas impacted by flooding in 2002: A 4-year survey. *Bulletin of Environmental Contamination and Toxicology, 81*(3), 299–304. https://doi.org/10.1007/s00128-008-9486-6

Rangasami, S., Murugaragavan, R., Rakesh, S., Prabu, P., Ramesh, A. R., & Elangovan, R. (2022). Chemistry of flooded soils: A review. *The Pharma Innovation Journal, 11*(2), 2534–2539.

Rassios, A. E., Ghikas, C., Dilek, Y., & Kostopoulos, D. (2019). "Rootless" Ophiolites above the Exhuming Pelagonian Core Complex, Northern Greece. *Deltio tis Ellinikis Geologikis Etaireias, 54*(1), 60–94.

Rozos, D. E., Chatzinakos, I., & Nikolaou, N. S. (1989). Recording and classification of landslide phenomena in the mountainous network of Karditsa Prefecture. *Proceedings of the 4th Conference. Bulletin of the Greek Geological Society, XXIII*(3), 95–109. https://www.geosociety.gr/images/arxeio-teuxwn/GSG_XLIII_1.pdf

Rudnick, R., & Gao, S. (2003). Composition of the Continental Crust. . *Treatise on geochemistry, 3*, 1-64. https://doi.org/10.1016/B0-08-043751-6/03016-4

Salminen, R., De Vos, W., & Tarvainen, T. (2006). *Geochemical atlas of Europe*. Geological survey of Finland Espoo, Finland.

Sappa, G., Barbieri, M., & Andrei, F. (2020). Assessment of trace elements natural enrichment in topsoil by some Italian case studies. *SN Applied Sciences, 2*(8), 1409. https://doi.org/10.1007/s42452-020-03214-y

Sentas, A., Psilovikos, A., Karamoutsou, L., & Charizopoulos, N. (2018). Monitoring, modeling, and assessment of water quality and quantity in River Pinios, using ARIMA models. *Desalination and Water Treatment, 133*, 336-347. https://doi.org/doi: DOI: 10.5004/dwt.2018.23239

Shurson, G. C., Urriola, P. E., & Hung, Y.-T. (2022). Too much of a good thing: Rethinking feed formulation and feeding practices for zinc in swine diets to achieve one health and environmental sustainability. *Animals (Basel)*, *12*(23), 3374. https://doi.org/https://doi.org/10.3390/ani12233374

Skordas, K. (2006). Environmental geochemical research on the bottom sediments of the old and the new bed of Pinios River (Thessaly, Greece): Assessment of the potential sources of pollution. *Hellenic Society of Mining and Metallurgical Engineers MMX*, *2*, 69–80.

Soil Survey Staff. (2024). Field book for describing and sampling soils. *Version 4.0*. USDA, Natural Resources Conservation Service. Government Printing Office.

Suchara, I., Sucharová, J., & Holá, M. (2021). Changes in selected physico-chemical properties of floodplain soils in three different land-use types after flooding. *Plant, Soil and Environment*, *67*(2), 99–109. https://doi.org/https://doi.org/10.17221/435/2020-PSE

Teou, X. (2014). *Pollution of the Pinios River by heavy metals* [Master's Thesis, National and Kapodistrian University of Athens]. Athens.

Thomas, G., Sparks, D. L., Page, A. L., Helmke, P. A., Loeppert, R. H., Soltanpour, P. N., Tabatabai, M. A., Johnston, C. T., & Sumner, M. E. (1996). Soil pH and Soil Acidity. In *Methods of Soil Analysis. Part 3. Chemical Methods* (Vol. SSSA Book Series no. 5, pp. 1390). Soil Science Society of America and American Society of Agronomy.

Tsinopoulos, T. (2007). *Ecological study and mapping of wetland vegetation in the Karditsa Prefecture wetlands*. National and Kapodistrian University of Athens.

Turekian, K. K., & Wedepohl, K. H. (1961). Distribution of the elements in some major units of the earth's crust. *Geological Society of America Bulletin*, *72*(2), 175–192.

U.S. Environmental Protection Agency. (1996). *Method 3050B: Acid digestion of sediments, sludges, and soils (Revision2)*. United States Environmental Protection Agency.

Varol, M. (2011). Assessment of heavy metal contamination in sediments of the Tigris River (Turkey) using pollution indices and multivariate statistical techniques. *Journal of Hazardous Materials*, *195*, 355–364. https://doi.org/https://doi.org/10.1016/j.jhazmat.2011.08.051

Vlček, V., & Pohanka, M. (2018). Adsorption of copper in soil and its dependence on physical and chemical properties. *Acta Universitatis Agriculturae et Silviculturae Mendelianae Brunensis, 66*(1), 219–224. https://doi.org/https://doi.org/10.11118/actaun201866010219

Wang, Z., Liu, X., Liang, X., Dai, L., Li, Z., Liu, R., & Zhao, Y. (2022). Flooding-drainage regulate the availability and mobility process of Fe, Mn, Cd, and As at paddy soil. *The Science of the Total Environment, 817*, 152898. https://doi.org/http://dx.doi.org/10.1016/j.scitotenv.2021.152898

Weber, A., Wolf, S., Becker, N., Märker-Neuhaus, L., Bellanova, P., Brüll, C., Hollert, H., Klopries, E.-M., Schüttrumpf, H., & Lehmkuhl, F. (2023). The risk may not be limited to flooding: Polluted flood sediments pose a human health threat to the unaware public. *Environmental Sciences Europe, 35*(1), 1–19. https://doi.org/10.1186/s12302-023-00765-w

Xia, F., Zhu, Y., Hu, B., Chen, X., Li, H., Shi, K., & Xu, L. (2021). Pollution characteristics, spatial patterns, and sources of toxic elements in soils from a typical industrial city of Eastern China. *Land, 10*(11), 1126. https://doi.org/ https://doi.org/10.3390/land10111126

Yaroshevsky, A. A. (2006). Abundances of chemical elements in the Earth's crust. *Geochemistry International, 44*(1), 48–55.

Zhang, J., & Liu, C. (2002). Riverine composition and estuarine geochemistry of particulate metals in China—Weathering features, anthropogenic impact and chemical fluxes. *Estuarine, Coastal and Shelf Science, 54*(6), 1051–1070. https://doi.org/https://doi.org/10.1006/ecss.2001.0879

Zhang, Z., Li, J., Tsui, C.-C., & Chen, Z.-S. (2020). The study of gaining more detailed variability information of soil organic carbon in surface soils and its significance to enriching the existing soil database. *Sustainability, 12*(12), 4866.

Chapter 6
Forest Fire Risk Assessment and Its Mitigation Strategies:
A Case Study of the 2021 Varimpompi Fire in Attica, Greece

Spyridon Kaloudis
https://orcid.org/0000-0002-5432-5236
Agricultural University Athens, Greece

Grigorios K. Koniavitis
Agricultural University of Athens, Greece

ABSTRACT

Forest fires are a critical challenge, posing risks to ecosystems, human life, and property, with their rising frequency linked to poor forest management and climate pattern changes. This work proposes a fire destruction danger index model and mitigation strategies, utilizing a Geographic Information Systems-based approach. This model identifies high-risk areas by synthesizing the variables, values in threat, values fire sensitivity, fire severity, and fire incidence probability. Based on the proposed model, the area burned from the 2021 Varimpompi fire in Attica, Greece, is studied. Findings highlight dense pine forests, a mix of settlement and forest areas, accumulated biomass, and insufficient firebreaks as key risk factors. Proposed strategies include integrated forest management and the construction of adapted firebreaks to increase forest resilience and provide valuable services to the nearby population of Athens. Leveraging Geographic Information Systems and up-to-date technologies, along with community-based prevention plans and centralized fire management policies, aims to mitigate fire risk.

DOI: 10.4018/979-8-3693-8237-0.ch006

INTRODUCTION

Forests cover a significant part of the land and play an important role from an ecological-environmental and socio-economic point of view (Zacharakis & Tsihrintzis, 2023). However, wildfires cause significant damage to forests and threaten human lives and property (Bowman et al., 2017; Ruffault et al., 2018; Samara et al., 2018; Trigo et al., 2006; Zagas et al., 2013). Poor forest management and rural abandonment, combined with declining land-use practices, have resulted in higher dead biomass accumulation and increased wildfire risk (Kaloudis, 2008; Kaloudis et al., 2008). The increasing frequency and intensity of extreme weather events, attributed to climate change, are also significant aggravating factors in wildfire risk. They generate hotter, drier, and windier conditions that facilitate ignition and accelerate fire spread and severity (Abatzoglou & Williams, 2016; Bowman et al., 2017; Jolly et al., 2015; Kirchmeier-Young et al., 2019; Ruffault et al., 2018; Turco et al., 2018).

The effects of forest fires can be categorized into both immediate and long-term consequences. Immediate impacts occur during the fire and immediately after it is extinguished, while long-term effects persist for a prolonged period, even until the ecosystem returns to its original state. More specifically, direct impacts concern threats to the lives and health of citizens and participating firefighting forces, as well as to public and private property, and the health of ecosystems. Long-term effects on human health arise from elevated concentrations of chemical compounds, such as heavy metals and hydrocarbons, in soil and water. At the landscape level, long-term effects include aesthetic degradation, which impacts people's psychological well-being, and an increased risk of flooding and soil erosion. The effects of fires usually diminish gradually over time, depending on the success of restoring the ecosystem to its original state.

Wildlife suffers a great burden of ecosystem destruction. Slow-moving animals, such as turtles and hedgehogs, are the most vulnerable, while others, like deer and hares, panic. Birds, although they can fly, leave newborns behind in their nests (Garcês & Pires, 2023). During or immediately after fires, increased concentrations of trace gases and specific pollutants, such as carbon monoxide, are detected (Duncan et al., 2003). Wildfires also release significant quantities of CO_2 into the atmosphere, further contributing to greenhouse gas emissions.

Forest fires have serious economic consequences. The sectors most economically affected in a devastated area are forestry, timber trade, livestock, agriculture, and possibly tourism. Economic losses also include the costs of fire suppression and mitigating pollution from forest fires. The total cost of wildfires should encompass multiple components, including the reduction in future forest production and the expenses associated with vegetation regeneration. It should also include the restoration of public utility networks and infrastructure, as well as the preservation and maintenance of cultural and archaeological heritage.

Climate change alters the functioning of ecosystem processes (Millar & Stephenson, 2015; Turner, 2010). These alterations may be exacerbated by improper land-use practices such as agricultural expansion and excessive logging (Hagmann et al., 2022). The combined influence of climatic stressors and human pressures undermines the capacity of forest ecosystems to absorb disturbance and maintain their overall resilience capacity (Stevens-Rumann et al., 2018). When ecological thresholds are overcome, ecosystems may undergo irreversible transformations, and in such cases, complete restoration to pre-disturbance conditions is rarely achievable. Instead, management must shift toward fostering adaptive capacity and promoting trajectories that sustain ecosystem functionality under novel environmental regimes.

In the context of wildfire risk, which is increasingly influenced by climate change, a diverse range of modeling approaches has been developed to assess potential impacts to support decision-making. Wildfire modeling approaches range from empirical and statistical models, which establish relationships between climatic variables and fire occurrence, to process-based models that simulate vegetation dynamics and fuel accumulation. More recently, integrated simulations have been developed to capture the complex interactions among climate, fuels, and human activities (Johnstone et al., 2016; Plucinski, 2012; Seidl et al., 2017).

Studies on the forests of the western Mediterranean, spanning 27 years from 1988 to 2015, confirm the above findings (Blanco-Rodríguez et al., 2023). The duration of drought is a crucial factor that negatively impacts ecosystem recovery after a fire. Adverse weather conditions, such as prolonged droughts, heatwaves, and strong winds, increase the frequency and intensity of wildfires. These exaggerations of burning conditions cause severe ecosystem disturbances and substantially delay ecosystem recovery and post-fire vegetation regeneration (Enright et al., 2015; Johnstone et al., 2016; Pausas & Keeley, 2014, 2019). Specifically, regeneration density is influenced by factors, including fire conditions, distance from surviving forest, topography, climate, and soil quality (Chambers et al., 2016). It is evident that prolonged droughts, coupled with more frequent and intense large fires, could significantly alter the structure and composition of forest ecosystems (Blanco-Rodríguez et al., 2023).

Forest ecosystems, in cases where environmental changes are gradual, can adapt satisfactorily to fluctuations by adapting to new conditions through the process of natural selection (Falk et al., 2022). However, the soil degrades after frequent fires, even in plant communities adapted to fire. The reduction or total elimination of mother-seeding trees makes it very difficult, in many cases, to regenerate the main forest species, ultimately leading to ecosystem modification (Falk et al., 2022).

In the context of mitigating the adverse effects of forest fires, the current state of the art focuses on enhancing forest resilience (Nikinmaa et al., 2024). Forest ecosystem resilience refers to the capacity to withstand and recover from disturbances—such as fires, insect outbreaks, and extreme weather events—while maintaining structural integrity, ecological functions, and biodiversity. Resilient forests are better equipped to adapt to gradually changing environmental conditions and recover from disturbances without undergoing significant alterations in their fundamental ecological characteristics. The concept of resilience further acknowledges that forests are complex, dynamic systems shaped by the interaction of biological, physical, and social factors (Forzieri et al., 2022).

In the above context, there are many definitions of resilience. According to the European Forest Institute, the three most used definitions refer to Engineering, Ecological, and Socio-ecological resilience. In forest science, these three basic concepts of resilience are used. Resilience can be described in three complementary ways.

1. Mechanical resilience is defined as the time required for variables to return to equilibrium after a disturbance (Pimm, 1984).
2. Ecological resilience refers to the ability of the system to absorb shocks without changing, as well as the ability to self-organize and build adaptive capacity (Holling, 1973).
3. Social-ecological resilience is understood as the ability of a socioecological system to absorb or withstand shocks and other stressors so that the system remains within the same regime; essentially maintaining its structure and functions.

Humans are considered to be one of the primary causes of most fires, with a small percentage resulting from natural causes, such as lightning (Sari, 2022). The statistical data, regarding fires with a burned area of more than 30 hectares for the period 2006-2023, show that a significant number of fire events occur in countries of the European Union, particularly in Mediterranean countries (Copernicus, 2024). Portugal, Italy, Greece, France, and Spain are the countries that suffer the most significant proportionate catastrophic effects of large-scale fires (Copernicus, 2024). Regarding the annual percentage of the country's total area that is burnt, Portugal holds the first place with 1.02%, followed by Greece with 0.38%, Cyprus with 0.30%, and Croatia with 0.24%. According to various sources, forests in Greece have generally improved in recent decades in terms of covered area, density, and composition of forest species in mountainous areas (Kaloudis et al., 2023). However, in the lowlands, where dry and hot conditions prevail during the summer, forests are affected by large forest fires. The number of annual forest fire incidents was relatively stable until 1979, while the average annual number of fires between 1980 and 2010 more than doubled. In recent decades, the burned areas have almost tripled compared to previous decades (Dimitrakopoulos et al., 2009). A report for the period 1983–2008 shows that the total burnt areas correspond to 10.3% of the country's total area (Tsangari et al., 2011).

Significant effort has been made in the field of forest fires in recent decades, both in assessing fire risk and implementing measures to enhance the resilience of forest ecosystems (Demestichas et al., 2025; Kaloudis, 2008; Kaloudis et al., 2005). However, implementing appropriate policies is hampered by the high cost. These policies and related measures have been implemented in a piecemeal manner and are not integrated into the overall management of forest ecosystems. The consequence of this practice is the reduced effectiveness of measures and the excessive consumption of available resources to combat forest fires (Arbinolo et al., 2024; Stougiannidou & Zafeiriou, 2022)

The complete exclusion of fire from forest ecosystems is neither ecologically necessary nor practically achievable. However, integrated forest management, including the utilization of excess biomass and controlled burning, offers a practical and cost-effective tool for reducing fuel loads. These practices help mitigate fire risk while enhancing the resilience and protection of forest ecosystems (Agee & Skinner, 2005; Fernandes, 2013; Kalapodis et al., 2025; Kaloudis, 2008; Kaloudis et al., 2008; North et al., 2012; Stephens et al., 2013)

The reduction of fire damage can be based on the modification of two factors, specifically: a) preventing the start of a fire and b) preventing its spread. Fire prevention and mitigation efforts should begin with reducing potential sources of ignition. This can be achieved through a combination of preventive actions. Public awareness campaigns play a key role, alongside the enforcement of regulations that restrict outdoor burning during high-risk periods. Additional precautionary measures include clearing vegetation along railway lines and roads that pass through forested areas. Additionally, the prevention of fire spread can be enhanced by establishing firebreaks in strategic locations. Fire characteristics can also be reduced through the targeted modification of forest fuels, integrated within holistic forest management planning.

Implementing integrated forest management plans for fire risk reduction requires the analytical determination of appropriate Fire Destruction Danger Indices (FDDI). A key distinction between these indices lies in the forecast time, which can be categorized into short-term and long-term forecasting.

This work focuses on several key objectives. It first defines and justifies a Fire Destruction Danger Index (FDDI) capable of assessing fire risk in both spatial and temporal dimensions, with particular emphasis on long-term patterns. It then establishes an integrated, multi-purpose forest management framework designed to reduce fire risk. Attention is also given to identifying types of firebreaks that can effectively mitigate fire hazards while minimizing environmental and aesthetic impacts. Finally, the proposed index and management framework are applied to the Varimpompi area of Attica, a peri-urban landscape with particularly high fire risk.

METHODS

This chapter describes the definition of FDDI and the measures for forest management aimed at improving forest resilience.

Fire Risk Indices

Fire hazard is defined as the possibility of causing a fire, which is determined by the presence and action of every cause (FAO, 1986). Another definition of fire hazard is the potential number of fire initiation sources (Hawkes et al., 1996). As part of a more integrated approach, it was defined that fire risk (FR) encompasses the probability of an adverse event and its effects (Bachmann & Allgöwer, 1998). Similar content can be found in the Dictionary of Forest Terms published by the American Forestry Association (Helms, 1998).

Short-Term Forecast Fire Risk Indices

Short-term forecasting indices provide forecasts for a few days and are used to inform the public and prepare firefighting forces for potential fire events. Comprehensive short-term forecast indices considered a) the Canadian Forest Fire Danger Rating System - CFFDRS (Forestry Canada Fire Danger Group, 1992) and b) the U.S.A. National Fire Danger Rating System – NFDRS (Bradshaw et al., 1984; Gimeno & San-Miguel-Ayanz, 2004). The NFDRS consists of the Spread and Energy Release Components (SC and ERC) taken from Rothermel (1972) as modified by Albini (1976). The Burning Index (BI) is based on Byram's flame-length model (Byram, 1959). The Ignition and two occurrence models were developed for the NFDRS, as was the Fire Load Index (FLI) (Cohen, 1985). The Canadian Forest Fire Weather Index (FWI) System, the Canadian Forest Fire Behavior Prediction (FBP), and the NFDRS are tested internationally alongside other indices (Dimitrakopoulos et al., 2011; Giannakopoulos et al., 2012). Much effort has been put into fire risk assessment at the international level, especially to match weather conditions with fire risk (Dimitrakopoulos & Bemmerzouk, 2003; Kambezidis & Kalliampakos, 2016; Karali et al., 2014; Mhawej et al., 2016; Mitsopoulos et al., 2020; Papagiannaki et al., 2020; Politi et al., 2023; Troumbis et al., 2023; Varela et al., 2020).

One subcategory of indices is the fire initiation risk indices (Catry et al., 2009; Duncan et al., 2010; Genton et al., 2006; Romero-Calcerrada et al., 2010; Romero-

Calcerrada et al., 2008; Sanchez Sanchez et al., 2018; Syphard et al., 2008; Wang & Anderson, 2010; Ye et al., 2017; Zhang et al., 2010). These indices specialize in assessing fire initiation probability, considering the reference period and environmental conditions in combination with spatial parameters, such as the distance of a location to the road network. The importance of these indices lies in providing early warning to apply preventive measures and minimize the start of a fire, as well as increasing the preparedness of firefighting forces.

Long-Term Forecast Fire Risk Indices

Reducing fire risk by modifying the structure of forest fuels, based on the modeling of fire behavior, requires accurate knowledge of the FDDI values at each forest site over a long period. Generally, the FDDI should consider the probability of a catastrophic fire event and the associated consequences.

Definition of the FDDI. Based on the above reasoning, a modified version of the Fire Destruction Danger Index (FDDI), initially proposed by Kaloudis et al. (2005), is applied. The framework is conceptually aligned with the fire risk assessment approach proposed by Chuvieco et al. (2023). This index meets the requirements of long-term fire risk management, considering the threatened values of the forest itself and its surrounding rural and urban areas.

The modified FDDI version incorporates four key variables:

- Values in threat (ViT): are the forest and non-forest values that are threatened by fire.
- Values fire sensitivity (VFS): is the susceptibility of threatened values to fire destruction.
- Fire severity (FS): represents the destructive capacity of fire and the difficulty of extinguishing it. This parameter depends on the characteristics of fire.
- Fire incidence probability (FIP): represents the probability of a fire event.

The determination of each variable value is as follows:

- Values in Threat (ViT) are assessed using land cover data from remote sensing, thematic maps, and field inspections.
- Values Fire Sensitivity (VFS) is derived from literature and empirical observations of fire consequences in comparable land cover types.
- Fire Severity (FS) is evaluated based on the potential for extreme yet plausible fire events in various land cover categories.
- Fire Incidence Probability (FIP) is estimated from two components:

(a) Fire history (Fh) that is calculated as the mean probability of an area being burned. It is defined as the ratio of the annual average burned area to the total study area, assuming that a given area cannot burn more than once within the same fire season. Fire history can also be stratified by land cover types (e.g., forested, agricultural) to capture differentiated risk.

(b) Fire risk ignition (FRI): determined by the variability of ignition sources across locations and the influence of land cover type.

The assumptions used in this work are: (a) all sites, or strata, of a study area are at the same risk, based on fire history; (b) the average annual number of fires and the average burned area per fire will remain stable during the next management period (5 to 10 years). According to the above, the value of the FIP is given by equation (1).

$$FIP = Fh \times FRI \quad (1)$$

The overall FDDI values result from the combined effect of all variables, weighted by proximity to high-value locations, and are given by Equation (2):

$$FDDI = R \times W_i \quad (2)$$

Where: $R = ViT \times VFS \times FS \times FIP$

Here, Wi is a weight coefficient that reflects both the distance from high-value sites and the amount of the protective values, ranging between 1 and 1.5. Since the variables of FDDI are heterogeneous, they must first be standardized on a standard arithmetic scale. For this purpose, a six-point Likert scale (Table 1) is used to match qualitative assessments with numerical values.

While ViT, VFS, FS, and FIP are individually normalized on this scale, their multiplicative product (R) is also normalized prior to calculating FDDI, ensuring consistency with the six-point scale. By contrast, Wi is not normalized and retains its original range [1–1.5]. As a result, FDDI values can exceed the maximum threshold of [5] solely due to the influence of Wi; when this occurs, the fire risk is classified as "particularly high."

Table 1. Matching qualitative with numerical values of FDDI on a six-point Likert scale

Qualitative Value	Numerical value
Very High (VH)	5
High (H)	4
Average (A)	3

continued on following page

Table 1. Continued

Qualitative Value	Numerical value
Low (L)	2
Very Low (VL)	1
Zero (Z)	0

Based on the six-point scale presented in Table 1, the values assigned to the FDDI variables for each land cover type in the study area are summarized in Table 2.

Table 2. Assigning values to FDDI variables based on the Corine land cover

Level 1	Level 2	Level 3	ViT*	VFS*	FS*	FRI*
1. Artificial surfaces	1.1 Urban fabric	1.1.1 Continuous urban fabric	VH	H	H	H
		1.1.2 Discontinuous urban fabric	VH	VH	H	VH
	1.2 Industrial, commercial, and transport units	1.2.1 Industrial or commercial units	VH	VH	VH	VH
		1.2.2 Road and rail networks and associated land	H	H	A	H
		1.2.4 Airports	VH	VH	A	A
	1.3 Mine, dump, and construction sites	1.3.1 Mineral extraction sites	H	L	A	H
	1.4 Artificial, non-agricultural vegetated areas	1.4.1 Green urban areas	VH	H	H	H
		1.4.2 Sport and leisure facilities	VH	H	H	H
2. Agricultural areas	2.1 Arable land	2.1.1 Non-irrigated arable land	H	VH	L	H
	2.2 Permanent crops	2.2.1 Vineyards	VH	VH	A	A
		2.2.3 Olive groves	VH	VH	H	A
	2.3 Pastures	2.3.1 Pastures	L	L	A	A
	2.4 Heterogeneous agricultural areas	2.4.2 Complex cultivation patterns	H	VH	A	A
		2.4.3 Land principally occupied by agriculture, with significant areas of natural vegetation	H	VH	A	A
3. Forest	3.1 Forests	3.1.1 Broad-leaved Forest	H	VH	H	A
		3.1.2 Coniferous Forest	H	VH	VH	H
		3.1.3 Mixed Forest	H	VH	H	H
	3.2 Scrub and/or herbaceous vegetation associations	3.2.1 Natural grasslands	L	VH	L	H
		3.2.3 Sclerophyllous vegetation	A	VH	A	H
		3.2.4 Transitional woodland-shrub	H	VH	H	H
	3.3 Open spaces with little or no vegetation	3.3.3 Sparsely vegetated areas	L	H	L	L
		3.3.4 Burnt areas	L	L	L	M
4. Wetlands	4.1 Inland wetlands	4.1.1 Inland marshes	VL	VL	L	VL
5. Water bodies	5.1 Inland waters	5.1.2 Water bodies	Z	Z	Z	Z

Forest Ecosystem Management Framework to Reduce FDDI

According to the extensive wildfire modeling literature, e.g., (Albini, 1976; Rothermel, 1972), the most common ecologically acceptable and cost-effective treatments of forest management for increasing forest resilience to forest fires are considered the following (Kaloudis, 2008):

Removal of logging residues: which reduces the surface fuel load and, therefore, the characteristics of the fire. However, it is noted that its intensive application may cause a lack of nutrients in the forest ecosystem (Baeza et al., 2002; Baeza et al., 2003; Carter & Darwin Foster, 2004; Fernandes & Botelho, 2003; Gibbons et al., 2000; Kalabokidis & Omi, 1998; Scherer et al., 2000; Zabowski et al., 2000).

Tree pruning: reduces the likelihood of surface fires escalating into canopy fires, while also improving the quality of the timber produced.

Removal of the surface fuel: which reduces the characteristics of a fire. In this way, in combination with tree pruning, the likelihood of a canopy fire is significantly reduced. Implementing this treatment can be facilitated by controlled grazing, which also generates additional income for local communities through livestock production.

Thinning of the forest: reduces fire characteristics and helps avoid catastrophic crown fires. In addition, it has a positive impact on forest production, enhancing the quality of timber and promoting forest regeneration. It is worth noting that excessive thinning can have several negative impacts on forests. It reduces timber yield and overall productivity, promotes the development of an undesirable dense understory, diminishes aesthetic value, and increases the risk of soil erosion (Baldwin et al., 2000; Graham et al., 1999).

Where the above vegetation treatments are applied, the cut fuel materials are considered not to participate in a possible future fire event. This hypothesis presupposes the appropriate handling of the cut biomass.

Encouraging species with high fire resistance: which refers to the support of native broadleaf tree species, thus increasing the resistance of the forest to fires (Dimitrakopoulos, 2001a, 2001b; Dimitrakopoulos & Panov, 2001; Dimitrakopoulos & Papaioannou, 2001; Liodakis et al., 2003). When mixed with coniferous and broadleaf species, it also enhances the resistance of these forests to insects and pathogens, while improving the landscape's aesthetics. Moreover, increasing the biodiversity of a forest can simultaneously contribute to its ecological resilience (Oliver et al., 2015; Thompson et al., 2009).

Grazing of domestic animals: which reduces the fuel load (bushes and herbs) and contributes to the increase of farmers' income through the production of livestock products (Bachelet et al., 2000; Liedloff et al., 2001; Ribeiro et al., 2023; Torrano & Valderrábano, 2005; Valderrábano & Torrano, 2000).

Firebreaks Construction

The construction of firebreaks can aid in reducing the destructive effects of fires because they provide prepared locations where it is easier to extinguish the fire. According to Butler and Cohen (Butler, 2014; Butler & Cohen, 1998a, 1998b), to adequately protect firefighting forces from fire, the distance of the fire front from the forces must exceed four times the estimated maximum height of the produced flame in the fire front. The width of the firebreaks resulting from the above rule is considered excessive from both ecological and aesthetic perspectives. To mitigate the above impacts of creating firebreaks, Kaloudis (2008) proposed composite firebreaks consisting of parallel subzones, along roads, covered by vegetation of adapted height, with only one of these subzones bare of vegetation.

For reasons of completeness, the structure of this type of composite bilateral firebreak is given in Table 3 (Kaloudis, 2008).

Specifically, for a typical composite firebreak constructed along a road for both sides, the following three subzones are proposed:

Bare Subzone: This subzone must remain completely free of vegetation or other flammable materials.

Herbaceous Vegetation Subzone: This subzone is characterized by herbaceous vegetation, and its form can be maintained with minimal intervention, aided by animal grazing.

Shrubland Subzone: This subzone comprises shrub vegetation or young trees. Based on the maximum flame height produced in this subzone, periodic vegetation treatments may be necessary, which can also be aided by animal grazing.

Table 3. Subzones of a bilateral composite firebreak.

Subzone No	Subzone cover
-	Natural forest
3	Shrubland subzone
2	Herbaceous vegetation subzone
1	Bare subzone
0	**Road deck**
1	Bare subzone
2	Herbaceous vegetation subzone
3	Shrubland subzone
-	Natural forest

Considering that the flame height is FH and the required width of the firebreak zone is D, then according to Kaloudis (2008) the following must hold:

$$D \geq 4FH \quad (3)$$

The subzones are arranged so that those with the lowest fire characteristics are positioned on the side closer to the firefighting forces, i.e., towards a road. Thus, the following holds:

$$\left(FH_{i+1} - FH_i\right) > 0 \quad (4)$$

Simultaneously, the sum of the subzones' widths towards the road or protected area must exceed four times the flame height in the subzone with the following higher i number (Table 3). Hence, if FH_i is the estimated flame height in subzone i and D_i is its width, then:

$$\sum_{i=1}^{n} D_i \geq 4FH_{n+1} \quad (5)$$

For n = 1, 2, 3.

Moreover, in locations with very high FDDI values, thinning the forest stands adjacent to the shrubland subzone, at an adequate depth within the forest, could minimize the probability of an active crown fire and consequently reduce fire characteristics, and specifically FH.

The proposed composite firebreaks, compared to the conventional bare ones of the same width, offer the following advantages:

- Reduces the ecological impact.
- Reduces the aesthetic degradation of the landscape.
- Lower maintenance effort, as complete surface clearing is not required, and partial vegetation control can be achieved through livestock grazing.
- Facilitates greater social acceptance of its creation.

CASE STUDY

In this section, the materials used and a description of the case study area are provided.

MATERIALS

For the implementation of the case study in this work, the following materials were used: a) the ArcMap Geographic Information Systems software; b) the orthophoto maps of the study area (photo shooting 2014-2016, with a spatial resolution of 25 by 25 cm) provided by the Public Legal Entity "Hellenic Cadaster"; c) the Thematic maps of the land cover 2018 from Copernicus, the road network from Greece open data and burned area boundaries from the European Forest Fire Information System. On-site inspections were also carried out.

STUDY AREA

Attica ranks among the country's most fire-affected regions, experiencing an annual loss of approximately 4,312 hectares of forests and pastures. The single fire event of August 3, 2021, devastated a total of 8,447 ha (Hellenic Fire Service). The study area encompasses the burned area and extends into the Municipalities of Acharnes, Dionysos, Oropos, Kifissia, and Marathon in the Attica Region, covering an area of 32,280 ha (Figure 1). This extension was designed to facilitate a more comprehensive planning approach from a forest perspective, encompassing unburned areas and part of the urban fabric. This choice is based on the fact that the fire risk of an area also depends on the fire risk in the neighboring locations.

The forests of the study area are a continuation of the residential fabric in the Attica basin, and they are considered peri-urban. The lack of green spaces in the residential fabric of the Attica basin burdens air quality. It also deprives residents of the opportunity to utilize public spaces for recreation, communication, and cultural events. This burden is disproportionately experienced by vulnerable population groups, including children, the elderly, and individuals of lower socioeconomic status, who frequently lack the capacity to travel long distances to meet their recreational needs.

The ecological value of peri-urban forests derives from their capacity to sustain diverse plant and animal communities, even under inhospitable conditions created by arid climates and increasing residential development pressures. They also beneficially affect the area's microclimate by regulating extreme temperatures, especially during summer. In addition to the above, they protect the urban area by regulating flooding phenomena and protecting mountainous soils from erosion.

Figure 1. Orientation map of the study area.

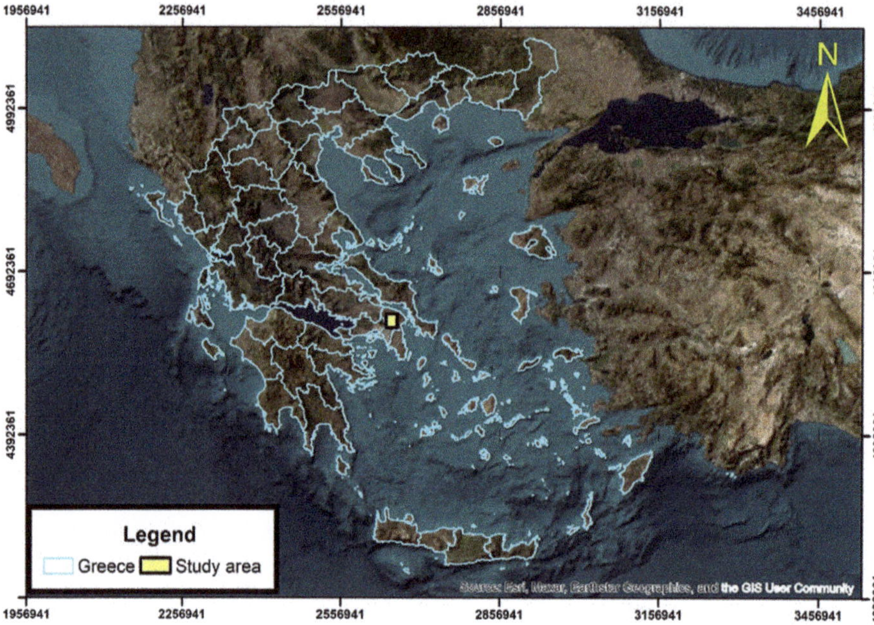

In the wider area, three main vegetation zones can be distinguished based on topography and flora: (a) the Eumediterranean zone (*Quercetalia ilicis*), covering coastal, hilly, and lower submontane areas; (b) the Sub-Mediterranean zone (*Quercetalia pubescentis*), extending across hilly and submontane sites; and (c) at higher altitudes, the Mountainous–Subalpine Conifer zone (*Vaccinio-Picetalia*) (Athanasiadis, 1986; Blondel et al., 2010; Dafis, 1986; Quézel & Médail, 2003; Rivas-Martínez et al., 2011). More specifically, the forest vegetation in Attica consists of the dominant forest species, which are evergreen broadleaves, Aleppo pine, and fir. In small areas, there are plane trees and oaks. Kefalonian fir (*Abies cephalonica*) grows at higher altitudes and is the only fir forest in Attica. In the *Abies cephalonica* zone, broadleaf species also appear, such as *Quercus coccifera, Quercus ilex, Phyllirea media, Juniperus oxycedrus, Berberis cretica, Rhus coriaria, Crataegus, Ostrya carpinifolia, Quercus pubescens, and Fraxinus ornus*, as well as a large number of herbaceous plant species.

In the study area, the dominant species is Aleppo pine (*Pinus halepensis* Mill*)*. The understory vegetation in the Aleppo pine forest is mainly composed of the species *Quercus coccifera* and *Quercus ithaburensis subsp. macrolepis, Pistacia lentiscus, Pistacia terebinthus, Juniperus phoenicea, Sarcopoterium spinosum (L.), Cistus spp., Juniperus oxycedrus, Arbutus unedo, Myrtus communis, Crataegus heldreichii, and Berberis cretica*. Despite the intense, dry, and hot climate, hydrophilic tree species are also found, primarily around springs or on the banks of streams that usually flow during winter. Hydrophilic species in the study area are represented mainly by the plane tree (*Platanus orientalis*), the white poplar (*Populus alba* L.), and several willow species (*Salix* spp.).

Figure 2 presents the land cover of the study area, based on CORINE Land Cover 2018 data. According to these data, the urban fabric accounts for about 20% of the total area, while industrial and commercial zones cover almost 4.4%. Agricultural land occupies 21.1%, whereas forest land, prior to the August 2021 fire, represented the largest share, with 49.7%. Other land cover types are present only to a minor extent (Figure 2). It is worth noting that, under current national legislation, the burned forest area has not lost its legal status as forest land.

The broader study area is home to a diverse fauna, comprising seven species of amphibians, three species of turtles, ten species of lizards, and nine species of snakes. The avifauna is also rich, with 132 recorded bird species (Annousis et al., 2021; Karaiskou et al., 2014). Parnitha supports 45 species of mammals, among which the red deer (*Cervus elaphus*) maintains a substantial population. The fauna of Parnitha remains among the richest in Attica, despite the pressures of human activities and environmental challenges.

In geological terms, the study area belongs to the Pelagonic zone and is structured by Alpine formations. The soil developed on these rocks is shallow, with intense erosion on steep slopes. The predominant soil types identified are Regosol, Luvisol, Agrisol, and Paleozem. Their texture is mainly loamy and, to a lesser extent, loamy and sandy loamy (FAO, 1998; IGME, 1983, 1993; Kosmas et al., 1993; Yassoglou et al., 1987).

Figure 2. Land cover of the study area

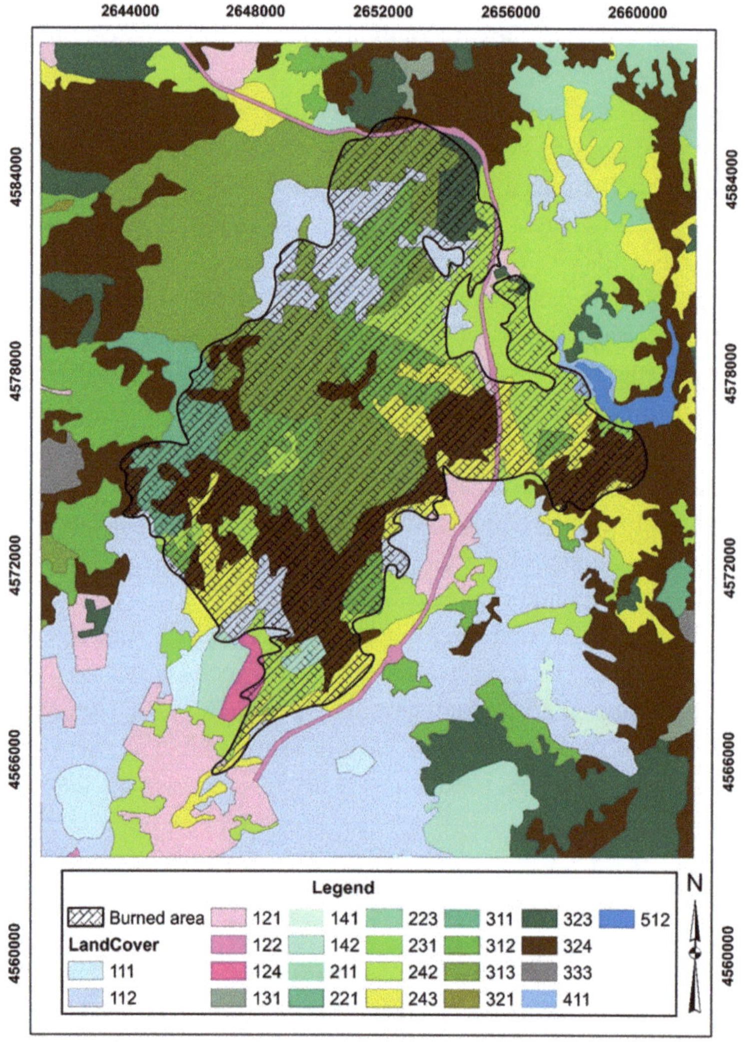

Note. Label numbers correspond to the Corine landcover codification listed in Table 2.

Climate is a crucial factor in the ecology of an area, and, in combination with topography and geological background, determines the growth environment of plants and, consequently, the type and composition of vegetation (Dafis, 1986). Climate and related environmental factors also affect: (a) the flammability of fuels, by favoring specific vegetation types and species, and (b) the combustion conditions, and therefore the characteristics of fire. Thus, climate contributes significantly to the vulnerability of forest ecosystems to wildfires.

The study area is characterized by a thermo-Mediterranean climate. The nearest meteorological station is that of the Hellenic National Meteorological Service at Tatoi (Lat. 38°07′ N, Lon. 23°47′ E, altitude 235 m). Meteorological records from this station cover the period 1956–2010 and provide the basis for the climate description of the area (Founda & Giannakopoulos, 2010; Giannakopoulos et al., 2011; Hellenic National Meteorological Service).

According to the weather station data above, the prevailing winds in the study area are typically from the north and northeast, followed in frequency by those from the southwest. The maximum monthly wind intensity is observed in July and August, coinciding with dry and hot atmospheric conditions that compose the most significant fire risk period. The average annual rainfall in the broader Attica area amounts to 452 mm. The low rainfall, especially during the summer period, creates an intense and prolonged drought that stresses vegetation and creates arid conditions, making it more susceptible to fire. Additionally, the average air temperature is generally high during the summer, and coupled with the intense drought, creates conditions that favor the start and spread of fires. The relative air humidity in the study area is generally low, and the climate of Attica is well known for its low humidity throughout the year.

Figure 3. Annual cycle of monthly average values of soil water deficit for the period 1951–1992.

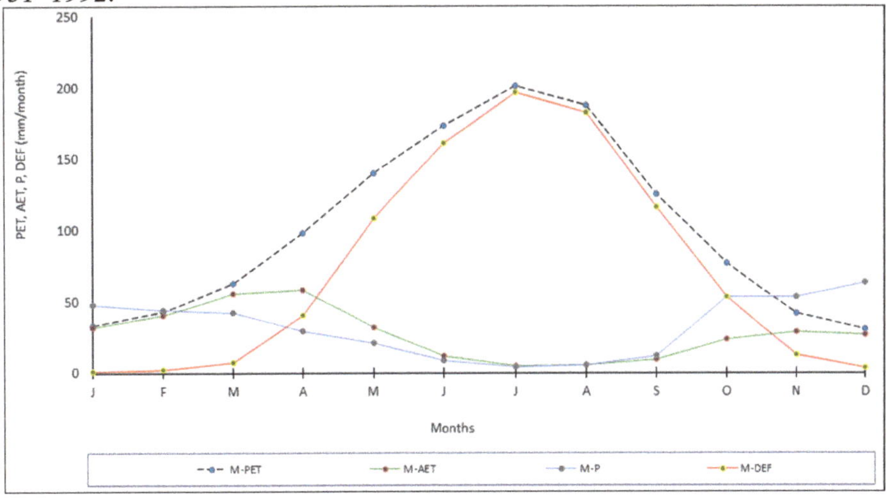

Note. M-PET = Monthly Potential Evapotranspiration; M-AET = Monthly Actual Evapotranspiration; M-P = Monthly Precipitation; M-DEF = Monthly Soil Moisture Deficit.

Figure 3 presents the seasonal evolution of soil water deficit components for the period 1951–1992, calculated using the Penman–Monteith method. The calculations were carried out within the framework of the study by Papaioannou et al. (1998), using data from the National Observatory of Athens meteorological station (located at an altitude of 197 m). The results indicate that water supply for vegetation is sufficient only during winter. At the same time, during the summer months, the monthly soil moisture deficit exceeds 160 mm, indicating the typical drought conditions of Mediterranean climates.

RESULTS

The data showed that the annual probability of destruction by fire for forest and grassland areas on the mainland of Attica, excluding Piraeus and the islands, is 0.0225. During a forest management planning period, typically carried out with a ten-year horizon, forest and grassland areas are expected to be burned at a rate of 0.225 of the total during a management cycle. The outcome is that, within approximately 45 years, all areas of these categories in Attica could be burned if no additional precautions are taken. The FDDI is considered a stable background of fire risk for all forest and grassland areas of the study area.

Based on the calculated FDDI values, we prioritized protecting settlements along the forest–urban interface. Where forests adjoin built-up areas, three consecutive buffers are measured outward from the urban boundary, as follows: Zone-100 (0 to 100 m), Zone-200 (100+ to 200 m), and Zone-300 (200+ to 300 m). A zone-specific weight factor W_i in equation (2) was applied using values of 1.25, 1.20, and 1.15 for the respective zones, with Wi = 1.50 assigned to urban areas and Wi = 1.00 to all other locations. This scheme captures (i) the elevated ignition likelihood near residences and (ii) the potential for fire transmission through these zones into the settlement.

Fire-behaviour inputs correspond to an extreme but possible scenario; for mature high-forest stands, we adopted a design flame height of 20 m. Figure 4 presents the resulting spatial distribution of FDDI across the study area computed using Equation (2).

MEASURES TO REDUCE FDDI

According to the FDDI calculation results (Figure 4), the study area is at significant fire risk, primarily due to its mixed coverage of forests and settlements.

Therefore, preventive measures are recommended to reduce future fire risk in the short and long term.

The short-term measures are applied seasonally or for a few days. They are mainly related to administrative measures, such as police provisions prohibiting the use of fire in open spaces, the dispersion of firefighting forces in timely locations, and the issuance of fire risk maps.

Figure 4. Spatial mapping of FDDI values across the study area

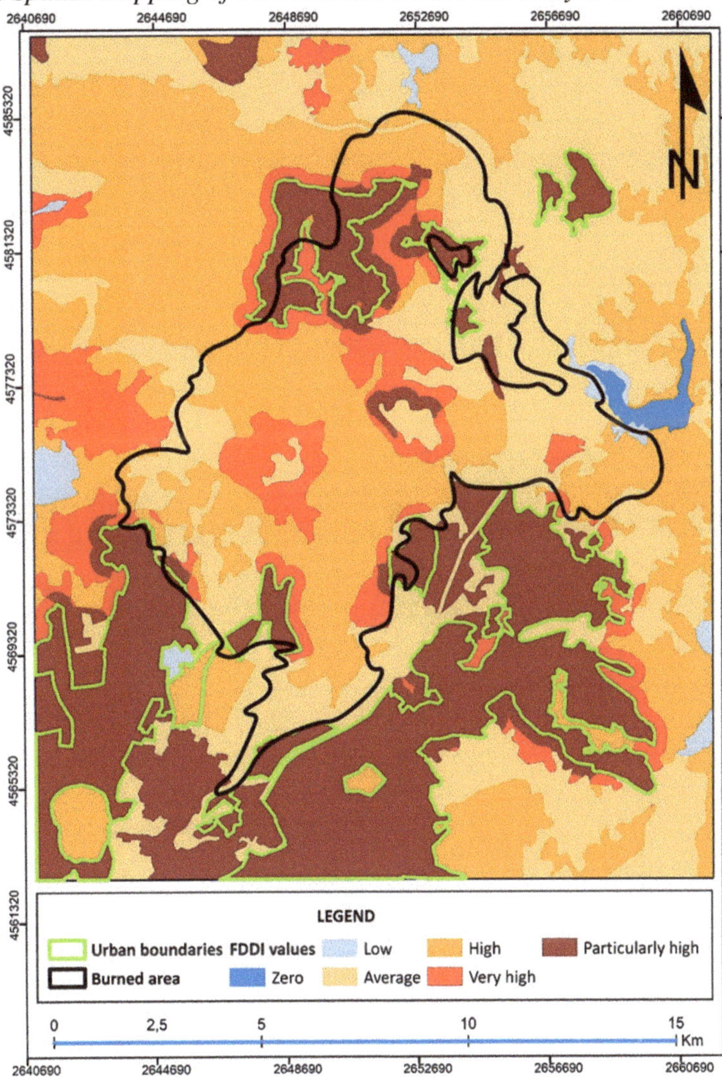

Long-term measures are implemented over extended timeframes and comprise two complementary and integrated components, as explained in the Methods section:

1. Integrated forest management planning. Silvicultural and fuel-management actions could be applied to lower fire risk. These include targeted thinning, understory thinning, pruning of ladder fuels, and diversification of species and stand structure. Such practices reduce potential fire-front intensity, rate of spread, and crown-fire potential, thereby improving the likelihood of successful containment.
2. Construction of firebreaks. Strategically site and maintain firebreaks to disrupt horizontal and vertical fuel continuity, creating pre-prepared control lines that facilitate suppression at safe, accessible locations. Properly designed firebreaks also reduce the likelihood that ignitions adjacent to assets will escalate into sustained runs.

Firebreaks Design

Firebreaks must provide continuous access for crews and heavy machinery and serve as pre-prepared control lines for direct or indirect fire attacks. Wherever feasible, they should be aligned with the existing road network to secure immediate access, minimize construction costs, and reduce ecological disturbance; new road segments are introduced only where necessary to ensure connectivity. The overall network should encompass assets and areas of interest, protecting them from fires approaching from any direction.

As explained in the Methods section, the width of fire breaks should be sufficient to protect the firefighting forces from the heat load of the fire front and effectively prevent its spread. Major firebreak construction is generally considered adequate only when it can support the intervention of firefighting forces. Implementing these zones is recommended to protect extended high-value areas along major roads, stopping the spread of fire and protecting significant areas with minimal intervention.

Alternatively, in locations with minor fire risk, it is possible to maintain only the bare vegetation zone along roads or at the boundaries of settlements, complemented by measures such as clearing the understory and pruning dead branches of trees deep in the forest. It is also recommended that the forest be thinned along the firebreaks to reduce the crown fuel load, thereby altering fire characteristics and the probability of active crown fire. Table 4 presents the types and descriptions of adapted firebreaks used in the study area.

Table 4. Features of Proposed Fire Breaks.

Firebreak type	Code	Firebreak description	Purpose
Rural	1	It is designed for use in rural environments, such as along roads or streams. It requires cleaning or ploughing crop residues or weeds that grow under tree crops along both sides of the road network or stream.	Protection is primarily targeting rural areas with low fire risk, preventing fires from spreading to high-risk areas.
Minor intervention	2	Firebreak of short length and width, depending on the vegetation type.	Protection of areas from relatively small possible fire passages in otherwise well-protected places.
Moderate intervention	3	This firebreak is relatively narrow in width.	Protection of areas with relatively low fire risk, mainly due to a low fuel load (sparse shrubby or herbaceous vegetation), far away from locations with high values.
One side zone	4	Firebreak along one side of a road.	Protecting important values with relatively low risk, such as residential areas, where building a firebreak on both sides of the road network is impossible.
On both sides of a highway	5	Firebreak along main highways on both sides.	To protect extensive areas—forested and non-forested. Highways typically have sufficient pavement widths, enabling the creation of an effective fire extinguishing zone along both sides.
Protection of settlements	6	Firebreak, single or double-sided, around settlements.	Protection of settlements and artisanal-industrial areas, or institutions and infrastructures of significant value.
On both sides of a simple road	7	Firebreak on both sides of a simple road.	Protection of forests and woodlands with heavy fuel loads. These zones protect large areas by providing adequate space for fire extinguishing.
On both sides of a new road	8	Firebreak on both sides of the new road construction.	Protection of important values and/or stopping the transmission of fire to other locations, and the construction of a new road.
Second-line defense within a settlement	9	Firebreak for second-line fire protection.	These zones aim to address fires in settlements with high vegetation and a risk of fire transmission, or on the boundaries of settlements where the fire is likely to overcome the outer firebreak.

The firebreaks were designed with an emphasis on protecting residential areas, particularly those adjacent to the forest. It is worth underlining that these types of firebreaks can be adapted and enriched in type, if necessary, to be applied in various environmental characteristics.

The forest area burned by the August 2021 fire will have low vegetation for several years after the fire, and thus, the flame height will be relatively small. For this reason, the proposed firebreaks should be adjusted according to the height growth of the regenerated forest's trees over time. The total length of the proposed fire breaks in the study αrea is approximately 247 km; while only about 5 km concerns the creation of new roads (Koniavitis, 2024).

FOREST MANAGEMENT PLANNING

Forest management in the study area is differentiated between burned and unburned forests. In burned forests, the primary objective is to secure successful regeneration, with particular emphasis on the recovery of tree species. Protecting areas under regeneration from subsequent fires or other disturbances is crucial, as repeated vegetation loss may lead to long-term and, in some cases, irreversible changes in ecosystem structure and function. Soil protection constitutes an additional priority. This can be achieved by promoting natural vegetation recovery, imposing temporary restrictions on domestic livestock grazing and, where necessary, implementing soil conservation measures in areas with a high risk of erosion. Further interventions, such as planting endemic species, may also be applied to mitigate soil degradation.

In unburned forests, the principal concern is to safeguard them against wildfire and to enhance their functionality, which is already compromised by the disturbance of adjacent burned areas.

Detailed forest management planning, along with protective measures for the urban–forest interface, is discussed in the following section.

ANALYSIS AND DISCUSSION

The study area consists of a complex of forest, agricultural, and residential areas (Figure 2). Potential fires pose a threat to the health and lives of urban populations due to the proximity of urban areas to forests. The spatially explicit FDDI map for the study area enabled the delineation of site-specific fuel management and rehabilitation treatments to reduce wildfire risk and support the recovery of burned landscapes. Although FDDI computation is data-intensive, it proved feasible using the multisource datasets described earlier. The index can be recalculated iteratively to test alternative intervention scenarios and optimize the type and intensity of measures. Moreover, periodic updates to the FDDI capture changes in risk driven by vegetation succession in the burned forest and by human modifications in residential and wildland–urban interface (WUI) areas, thereby supporting targeted and adaptive management.

INTEGRATED FIRE RISK MANAGEMENT

These forests hold significant value because of their proximity to large urban areas. However, the high cost of implementing fire risk reduction measures outside the forest management plan highlights the need for a systematic, multipurpose management approach that integrates fire risk reduction strategies and measures.

In a broader planning context, the site's proximity to the Athens metropolitan area shapes both primary and secondary objectives. The primary objective is to provide safe, high-quality public recreation within easy reach of the urban population. Secondary objectives include (i) hydrological regulation and flood-risk mitigation, (ii) conservation of habitats and species, (iii) sustainable timber production where appropriate, and (iv) controlled livestock grazing to reduce understory fuel loads. Across all objectives, systematic fire-risk reduction through integrated forest management is a prerequisite for maintaining ecosystem health and sustaining the ecosystem services.

The appropriate structure of the forest is considered the sub-all-aged or same-aged. The above forest structures provide flexibility for more effective forest management aimed at reducing fire risk. Removing "ladder" fuel treatments over extensive areas in such types of forests is more straightforward than in all-aged forests, where young trees are mixed with old trees and thus can act as "ladder" fuels. As a result, a more fire-resistant forest is built, especially when the conifers are mixed with broad-leaf trees. In the above context, forest resilience to fire improves when the treatments mentioned in the chapter on Methods are applied. In designated remote locations adequately protected by firebreaks, creating game reserves suitable for hosting species of domestic wildlife is possible through adaptation in the forest structure. Developing recreational facilities within peri-urban forests and ensuring systematic guarding during the fire season are key measures to enhance public benefits while reducing fire risk.

The implementation of these measures within a well-structured forest management plan is anticipated to generate substantial benefits, provided that potential drawbacks are systematically considered during the planning phase. In this case, the cost of applying forest treatments can be rationalized and offset by the increased income derived from the qualitative improvement of forest products and recreational facilities. For example, the improvement of timber quality through pruning is applied in the context of both fire protection and timber quality enhancement. Additionally, income from other ecosystem services, such as livestock products from grazing, resin tapping (where applicable), and beekeeping, helps offset the costs of protection. Especially for public forests, even though no fee is usually paid for most of the ecosystem services provided, the state benefits indirectly, as increased turnover generates higher tax revenues and ultimately contributes to the growth of Gross Domestic Product (GDP).

As forest management practices progress, the risk of fire is expected to decrease. Accounts should also be considered by reducing extinguishing costs and losses due to fire damage, as well as compensation and aid paid by reducing burnt areas.

STREAM MANAGEMENT

Streams are an integral part of an area's relief and play a crucial role in soil erosion, draining rainwater, and protecting the area from flooding. They also enrich an ecosystem's biodiversity, as aquatic vegetation grows along them, which in turn favors the survival of species of fauna adapted to those ecosystems.

In the study area, a dense network of streams flows primarily during the rainy season, traversing the low-lying agricultural areas and settlements. In the context of fire risk reduction, special measures should be taken to reduce the risk posed by streams in the study area. More specifically, the following treatments are proposed:

- Reinforcement of broad-leaved tree species within the bed and on the banks of streams.
- Interrupting the continuity of vegetation within the stream helps prevent possible fires from spreading along the stream. Suitable places for interrupting the continuity of vegetation are at technical works, such as bridges, where there is usually a floor of inflammable material (cement or stones) and where, due to shading, there is no growing rich vegetation.
- In rural areas along stream banks, plowing agricultural fields to create an annual fire zone of bare vegetation is suggested to prevent the spread of fire to nearby agricultural areas.
- Through vegetation management, strengthen broadleaf species that are considered more resistant to fire in forested areas along the streams on both banks. This would create a zone with a low potential fire intensity and facilitate its extinguishment.

FOREST ECOSYSTEM REHABILITATION

Ecologically centered forest rehabilitation could be employed as a drastic measure to counter the widespread loss and degradation of forests. As with other management interventions, rehabilitation efforts should be directed first to areas where maximum benefits are likely to be achieved.

A key goal for rehabilitation is prioritizing appropriate criteria and indices, an issue that has not been adequately addressed (Orsi et al., 2011). Regarding forests, criteria related to management objectives can be helpful in sustainable forest management (Nussbaum & Simula, 2013). The criteria can be defined as the essential elements or main components that define sustainable forest management. Indices are qualitative or quantitative parameters of a criterion that provide a basis for assessing the state and trends of forests and forest management. Nevertheless, forest protection and rehabilitation can be seen as one of the management options that contribute to the broader objectives of sustainable forest management (Orsi et al., 2011). The rehabilitation method should be consistent with the assessed damage, not only to the vegetative cover but also to the various components of the soil, such as the soil organic matter content (Soto & Valencia, 2021).

Rehabilitation can contribute to biodiversity conservation while providing economic and livelihood benefits to local communities and should not be overlooked (Mansourian et al., 2005; Orsi et al., 2011). Different objectives may lead to the identification of distinct priority areas, the establishment of different tree species, and the selection of various rehabilitation methods. Therefore, approaches should consider the multiple objectives and the possible consequences to be explored.

However, the benefits of forest rehabilitation depend on the scale and spatial distribution of the rehabilitation process. Forest rehabilitation on a landscape scale has a positive impact on the environment by restoring biodiversity, protecting soil, reducing flood risk, absorbing carbon (Lewis et al., 2019), and enhancing water retention and underground reservoir enrichment (Huang et al., 2019).

Forest rehabilitation, when properly planned and carefully implemented, is expected to be one of the key pathways for advancing biodiversity conservation (Egoh et al., 2014; Fernandez et al., 2017; Huang et al., 2019). Also, rehabilitation promoting livelihoods and prosperity (Erbaugh & Oldekop, 2018) as well as landscape regeneration and restoration of degraded ecosystems (Carvalho Ribeiro et al., 2020; Huang et al., 2019; Keesstra et al., 2018; Lewis et al., 2019; Luo et al., 2019).

The burned part of the study area is undergoing natural regeneration with native species. The regeneration of Aleppo pine (*Pinus halepensis*) is satisfactory, while the recovery of shrubs and oaks (*Quercus spp.*) is consistent with pre-fire conditions (Figure 5). Based on these outcomes, artificial plantations are not generally required, except in sites where natural regeneration fails, particularly in areas that have been repeatedly burned or otherwise degraded.

Figure 5. Natural Regeneration of Pine in the Study Area.

Nevertheless, it is advisable to reinforce the broadleaf component, primarily oaks, actively, and to guide stand development toward mixed conifer–broadleaf mosaics. Such mixtures typically enhance fire resilience and increase biotic resistance, including resistance to insects and pathogens, by promoting higher structural and species diversity. Where enrichment planting is implemented, locally adapted species should be used, and basic establishment care should be provided until successful survival is ensured (Gómez et al., 2020; Jactel et al., 2021; MacLean & Clark, 2021).

Mixed stands can be developed over time through adaptive silviculture. In both the initial and subsequent phases, the broadleaf component of lowland pine forests can be promoted by releasing existing broadleaf individuals. This may involve selectively removing adjacent pines to create small canopy gaps and protecting regenerating broadleaves from competing vegetation. Although artificial oak planting is theoretically desirable in the study area, it is not recommended as a general practice owing to the high costs of establishment and the irrigation requirements for some years during the hot and dry summer period. Where sufficient resources are available, enrichment planting with locally adapted oak species should be selectively applied to high-quality sites and to locations where natural regeneration has not succeeded.

Extensive post-fire erosion-control works are no longer necessary. Exceptions include steep or convergent terrain and locations lacking adequate ground cover, where localized measures may still be justified until vegetation fully stabilizes the soil.

The cost of fighting and extinguishing fires is generally high (Birot, 2009). Integrated forest management can effectively reduce the size of burned areas, thereby reducing the impact and costs associated with fighting forest fires.

Implementing rational and integrated management of forest ecosystems primarily depends on clearly defining management objectives. In this work, the value of peri-urban forests of the Attica region is primarily viewed from an aesthetic and protective perspective. More specifically, these forests, alongside their recreational function, protect residential areas from floods and landslides and improve the local climate. They are regarded as having an exceptional value that significantly surpasses their traditional management purpose, as it is timber production.

These forests play a vital role in the quality of life of the population of neighboring urban areas. The direct economic valuation of ecosystem services is often not feasible, as important services, such as recreation, do not have a readily identifiable market value. However, there are methods by which their value can be estimated, such as the contingent valuation method (Kaloudis et al., 2021).

The most economical and practical approach to protecting the forest is considered to be integrated multi-purpose forest management. This approach has significant advantages, including the reintroduction of traditional forest uses, such as timber production, controlled grazing of livestock herds, and beekeeping. These uses will bring workers back to the forest, who will help guard and remove excess biomass. At the same time, valuable goods will be produced, such as livestock products from grazing livestock herds and woody products from forest management.

Firebreaks, when appropriately designed and adapted to site-specific ecological and topographical conditions, constitute a critical component of integrated at landscape-level forest management. Although firebreaks alone are insufficient to suppress wildfires, their layout should ensure continuous and reliable access for firefighting personnel and equipment, thereby enhancing the overall effectiveness of wildfire response efforts.

Investments in well-managed forestry and outdoor-recreation infrastructure (e.g., hunting grounds, game reserves, way-marked hiking trails, lookouts) can increase legitimate human presence in forests. When combined with effective fire prevention and safety measures, these activities can reduce wildfire risk by facilitating early detection and enabling rapid response.

Complementary Measures

Citizen Engagement. Protection of individual dwellings: Individual dwellings within forests or wooded areas should be protected and fire-proofed under the responsibility of their owners. Homeowners should clear vegetation around houses. If dense forest is present nearby, they must obtain local authority approval to reduce fuel loads and improve fire protection (Konstantinidis, 2003).

Educating citizens. Proper training and information for citizens, provided through school education, mass media, and specialized seminars, could reduce human carelessness and negligence, which are the primary causes of most forest fire ignitions. Public awareness campaigns and training programs for local volunteers could also improve fire preparation, forest surveillance, and response to fire. Moreover, engaging local communities in self-fire prevention efforts, such as maintaining a safe home environment, is crucial for reducing the risk of fire.

Enhancing the Resilience of Residential Areas. Improving the resilience of residential and industrial zones to wildfire is a critical priority. Beyond the measures detailed above, additional interventions can enhance the resilience of residential areas, such as:

Vegetation replacement inside the urban fabric. Gradual replacement of highly flammable species—especially resinous pines—with site-adapted, low-flammability native broad-leaved trees (e.g., oaks and plane trees). Pair replacement with routine maintenance (pruning, litter removal, irrigation where appropriate).

Evacuation Planning. Designate and maintain primary and alternate evacuation routes to account for potential blockages of main roads. Include wayfinding, night-time visibility, and provisions for persons with reduced mobility.

Safe assembly/temporary areas. Establish and equip clearly identified gathering points with first-aid supplies, water, communication equipment, and shelter provisions, with specific arrangements in place for vulnerable populations.

Protected approach and egress corridors. Provide fuel-reduced approach and escape routes to and from settlements to facilitate safe evacuation and emergency access under smoke conditions.

CENTRALIZED FIRE RISK MANAGEMENT

The growing frequency and severity of wildfires make fire risk reduction and integrated forest management increasingly demanding. Local authorities often face resource and expertise shortages, which limit their ability to design and implement effective prevention plans. Covering these gaps exclusively through large-scale recruitment and funding is very demanding in terms of financing and time spent;

Alternatively, establishing a centralized forest and fire management agency could represent a cost-effective solution.

Such an agency would combine advanced technologies, scientific knowledge, and specialized personnel to design and implement pilot fire risk reduction plans through forest management, supported by monitoring systems that evaluate the effectiveness of these plans. A key role would be to transform knowledge and best practices into applicable guidelines and specifications. Regional and local forest services would remain essential partners, ensuring that national frameworks are adapted to local ecological and socio-economic conditions through adapted planning, implementation, and monitoring. The advantages of such a centralized structure include enhanced coordination across regions, consistent application of prevention policies, reduced fragmentation, and more efficient use of resources. International experience supports this approach as in Portugal, the Agência para a Gestão Integrada de Fogos Rurais provides nationwide coordination; while in the United States and Canada, the U.S. Forest Service, the National Interagency Fire Center, and the Canadian Interagency Forest Fire Centre ensure technological support, monitoring, and resource-sharing across local services.

These examples demonstrate that centralized frameworks, when complemented by strong local-level collaboration, can significantly enhance wildfire risk reduction through integrated forest management.

LIMITATIONS AND FUTURE WORK

The Fire Destruction Danger Index (FDDI) relies on recent historical fire data and assumes relative stability in ignition frequency and average burned area within the forthcoming forest management period. Given that the projection horizon is relatively short (usually five to ten years), this assumption is considered reasonable, as the datasets already reflect the most recent conditions and trends. Nevertheless, uncertainties related to potential shifts in climate extremes, land-use practices, and human activity cannot be entirely excluded. Furthermore, scenario analysis can be implemented to account for possible deviations from recent historical patterns.

Future research should focus on benchmarking the FDDI against other long-term or landscape-level fire risk assessment approaches, particularly those that integrate vegetation dynamics, climate projections, and land-use change. Comparative testing with such frameworks, especially those developed for Mediterranean and other fire-prone regions, would help evaluate the relative strengths and weaknesses of the index. Applying the FDDI in diverse ecological and socio-economic contexts would also be highly valuable to test its robustness, adaptability, and transferability beyond the present case study. Such applications could highlight specific advantages and reveal aspects requiring refinement. While socio-economic cost–benefit evaluations of mitigation measures are of great importance for forest management and policy, they extend beyond the core scope of fire risk assessment.

CONCLUSION

Forest fires have consistently increased in frequency and intensity in recent decades, resulting in the loss of human lives and the destruction of natural ecosystems, private properties, and public infrastructure. The increase in fires is primarily attributed to poor forest management, the abandonment of rural areas, and extreme weather conditions. Areas of the planet located in temperate zones and near the equatorial zone, such as the Mediterranean region, are most affected.

The costs associated with wildfire suppression, together with their environmental and economic impacts, are substantial. Consequently, forest fires should be addressed in the same manner as natural disasters, requiring comprehensive legislative frameworks, effective organizational structures, and adequate financial resources. Policymakers, therefore, need to recognize the magnitude of the problem and engage all relevant stakeholders in coordinated strategies.

The defined adapted FDDI proved to be applicable over extended areas and provided a spatial spread of the fire risk in the study area. Based on the calculated FDDI values, integrating fire risk reduction into multipurpose forest management planning enhances forest resilience while promoting economic viability and ecological sustainability. This approach is particularly crucial for peri-urban forests, which hold high ecological and social value by providing essential ecosystem services to adjacent residential areas. At the same time, sustainable forest exploitation contributes to the country's gross domestic product by producing goods and services and increases the income of local communities.

The protection of residential areas can be further advanced through the design and creation of appropriate fire breaks, the management of vegetation within settlements, and the preparation of gathering places and safe escape routes in case of fire.

The establishment of a centralized forest and fire management agency emerges as a promising strategy to address the escalating fire risk. By consolidating technological innovation, scientific expertise, and specialized personnel, such an agency can ensure the consistent implementation of fire risk reduction plans while maintaining the indispensable contribution of regional and local forest services. The experiences of countries such as Portugal, the United States, and Canada provide evidence that centralized planning and coordination, when coupled with strong local collaboration, enhance efficiency, reduce fragmentation, and increase resilience. Therefore, adopting a similar approach could provide a sustainable, cost-effective, and adaptive framework for improving wildfire prevention and integrated forest management at the national level.

A positive development is that both the European Union and national administrations have recognized the current state of forest fire risk, the consequent legislative regulations required, and the need for effective implementation of forest management plans to adequately address wildfire challenges.

ACKNOWLEDGMENTS

We want to thank the "Hellenic Cadaster" for providing the orthophotography maps of the study area.

REFERENCES

Abatzoglou, J. T., & Williams, A. P. (2016). Impact of anthropogenic climate change on wildfire across western US forests. *Proceedings of the National Academy of Sciences of the United States of America, 113*(42), 11770–11775. DOI: 10.1073/pnas.1607171113

Agee, J. K., & Skinner, C. N. (2005). Basic principles of forest fuel reduction treatments. *Forest Ecology and Management, 211*(1-2), 83–96. DOI: 10.1016/j.foreco.2005.01.034

Albini, F. A. (1976). *Estimating Wildfire Behavior and Effects.* (INT-30).

Arbinolo, M., Patimo, G., Rey, E., Stokkeland, O., Verde, J. C., Casartelli, V., Marengo, A., Melinato, S., Mysiak, J., Salpina, D., Afentoulidis, S., Brăilescu, C., & Sørensen, J. (2024). *UCPM Wildfire Peer Review Report: Greece 2024.* https://doi.org/DOI: 10.25424/CMCC-79TS-VV91

Athanasiadis, N. (1986). *Forest Botany (Trees and Shrubs of the Forests of Greece) part II.* Giahoudi.

Bachelet, D., Lenihan, J. M., Daly, C., & Neilson, R. P. (2000). Interactions between fire, grazing and climate change at Wind Cave National Park, SD. *Ecological Modelling, 134*(2-3), 229–244. DOI: 10.1016/s0304-3800(00)00343-4

Bachmann, A., & Allgöwer, B. (1998). Framework for Wildfire Risk Analysis. In Viegas, D. X. (Ed.), *Forest Fire Research.* University of Coimbra.

Baeza, M. J., De Luis, M., Raventos, J., & Escarre, A. (2002). Factors influencing fire behaviour in shrublands of different stand ages and the implications for using prescribed burning to reduce wildfire risk. *Journal of Environmental Management, 65*(2), 199–208. DOI: 10.1006/jema.2002.0545

Baeza, M. J., Raventós, J., Escarré, A., & Vallejo, V. R. (2003). The effect of shrub clearing on the control of the fire-prone species Ulex parviflorus. *Forest Ecology and Management, 186*(1-3), 47–59. DOI: 10.1016/s0378-1127(03)00237-8

Baldwin, V. C., Peterson, K. D., Clark, A., Ferguson, R. B., Strub, M. R., & Bower, D. R. (2000). The effects of spacing and thinning on stand and tree characteristics of 38-year-old Loblolly Pine. *Forest Ecology and Management, 137*(1-3), 91–102. DOI: 10.1016/s0378-1127(99)00340-0

Birot, Y. (2009). *Living with wildfires: What science can tell us - A contribution to the science-policy dialogue.*

Blanco-Rodríguez, M. Á., Ameztegui, A., Gelabert, P., Rodrigues, M., & Coll, L. (2023). Short-term recovery of post-fire vegetation is primarily limited by drought in Mediterranean forest ecosystems. *Fire Ecology, 19*(1), 68–68. DOI: 10.1186/s42408-023-00228-w

Blondel, J., Aronson, J., Bodiou, J. Y., & Boeuf, G. (2010). *The Mediterranean Region: Biological Diversity in Space and Time* (2 ed.). Oxford University Press.

Bowman, D. M. J. S., Williamson, G. J., Abatzoglou, J. T., Kolden, C. A., Cochrane, M. A., & Smith, A. M. S. (2017). Human exposure and sensitivity to globally extreme wildfire events. *Nature Ecology & Evolution, 1*(3), 0058. https://doi.org/DOI: 10.1038/s41559-016-0058

Bradshaw, L. S., Deeming, J. E., & Burgan, R. E. (1984). The 1978 national fire-danger rating system: technical documentation. *General Technical Report INT-169. Ogden, UT: US Department of Agriculture, Forest Service, Intermountain Forest and Range Experiment Station. 44 p., 169.*

Butler, B. W. (2014). Wildland firefighter safety zones: A review of past science and summary of future needs. *International Journal of Wildland Fire, 23*(3), 295–308. DOI: 10.1071/WF13021

Butler, B. W., & Cohen, J. D. (1998a). Firefighter safety zones: A theoretical model based on radiative heating. *International Journal of Wildland Fire, 8*(2), 73–77. DOI: 10.1071/WF9980073

Butler, B. W., & Cohen, J. D. (1998b). Firefighter Safety Zones: How Big Is Big Enough? *Fire Management Notes, 58*(1), 13–16.

Byram, G. M. (1959). Combustion of Forest Fuels. In Use, C. a. (Ed.), *Forest Fire*. McGraw-Hill.

Carter, M. C., & Darwin Foster, C. (2004). Prescribed burning and productivity in southern pine forests: A review. *Forest Ecology and Management, 191*(1-3), 93–109. DOI: 10.1016/j.foreco.2003.11.006

Carvalho Ribeiro, S. M., Rajao, R., Nunes, F., Assis, D., Neto, J. A., Marcolino, C., Lima, L., Rickard, T., Salomao, C., & Filho, B. S. (2020). A spatially explicit index for mapping Forest Restoration Vocation (FRV) at the landscape scale: Application in the Rio Doce basin, Brazil. *The Science of the Total Environment, 744*, 140647. DOI: 10.1016/j.scitotenv.2020.140647

Catry, F. X., Rego, F. C., Bação, F. L., & Moreira, F. (2009). Modeling and mapping wildfire ignition risk in Portugal. *International Journal of Wildland Fire, 18*(8), 921–931. DOI: 10.1071/wf07123

Chambers, M. E., Fornwalt, P. J., Malone, S. L., & Battaglia, M. A. (2016). Patterns of conifer regeneration following high severity wildfire in ponderosa pine – dominated forests of the Colorado Front Range. *Forest Ecology and Management, 378*, 57–67. DOI: 10.1016/j.foreco.2016.07.001

Chuvieco, E., Yebra, M., Martino, S., Thonicke, K., Gómez-Giménez, M., San-Miguel, J., Oom, D., Velea, R., Mouillot, F., Molina, J. R., Miranda, A. I., Lopes, D., Salis, M., Bugaric, M., Sofiev, M., Kadantsev, E., Gitas, I. Z., Stavrakoudis, D., Eftychidis, G., & Viegas, D. (2023). Towards an Integrated Approach to Wildfire Risk Assessment: When, Where, What and How May the Landscapes Burn. *Fire (Basel, Switzerland), 6*(5). Advance online publication. DOI: 10.3390/fire6050215

Cohen, D. J., Deeming, E.J., . (1985). *The National Fire Danger Rating System: Basic Equations* (PSW-82).

Copernicus. (2024). EFFIS statistics – Estimates. https://forest-fire.emergency.copernicus.eu/apps/effis.statistics/estimates

Dafis, S. (1986). *Forest Ecology*. Giahoudi.

Demestichas, K., Sykas, D., Zografakis, D., Kaloudis, S., Kalapodis, N., Sakkas, G., Athanasiou, M., & Costopoulou, C. (2025). Measuring Forest Resilience Against Wildfires and Climate Change: Methods and Technical Approaches. In *Paradigms on Technology Development for Security Practitioners* (pp. 53-63). Springer. https://doi.org/DOI: 10.1007/978-3-031-62083-6_5

Dimitrakopoulos, A., Baxevanis, K. E., & Mitsopoulos, I. D. (2009). Time-course correlation of forest fires and socioeconomic factors in Greece, during the period 1960-2000. 19th Panhellenic Forestry Conference, 1-4 November 2009, Patra.

Dimitrakopoulos, A., Bemmerzouk, A., & Mitsopoulos, I. (2011). Evaluation of the Canadian fire weather index system in an eastern Mediterranean environment. *Meteorological Applications, 18*(1), 83–93. https://doi.org/https://doi.org/10.1002/met.214

Dimitrakopoulos, A. P. (2001a). A statistical classification of Mediterranean species based on their flammability components. *International Journal of Wildland Fire, 10*(2), 113–113. DOI: 10.1071/wf01004

Dimitrakopoulos, A. P. (2001b). Thermogravimetric analysis of Mediterranean plant species. *Journal of Analytical and Applied Pyrolysis, 60*(2), 123–130. DOI: 10.1016/s0165-2370(00)00164-9

Dimitrakopoulos, A. P., & Bemmerzouk, A. M. (2003). Predicting live herbaceous moisture content from a seasonal drought index. *International Journal of Biometeorology, 47*(2), 73–79. DOI: 10.1007/s00484-002-0151-1

Dimitrakopoulos, A. P., & Panov, P. I. (2001). Pyric properties of some dominant Mediterranean vegetation species. *International Journal of Wildland Fire, 10*(1), 23–23. DOI: 10.1071/wf01003

Dimitrakopoulos, A. P., & Papaioannou, K. K. (2001). Flammability Assessment of Mediterranean Forest Fuels. *Fire Technology, 37*(2), 143–152. DOI: 10.1023/a:1011641601076

Duncan, B. N., Martin, R. V., Staudt, A. C., Yevich, R., & Logan, J. A. (2003). Interannual and seasonal variability of biomass burning emissions constrained by satellite observations. *Journal of Geophysical Research, D, Atmospheres, 108*(D2). Advance online publication. DOI: 10.1029/2002jd002378

Duncan, B. W., Adrian, F. W., & Stolen, E. D. (2010). Isolating the lightning ignition regime from a contemporary background fire regime in east-central Florida, USA. *Canadian Journal of Forest Research, 40*(2), 286–297. DOI: 10.1139/x09-193

Egoh, B. N., Paracchini, M. L., Zulian, G., Schägner, J. P., Bidoglio, G., & Jones, J. (2014). Exploring restoration options for habitats, species and ecosystem services in the European Union. *Journal of Applied Ecology, 51*(4), 899–908. DOI: 10.1111/1365-2664.12251

Enright, N. J., Fontaine, J. B., Bowman, D. M. J. S., Bradstock, R. A., & Williams, R. J. (2015). Interval squeeze: Altered fire regimes and demographic responses interact to threaten woody species persistence as climate changes. *Frontiers in Ecology and the Environment, 13*(5), 265–272. DOI: 10.1890/140231

Erbaugh, J. T., & Oldekop, J. A. (2018). Forest landscape restoration for livelihoods and well-being. *Current Opinion in Environmental Sustainability, 32*, 76–83. DOI: 10.1016/j.cosust.2018.05.007

Falk, D. A., van Mantgem, P. J., Keeley, J. E., Gregg, R. M., Guiterman, C. H., Tepley, A. J., Jn Young, D., & Marshall, L. A. (2022). Mechanisms of forest resilience. *Forest Ecology and Management*, *512*, 120129–120129. DOI: 10.1016/j.foreco.2022.120129

FAO. (1986). *Wildland fire management terminology.*

FAO. (1998). *World Reference Base for Soil Resources.* Food and Agriculture Organization of the United Nations.

Fernandes, P. M. (2013). Fire-smart management of forest landscapes in the Mediterranean basin under global change. *Landscape and Urban Planning*, *110*, 175–182. DOI: 10.1016/j.landurbplan.2012.10.014

Fernandes, P. M., & Botelho, H. S. (2003). A review of prescribed burning effectiveness in fire hazard reduction. *International Journal of Wildland Fire*, *12*(2), 117–117. DOI: 10.1071/wf02042

Fernandez, F. A. S., Rheingantz, M. L., Genes, L., Kenup, C. F., Galliez, M., Cezimbra, T., Cid, B., Macedo, L., Araujo, B. B. A., Moraes, B. S., Monjeau, A., & Pires, A. S. (2017). Rewilding the Atlantic Forest: Restoring the fauna and ecological interactions of a protected area. *Perspectives in Ecology and Conservation*, *15*(4), 308–314. DOI: 10.1016/j.pecon.2017.09.004

Forzieri, G., Dakos, V., McDowell, N. G., Ramdane, A., & Cescatti, A. (2022). Emerging signals of declining forest resilience under climate change. *Nature*, *608*(7923), 534–539. DOI: 10.1038/s41586-022-04959-9

Founda, D., & Giannakopoulos, C. (2010). The exceptionally hot summer of 2007 in Athens, Greece — A typical summer in the future climate? *Atmospheric Research*, *98*(2-4), 245–252. DOI: 10.1016/j.atmosres.2010.07.002

Garcês, A., & Pires, I. (2023). The Hell of Wildfires: The Impact on Wildlife and Its Conservation and the Role of the Veterinarian. *Conservation*, *3*(1), 96–108. DOI: 10.3390/conservation3010009

Genton, M. G., Butry, D. T., Gumpertz, M. L., & Prestemon, J. P. (2006). Spatio-temporal analysis of wildfire ignitions in the St Johns River Water Management District, Florida. *International Journal of Wildland Fire*, *15*(1), 87–97. DOI: 10.1071/wf04034

Giannakopoulos, C., Bindi, M., Moriondo, M., LeSager, P., & Tin, T. (2011). An integrated assessment of climate change impacts for Greece in the near future. *Regional Environmental Change*, *11*(4), 829–843. DOI: 10.1007/s10113-011-0219-8

Giannakopoulos, C., LeSager, P., Moriondo, M., Bindi, M., Karali, A., Hatzaki, M., & Kostopoulou, E. (2012). Comparison of fire danger indices in the Mediterranean for present day conditions. *iForest-Biogeosciences and Forestry, 5*(4), 197-203. https://doi.org/https://doi.org/10.3832/ifor0622-005

Gibbons, P., Lindenmayer, D. B., Barry, S. C., & Tanton, M. T. (2000). The effects of slash burning on the mortality and collapse of trees retained on logged sites in south-eastern Australia. *Forest Ecology and Management, 139*(1-3), 51–61. DOI: 10.1016/s0378-1127(99)00333-3

Gimeno, M., & San-Miguel-Ayanz, J. (2004). Evaluation of RADARSAT-1 data for identification of burnt areas in Southern Europe. *Remote Sensing of Environment, 92*(3), 370–375. http://www.sciencedirect.com/science/article/B6V6V-4D7BNR7-3/2/d69d549ec6ce3112965fec443b49b287

Gómez, M. C., González, M., Gómez-Aparicio, L., & Serrano, M. S. (2020). Coexistent Mediterranean woody species as a driving factor of Phytophthora cinnamomi infectivity and survival. *Annals of Applied Biology, 177*(1), 41–50. DOI: 10.1111/aab.12599

Graham, R. T., Alan E., H., Threasa B. Jain, & Tonn, J. R. (1999). *The effects of thinning and similar stand treatments on fire behavior in western forests.*

Hagmann, R. K., Hessburg, P. F., Salter, R. B., Merschel, A. G., & Reilly, M. J. (2022). Contemporary wildfires further degrade resistance and resilience of fire-excluded forests. *Forest Ecology and Management, 506*, 119975–119975. DOI: 10.1016/j.foreco.2021.119975

Hawkes, B., Beck, J., & Sahle, W. (1996). A wildfire threat rating system for the McGregor Model Forest. Hellenic Fire Service. (Retrieved September 27, 2024). *Datasets.* https://www.fireservice.gr/el_GR/synola-dedomenon

Hellenic National Meteorological Service. (Retrieved September 26, 2024). *Climatic data per month.* https://www.emy.gr/en/climatic-data?tab=statistics-tab

Helms, J. A. (1998). *The Dictionary of Forestry.* Society of American Foresters.

Holling, C. S. (1973). Resilience and Stability of Ecological Systems. *Annual Review of Ecology and Systematics, 4*(1), 1–23. DOI: 10.1146/annurev.es.04.110173.000245

Huang, C., Zhou, Z., Peng, C., Teng, M., & Wang, P. (2019). How is biodiversity changing in response to ecological restoration in terrestrial ecosystems? A meta-analysis in China. *The Science of the Total Environment, 650*(Pt 1), 1–9. DOI: 10.1016/j.scitotenv.2018.08.320

IGME. (1983). *Geological Map of Greece, Scale 1:50,000, Sheet Athens*. Athens, Greece, IGME.

IGME. (1993). *Explanatory Notes on the Geological Map of Greece, Sheet Athens*.

Jactel, H., Moreira, X., & Castagneyrol, B. (2021). Tree Diversity and Forest Resistance to Insect Pests: Patterns, Mechanisms, and Prospects. *Annual Review of Entomology*, *66*(1), 277–296. DOI: 10.1146/annurev-ento-041720-075234

Johnstone, J. F., Allen, C. D., Franklin, J. F., Frelich, L. E., Harvey, B. J., Higuera, P. E., Mack, M. C., Meentemeyer, R. K., Metz, M. R., Perry, G. L. W., Schoennagel, T., & Turner, M. G. (2016). Changing disturbance regimes, ecological memory, and forest resilience. *Frontiers in Ecology and the Environment*, *14*(7), 369–378. DOI: 10.1002/fee.1311

Jolly, W. M., Cochrane, M. A., Freeborn, P. H., Holden, Z. A., Brown, T. J., Williamson, G. J., & Bowman, D. M. J. S. (2015). Climate-induced variations in global wildfire danger from 1979 to 2013. *Nature Communications*, *6*, 7537. DOI: 10.1038/ncomms8537

Kalabokidis, K. D., & Omi, P. N. (1998). Reduction of Fire Hazard Through Thinning/Residue Disposal in the Urban Interface. *International Journal of Wildland Fire*, *8*(1), 29–29. DOI: 10.1071/wf9980029

Kalapodis, N., Sakkas, G., Lazarou, A., Casciano, D., Demestichas, K., Athanasiou, M., Kaloudis, S., & Sykas, D. (2025). EU-Integrated Multifunctional Forest and Fire Management, Policies, and Practices: Challenges Between "As-Is" and "To-Be" State. In *Paradigms on Technology Development for Security Practitioners* (pp. 65-77). Springer. https://doi.org/DOI: 10.1007/978-3-031-62083-6_6

Kaloudis, S. (2008). *Decision Support System for Forest Ecosystem Management Based on Geographic Information Systems and Knowledge Systems*. [Doctoral dissertation, Agricultural University of Athens]. Athens, Greece.

Kaloudis, S., Costopoulou, C. I., Lorentzos, N. A., Sideridis, A. B., & Karteris, M. (2008). Design of forest management planning DSS for wildfire risk reduction. *Ecological Informatics*, *3*(1), 122–133. DOI: 10.1016/j.ecoinf.2007.07.008

Kaloudis, S., Glykou, M., Galanopoulou, S., Fotiadis, G., Yialouris, C., & Raptis, D. (2023). Land Cover Changes in Evrytania Prefecture (Greece). *Forests*, *14*(7), 1462–1462. DOI: 10.3390/f14071462

Kaloudis, S., Platia, K., Pantera, A., & Galanopoulou, S. (2021). Investigation on the importance of ecosystem services based on the contingent valuation method. *MIBES Transactions, 15*(1), 87-100. https://mibes.uth.gr/vol15_issue1_2021/Kaloudis%2C%20Platia%2C%20Pantera%20%26%20Galanopoulou.pdf

Kaloudis, S., Tocatlidou, A., Lorentzos, N. A., Sideridis, A. B., & Karteris, M. (2005). Assessing Wildfire Destruction Danger: A Decision Support System Incorporating Uncertainty. *Ecological Modelling, 181*(1), 25–38. DOI: 10.1016/j.ecolmodel.2004.06.021

Kambezidis, H. D., & Kalliampakos, G. K. (2016). Fire-Risk Assessment in Northern Greece Using a Modified Fosberg Fire-Weather Index That Includes Forest Coverage. *International Journal of Atmospheric Sciences, 2016*(1), 8108691. DOI: 10.1155/2016/8108691

Karali, A., Hatzaki, M., Giannakopoulos, C., Roussos, A., Xanthopoulos, G., & Tenentes, V. (2014). Sensitivity and evaluation of current fire risk and future projections due to climate change: The case study of Greece. *Natural Hazards and Earth System Sciences, 14*(1), 143–153. DOI: 10.5194/nhess-14-143-2014

Keesstra, S., Nunes, J., Novara, A., Finger, D., Avelar, D., Kalantari, Z., & Cerda, A. (2018). The superior effect of nature based solutions in land management for enhancing ecosystem services. *The Science of the Total Environment, 610-611*, 997–1009. DOI: 10.1016/j.scitotenv.2017.08.077

Kirchmeier-Young, M. C., Gillett, N. P., Zwiers, F. W., Cannon, A. J., & Anslow, F. S. (2019). Attribution of the influence of human-induced climate change on an extreme fire season. *Earth's Future, 7*(1), 2–10. DOI: 10.1029/2018EF001050

Koniavitis, G. (2024). *Forest fire risk assessment and mitigation measures: Case study of Varibombi*. Attica. [Master's thesis, Agricultural University of Athens, Department of Forestry and Natural Environment Management]

Konstantinidis, P. (2003). *Learning to Live with Forest Fires*. Christodoulides.

Kosmas, C., Danalatos, N., & Moustakas, N. (1993). The soils of the broader area of Attica: Properties, classification, and land use potential. *Geoderma, 56*(1-4), 401–419. DOI: 10.1016/0016-7061(93)90131-I

Lewis, T., Verstraten, L., Hogg, B., Wehr, B. J., Swift, S., Tindale, N., Menzies, N. W., Dalal, R. C., Bryant, P., Francis, B., & Smith, T. E. (2019). Reforestation of agricultural land in the tropics: The relative contribution of soil, living biomass and debris pools to carbon sequestration. *The Science of the Total Environment, 649*, 1502–1513. DOI: 10.1016/j.scitotenv.2018.08.351

Liedloff, A. C., Coughenour, M. B., Ludwig, J. A., & Dyer, R. (2001). Modelling the trade-off between fire and grazing in a tropical savanna landscape, northern Australia. *Environment International*, *27*(2-3), 173–180. DOI: 10.1016/s0160-4120(01)00079-4

Liodakis, S., Bakirtzis, D., & Dimitrakopoulos, A. P. (2003). Autoignition and thermogravimetric analysis of forest species treated with fire retardants. *Thermochimica Acta*, *399*(1-2), 31–42. DOI: 10.1016/s0040-6031(02)00400-8

Luo, Y., Lu, Y., Fu, B., Zhang, Q., Li, T., Hu, W., & Comber, A. (2019). Half century change of interactions among ecosystem services driven by ecological restoration: Quantification and policy implications at a watershed scale in the Chinese Loess Plateau. *The Science of the Total Environment*, *651*(Pt 2), 2546–2557. DOI: 10.1016/j.scitotenv.2018.10.116

MacLean, D. A., & Clark, K. L. (2021). Mixedwood management positively affects forest health during insect infestations in eastern North America[1]. *Canadian Journal of Forest Research*, *51*(7), 910–920.

Mansourian, S., Vallauri, D., & Dudley, N. (2005). *Forest Restoration in Landscapes*. Springer New York., DOI: 10.1007/0-387-29112-1

Mhawej, M., Faour, G., Abdallah, C., & Adjizian-Gerard, J. (2016). Towards an establishment of a wildfire risk system in a Mediterranean country. *Ecological Informatics*, *32*, 167–184.

Millar, C. I., & Stephenson, N. L. (2015). Temperate forest health in an era of emerging megadisturbance. *Science*, *349*(6250), 823–826. DOI: 10.1126/science.aaa9933

Mitsopoulos, I., Mallinis, G., Dimitrakopoulos, A., Xanthopoulos, G., Eftychidis, G., & Goldammer, J. G. (2020). Vulnerability of peri-urban and residential areas to landscape fires in Greece: Evidence by wildland-urban interface data. *Data in Brief*, *31*, 106025. DOI: 10.1016/j.dib.2020.106025

Nikinmaa, L., de Koning, J. H. C., Derks, J., Grabska-Szwagrzyk, E., Konczal, A. A., Lindner, M., Socha, J., & Muys, B. (2024). The priorities in managing forest disturbances to enhance forest resilience: A comparison of a literature analysis and perceptions of forest professionals. *Forest Policy and Economics*, *158*, 103119–103119. DOI: 10.1016/j.forpol.2023.103119

North, M., Collins, B. M., & Stephens, S. L. (2012). Using fire to increase the scale, benefits, and future maintenance of fuels treatments. *Journal of Forestry*, *110*(7), 392–401. DOI: 10.5849/jof.12-021

Nussbaum, R., & Simula, M. (2013). *The Forest Certification Handbook*. Routledge., DOI: 10.4324/9781849773003

Oliver, T. H., Heard, M. S., Isaac, N. J. B., Roy, D. B., Procter, D., Eigenbrod, F., Freckleton, R., Hector, A., Orme, C. D. L., Petchey, O. L., Proenca, V., Raffaelli, D., Suttle, K. B., Mace, G. M., Martin-Lopez, B., Woodcock, B. A., & Bullock, J. M. (2015). Biodiversity and Resilience of Ecosystem Functions. *Trends in Ecology & Evolution*, *30*(11), 673–684. DOI: 10.1016/j.tree.2015.08.009

Orsi, F., Geneletti, D., & Newton, A. C. (2011). Towards a common set of criteria and indicators to identify forest restoration priorities: An expert panel-based approach. *Ecological Indicators*, *11*(2), 337–347. DOI: 10.1016/j.ecolind.2010.06.001

Papagiannaki, K., Giannaros, T. M., Lykoudis, S., Kotroni, V., & Lagouvardos, K. (2020). Weather-related thresholds for wildfire danger in a Mediterranean region: The case of Greece. *Agricultural and Forest Meteorology*, *291*, 108076. DOI: 10.1016/j.agrformet.2020.108076

Papaioannou, G., Kaloudis, S., & Kerkides, P. (1998). On the proper employment of Piche evaporimeters in estimating evapotranspiration. *International Journal of Climatology*, *18*(11), 1247–1260. DOI: 10.1002/(SICI)1097-0088(199809)18:11<1247:AID-JOC310>3.0.CO;2-U

Pausas, J. G., & Keeley, J. E. (2014). Abrupt climate-independent fire regime changes. *Ecosystems (New York, N.Y.)*, *17*(6), 1109–1120. DOI: 10.1007/s10021-014-9773-5

Pausas, J. G., & Keeley, J. E. (2019). Wildfires as an ecosystem service. *Frontiers in Ecology and the Environment*, *17*(5), 289–295. DOI: 10.1002/fee.2044

Pimm, S. L. (1984). The complexity and stability of ecosystems. *Nature*, *307*(5949), 321–326. DOI: 10.1038/307321a0

Plucinski, M. P. (2012). *A review of wildfire occurrence research*. Bushfire Cooperative Research Centre.

Politi, N., Vlachogiannis, D., Sfetsos, A., Gounaris, N., & Varela, V. (2023). Investigation of Fire Weather Danger under a Changing Climate at High Resolution in Greece. *Sustainability*, *15*(3), 2498. https://www.mdpi.com/2071-1050/15/3/2498

Quézel, P., & Médail, F. (2003). *Ecology and Biogeography of Mediterranean Basin Forests*. Springer., DOI: 10.1007/978-3-662-05238-9

Ribeiro, I., Domingos, T., McCracken, D., & Proença, V. (2023). The use of domestic herbivores for ecosystem management in Mediterranean landscapes. *Global Ecology and Conservation*, *46*, e02577–e02577. DOI: 10.1016/j.gecco.2023.e02577

Rivas-Martínez, S., Sáenz, S. R., & Penas, A. (2011). Worldwide Bioclimatic Classification System. *Global Geobotany*, *1*, 1–634. DOI: 10.5616/gg110001

Romero-Calcerrada, R., Barrio-Parra, F., Millington, J. D. A., & Novillo, C. J. (2010). Spatial modelling of socioeconomic data to understand patterns of human-caused wildfire ignition risk in the SW of Madrid (central Spain). *Ecological Modelling*, *221*(1), 34–45. DOI: 10.1016/j.ecolmodel.2009.08.008

Romero-Calcerrada, R., Novillo, C. J., Millington, J. D. A., & Gomez-Jimenez, I. (2008). GIS analysis of spatial patterns of human-caused wildfire ignition risk in the SW of Madrid (Central Spain). *Landscape Ecology*, *23*(3), 341–354. DOI: 10.1007/s10980-008-9190-2

Rothermel, R. C. (1972). *A mathematical model for predicting fire spread in wildland fuels*. F. S. Intermountain Forest & Range Experiment Station, US Department of Agriculture.

Ruffault, J., Curt, T., Martin-StPaul, N. K., Moron, V., & Trigo, R. M. (2018). Extreme wildfire events are linked to global-change–type droughts in the northern Mediterranean. *Nature Ecology & Evolution*, *2*(2), 279–285. DOI: 10.1038/s41559-017-0421-7

Samara, T., Raptis, D., & Spanos, I. (2018). Fuel treatments and potential fire behavior in peri-urban forests in Northern Greece. *Environments*, *5*(7), 79. DOI: 10.3390/environments5070079

Sanchez Sanchez, Y., Martinez-Grana, A., Santos Frances, F., & Mateos Picado, M. (2018). Mapping Wildfire Ignition Probability Using Sentinel 2 and LiDAR (Jerte Valley, Caceres, Spain). *Sensors (Basel)*, *18*(3). Advance online publication. DOI: 10.3390/s18030826

Sari, F. (2022). Identifying anthropogenic and natural causes of wildfires by maximum entropy method-based ignition susceptibility distribution models. *Journal of Forestry Research*, *34*(2), 355–371. DOI: 10.1007/s11676-022-01502-4

Scherer, G., Zabowski, D., Java, B., & Everett, R. (2000). Timber harvesting residue treatment. Part II. Understory vegetation response. *Forest Ecology and Management*, *126*(1), 35–50. DOI: 10.1016/s0378-1127(99)00080-8

Seidl, R., Thom, D., Kautz, M., Martin-Benito, D., Peltoniemi, M., Vacchiano, G., Wild, J., Ascoli, D., Petr, M., Honkaniemi, J., Lexer, M. J., Trotsiuk, V., Mairota, P., Svoboda, M., Fabrika, M., Nagel, T. A., & Reyer, C. P. O. (2017). Forest disturbances under climate change. *Nature Climate Change*, *7*(6), 395–402. DOI: 10.1038/nclimate3303

Soto, M. C., & Valencia, A. P. (2021). Proposals of Rehabilitation and Management Actions for the Protection of Sclerophyllous Forests Affected by Forest Fires. *Forestist, 72*(2), 112–119. DOI: 10.5152/forestist.2021.21029

Stephens, S. L., Agee, J. K., Fulé, P. Z., North, M. P., Romme, W. H., Swetnam, T. W., & Turner, M. G. (2013). Managing forests and fire in changing climates. *Science, 342*(6154), 41–42. DOI: 10.1126/science.1240294

Stevens-Rumann, C. S., Kemp, K. B., Higuera, P. E., Harvey, B. J., Rother, M. T., Donato, D. C., Morgan, P., & Veblen, T. T. (2018). Evidence for declining forest resilience to wildfires under climate change. *Ecology Letters, 21*(2), 243–252. DOI: 10.1111/ele.12889

Stougiannidou, D., & Zafeiriou, E. (2022). Wildfire economic impact assessment: An empirical model-based investigation for Greek agriculture. *Modeling Earth Systems and Environment, 8*(3), 3357–3371.

Syphard, A. D., Radeloff, V. C., Keuler, N. S., Taylor, R. S., Hawbaker, T. J., Stewart, S. I., & Clayton, M. K. (2008). Predicting spatial patterns of fire on a southern California landscape. *International Journal of Wildland Fire, 17*(5), 602–613. DOI: 10.1071/wf07087

Thompson, I., Mackey, B., & McNulty, S. G. (2009). *Forest Resilience, Biodiversity, and Climate Change. A Synthesis of the Biodiversity/Resilience/Stability Relationship in Forest Ecosystems*. https://www.researchgate.net/publication/232129022

Torrano, L., & Valderrábano, J. (2005). Grazing ability of European black pine understorey vegetation by goats. *Small Ruminant Research, 58*(3), 253–263. DOI: 10.1016/j.smallrumres.2004.11.001

Trigo, R. M., Pereira, J. M., Pereira, M. G., Mota, B., Calado, T. J., Dacamara, C. C., & Santo, F. E. (2006). Atmospheric conditions associated with the exceptional fire season of 2003 in Portugal. *International Journal of Climatology, 26*(13), 1741–1758.

Troumbis, A. Y., Gaganis, C. M., & Sideropoulos, H. (2023). Probabilistic Wildfire Risk Assessment and Modernization Transitions: The Case of Greece. *Fire (Basel, Switzerland), 6*(4), 158. https://www.mdpi.com/2571-6255/6/4/158

Tsangari, K., Karetsos, G., & Proutsos, N. (2011). *Forest fires in Greece 1983-2008* (978-960-88241-3-3). https://www.researchgate.net/publication/351330520_Forest_fires_in_Greece_1983-2008_in_greek

Turco, M., Rosa-Cánovas, J. J., Bedia, J., Jerez, S., Montávez, J. P., Llasat, M. C., & Provenzale, A. (2018). Exacerbated fires in Mediterranean Europe due to anthropogenic warming projected with non-stationary climate-fire models. *Nature Communications*, *9*(1), 3821. DOI: 10.1038/s41467-018-06358-z

Turner, M. G. (2010). Disturbance and landscape dynamics in a changing world. *Ecology*, *91*(10), 2833–2849. DOI: 10.1890/10-0097.1

Valderrábano, J., & Torrano, L. (2000). The potential for using goats to control Genista scorpius shrubs in European black pine stands. *Forest Ecology and Management*, *126*(3), 377–383. DOI: 10.1016/s0378-1127(99)00108-5

Varela, V., Vlachogiannis, D., Sfetsos, A., Politi, N., & Karozis, S. (2020). Methodology for the Study of Near-Future Changes of Fire Weather Patterns with Emphasis on Archaeological and Protected Touristic Areas in Greece. *Forests*, *11*(11), 1168. https://www.mdpi.com/1999-4907/11/11/1168

Wang, Y., & Anderson, K. R. (2010). An evaluation of spatial and temporal patterns of lightning- and human-caused forest fires in Alberta, Canada, 1980 - 2007. *International Journal of Wildland Fire*, *19*(8), 1059–1072. DOI: 10.1071/wf09085

Yassoglou, N., Kosmas, C., Moustakas, N., & Detsis, V. (1987). *The Soils of Greece*. Soil Survey Division, Ministry of Agriculture.

Ye, J., Wu, M., Deng, Z., Xu, S., Zhou, R., & Clarke, K. C. (2017). Modeling the spatial patterns of human wildfire ignition in Yunnan province, China. *Applied Geography (Sevenoaks, England)*, *89*, 150–162. DOI: 10.1016/j.apgeog.2017.09.012

Zabowski, D., Java, B., Scherer, G., Everett, R. L., & Ottmar, R. (2000). Timber Harvesting Residue Treatment: Part I. Responses of Conifer Seedlings, Soils and Microclimate. *Forest Ecology and Management*, *126*(1), 25–34. http://www.sciencedirect.com/science/article/B6T6X-3Y9H9FT-3/2/7b40d498ed732b1e6a8ab8d9a09f2149

Zacharakis, I., & Tsihrintzis, V. A. (2023). Environmental Forest Fire Danger Rating Systems and Indices around the Globe: A Review. *Land (Basel)*, *12*. Advance online publication. DOI: 10.3390/land12010194

Zagas, T., Raptis, D., Zagas, D., & Karamanolis, D. (2013). Planning and assessing the effectiveness of traditional silvicultural treatments for mitigating wildfire hazard in pine woodlands of Greece. *Natural Hazards*, *65*(1), 545–561. DOI: 10.1007/s11069-012-0380-5

Zhang, Z. X., Zhang, H. Y., & Zhou, D. W. (2010). Using GIS spatial analysis and logistic regression to predict the probabilities of human-caused grassland fires. *Journal of Arid Environments*, *74*(3), 386–393. DOI: 10.1016/j.jaridenv.2009.09.024

Chapter 7
Bridging Gaps in Digital Agriculture Education in Europe

Konstantinos Demestichas
Department of Agricultural Economics and Rural Development, Agricultural University of Athens, Greece

Maria Ntaliani
https://orcid.org/0000-0001-7643-5009
Department of Agricultural Economics and Rural Development, Agricultural University of Athens, Greece

Effrosyni Bitakou
https://orcid.org/0009-0003-0206-8420
Department of Agricultural Economics and Rural Development, Agricultural University of Athens, Greece

Constantina Costopoulou
https://orcid.org/0000-0003-0151-5586
Agricultural University of Athens, Greece

Caterina Constantinou
https://orcid.org/0000-0003-3589-0739
Rezos Brands S.A., Greece

Angeliki Milioti
Smart Agro Hub, Greece

Branislava Lalić
University of Novi Sad, Serbia

Nenad Magazin
University of Novi Sad, Serbia

Mladen Radišić
University of Novi Sad, Serbia

Olivera Stojilović Trivunić
https://orcid.org/0009-0006-8869-4765
Food Scale Hub, Serbia

Jürgen Vangeyte
ILVO, Belgium

Charlotte Niyomizero
ILVO, Belgium

Reinout Godaert
ILVO, Belgium

Koen Uyttenhove
ODYC, Belgium

Jan Pieters
Ghent University, Belgium

DOI: 10.4018/979-8-3693-8237-0.ch007

ABSTRACT

The European Union has given special attention for supporting higher education institutions in widening countries that lag behind in research and innovation. Till now, non-widening countries have successfully fostered ecosystems for promoting collaboration between universities, leveraging advanced digital technologies. Digital agriculture comprises a particular area of concern for widening countries that requires adequate support to bridge the gap and achieve scientific excellence. Higher education institutions that provide agricultural courses can foster research and innovation while equipping the future agricultural workforce with essential skills in digital agriculture. The objective of this article is to study the current state and future needs of digital agriculture education in universities. A questionnaire-based survey was conducted in a sample of fifty faculty members in Greece, Serbia, and Belgium. Valuable remarks for the development of digital agriculture in higher education were provided for benefiting the stakeholders of the agricultural sector.

INTRODUCTION

In recent years, in the context of the digital transformation of the agricultural sector several concepts have emerged to express this evolution, such as Agriculture 4.0, smart farming, precision agriculture, and Digital Agriculture (DA) (Klerkx et al., 2019). As highlighted in the literature, DA is defined in various ways and not with a single agreed-upon definition (Saçtı et al., 2023; An, 2024). DA involves guiding decision-making in the agriculture value chain by using detailed digital information (Shepherd et al., 2018). From another perspective, it can be considered as a comprehensive set of tools that collect, store, analyze, and share digital information across the entire food value chain, from production to consumption. It integrates a rapidly evolving and diverse range of digital technologies across various segments of the value chain, with a strong foundation in data generation and management systems (OVER et al., 2021). The utilization of advanced sensor capabilities, enhanced data connectivity and data analytics, and the integration of computer-based artificial or augmented intelligence can enable decision support and self-learning systems (Shepherd et al., 2020), that facilitate connections between farmers and the broader value chain (Daum et al., 2022).

According to policies and industry perspectives, DA is seen as the key to enhancing the productivity, profitability, and sustainability of agricultural enterprises (Muzamil et al., 2025). Among the reported benefits of DA are higher yields with reduced inputs, comprising a solution to meeting the food demands of a growing global population. Also, DA enhances environmental stewardship by simultaneously addressing the adverse environmental and climate impacts associated with industrial agriculture, and social advantages such as decreased manual labor on farms (Edwards et al., 2020).

The adoption of digital solutions in agriculture can enhance both technical and allocative efficiency by optimizing the use of capital, such as machinery and equipment, and improving labor productivity and input utilization through better access to skills and knowledge. It also supports informed decision-making on farms by enabling the collection, processing, and dissemination of precise, timely, and geo-spatially tailored agronomic, weather, and market data. Furthermore, it reduces transaction costs by strengthening linkages between producers and consumers, streamlining market interactions, and improving efficiency across the value chain. By leveraging digital innovations, agriculture presents a significant opportunity to drive sustainable economic growth and development. In conclusion, DA is a multidisciplinary field that encompasses agriculture, mathematics, engineering, computer science, and other related disciplines (Soma & Nuckchady, 2021), and can have various effects on agriculture, economy, and society. However, its efficient exploitation requires digital readiness from agricultural stakeholders.

Studies in different countries around the world underline that there is an imbalance in the rapid evolution of DA technologies and the provision of relevant education and training (Pogorelskaia & Várallyai, 2020; Wu et al., 2023). This creates an educational and market gap that hinders the evolution of DA. It has been acknowledged that higher education has a significant role not only in bridging digital technology gaps, but also technology reskilling and upskilling future's agricultural stakeholders. Universities can speed up digital transformation in agriculture by enriching their curriculum with DA, transferring accumulated scientific knowledge into the market and fostering innovative ecosystems (Ozdogan et al., 2017). Nonetheless, more emphasis has to be given to research that focuses on closing this digital divide by leveraging higher education programs (Bambasiadou et al., 2024). Moreover, it is unclear where Higher Education Institutions' (HEIs) educators can fully integrate the opportunities and risks of DA in their courses and implement sufficient approaches in the process of transforming knowledge (Wu & Takács-György, 2023). The research for examining DA from HEIs educators' point of view is nascent, but crucial to investigate their approach and identify potential gaps that must be tackled (Soma & Nuckchady, 2021).

In this light, the objective of this article is to study the current state-of-the-art and future requirements of DA education in HEIs. Specifically, this study examines DA in HEIs from faculty members' side of view on the widening country cases of Greece and Serbia and one non-widening country, namely Belgium.

The concept of widening and non-widening countries is referred by the European Commission (EC) in the context of the Widening Participation and Spreading Excellence actions under the Horizon Europe Program. It focuses on building Research and Innovation (R&I) capacity for countries lagging behind, characterized as widening countries. Also, it aims to enhance and expand participation from countries that face disadvantages in R&I, such as insufficient infrastructure, limited funding, and challenges in building sustainable networks or retaining talent. Currently, fifteen member states are considered as widening countries along with all outermost regions, and the associated countries that perform low in R&I.

According to the "Science, Research and Innovation performance" (SRIP) report (European Commission, 2024) that analyses R&I performance and dynamics, Greece and Serbia have notably low expenditure levels compared to other European Union (EU) and neighboring countries. This limited investment places them well below the EU average, highlighting substantial gaps in research funding that may affect their capacity for scientific advancement, innovation, and competitiveness. Moreover, it can constrain opportunities for technological growth, reduce collaboration potential within the EU, and hinder economic diversification in these countries, making it challenging to catch up with heavily investing countries.

For both widening countries, one of the critical sectors for R&I investment is agriculture, given the importance of agricultural production to their economies. In Greece, the agricultural sector is predominantly composed of small, family-run farms that are widely distributed across the country. This sector employs about 12% of the workforce and accounts for roughly 3.9% of the national Gross Domestic Product (GDP) (World Bank, 2024a). However, despite its vital role in the economy, the sector is struggling with the establishment of digital transformation. Regarding the Digital Maturity Index (DMI), Greece ranks 27th out of 28 European countries, indicating significant challenges in advancing its digital capabilities within the economy and the business landscape (Deloitte, 2020).

Serbia stands out as the largest agricultural market in the Western Balkans, with agricultural production contributing around 10% to the national GDP (International Trade Administration, 2024). The country is largely self-sufficient in food production and is a significant exporter of grains and fruits. However, Serbia's agricultural sector still faces substantial issues, such as small farms with outdated farming tehniques, which hinder its ability to reach the level of highly self-sufficient countries (World Bank, 2024b); and low level of digitalization, such as lack of findable, accessible, interoperable and reusable (FAIR) agriculture-related data and limited dissemination and implementation of DA. Conversely, this sector has also demonstrated considerable innovation potential, with companies from both Greece and Serbia making noteworthy strides. There are agrifood enterprises that have gained substantial market shares both domestically and internationally by harnessing DA technologies, utilizing traditional local ingredients and implementing cutting-edge marketing and packaging strategies (Tsiforou, 2020).

However, both widening countries face challenges in fostering a skilled workforce capable of navigating the complexities of DA, which impedes their overall competitiveness and growth potential in the agricultural market. This disparity not only affects their economic development but also poses risks to food security and sustainability in the long term. In order to enhance the ability of agronomists, farmers and agribusinesses in these countries to fully exploit technological advancements, efforts should be made to support DA education and training. Towards this direction, HEIs that provide agricultural courses can play a vital role since they comprise the primary organizations for R&I and can provide future agriculture workforce with the necessary technology assets and DA skills. A study with policymakers in Greece highlights that the transition to a new technological era for the agri-food sector necessitates highly skilled workforce with specialized knowledge. Additionally, they state that HEIs' curricula need to be revised and bring together research and practice in agricultural production (Apostolopoulos et al., 2024). On the other hand, the lack of digital skills and skilled labor, as well as information asymmetry between agricultural stakeholders in the production chain and education institutions are among the obstacles for introducing DA, which are more pronounced in Serbia in comparison to other countries (Dimitrijević, 2023).

In sharp contrast to the limited R&D expenditure observed in Serbia and Greece, Belgium demonstrates a considerable investment to R&I, placing well above the EU average and also signifying its strategic prioritization of research-intensive sectors as key drivers of economic growth and competitiveness. This fact can possibly enable the cultivation of a robust knowledge-based economy, attract highly skilled researchers, thus strengthening its position within global innovation networks. Belgium's agricultural sector although relatively small, is well-developed and efficient. It focuses on high-value, intensive farming practices due to limited arable land and high population density. Major agricultural products include potatoes, sugar beets, and wheat. All three countries-Belgium, Greece, and Serbia- have a prevalence of small and medium-sized farms rather than large, industrial-scale operations, with many farms being family-owned. Fragmented land holdings are common, making it challenging to achieve economies of scale.

Particularly, this paper seeks to answer the specific research questions:

- **RQ1:** What are faculty members' views on the current integration of DA in their courses?
- **RQ2:** What do faculty members consider to be the most significant challenges in integrating DA into the curriculum from an educational perspective?

The novelty of this paper lies in revealing the existing gaps in DA education within HEIs. For comparison reasons, a questionnaire-based survey was conducted among the faculty of the two widening countries (Greece and Serbia) and the non-widening country (Belgium). This is crucial for addressing the growing disparities between widening and non-widening countries in this sector. Understanding the existing gaps in educational offerings will not only highlight areas for improvement but also encourage the development of targeted curricula that align with industry demands and technological advancements.

The structure of this paper is as follows: The background section presents the Transforming Access to Excellence with Successful Alliances of Higher Education in Digital Agriculture" (TALLHEDA) initiative that served as the springboard of this work and the research methodology. Then, the results and discussion on them are presented. Finally, the conclusion section argues on the interpretation of the findings, the limitations of this study, and avenues for future research.

BACKGROUND

TALLHEDA Project

TALLHEDA is an Horizon Widera project that aims to demonstrate a successful path to excellence in research and education for the widening of agricultural HEIs. It attempts to build a joint and long-term alliance between them and the surrounding ecosystems, allowing thus to share best pedagogical and research practices. Thus, it contributes to increase scientific, technological and staff capacity in DA. TALLHEDA alliance, walking on the footsteps of beacon non-widening countries like Belgium, aspires to improve access to DA excellence for two widening agricultural HEIs in Greece and Serbia.

The main objective of TALLHEDA is to raise a critical mass of highly skilled young scientists who will be able to address the challenges of present and future agricultural education and research by applying the principles of DA not only within the confines of academia but also across society. To support its main objective, TALLHEDA engages a bottom-up methodological approach of multi-actor, inter-sector and international complementary actions (Figure 1). Green arrows represent input knowledge and red arrows represent output knowledge. It tries to achieve the following:

- Setting up: i) sustainable, innovation-driven ecosystems in Widening countries, according to their R&I needs and barriers by utilizing a multi-actor engagement methodology and ii) a Virtual Innovation Hub (VIH), which will mainstream the creation of conducive ecosystems strengthening R&I excellence, skills, training, education, and research know-how.

- Achieving a balanced circulation of talents via several mobility activities to introduce academic and research staff from the Widening HEIs to conducive educational, research, management, and administrative environments in the field of DA, to promote better quality and capacity in R&I.
- Building transnational research bridges through joint courses and research activities, to strengthen the quality of higher education in widening countries and Europe's R&I. These activities will build long-lasting collaboration bridges between participating HEIs, and equip young researchers with the right knowledge, experience and skills for their future professional and personal development.
- Mainstreaming practices and tools for open sharing of knowledge and data. Tools for open sharing of knowledge and data will be developed to be the basis supporting educational and research reform of widening HEIs.
- Strengthening careers of researchers from Widening HEIs. Academic staff, early and experienced researchers from Widening HEIs will have access to non-widening R&I excellence in DA, mobility opportunities, brain circulation, joint educational and research activities, open science practices and inclusiveness.
- Bridging the widening HEIs with local surrounding ecosystems. This will involve introducing them to external, well-organized ecosystem(s) of knowledge transfer among stakeholders (e.g., HEIs-competence centers, businesses, -farmers). The TALLHEDA Innovation Hub will foster extroversion of widening HEIs and reinforce their role in local innovation ecosystems.

Figure 1. TALLHEDA bottom-up strategy

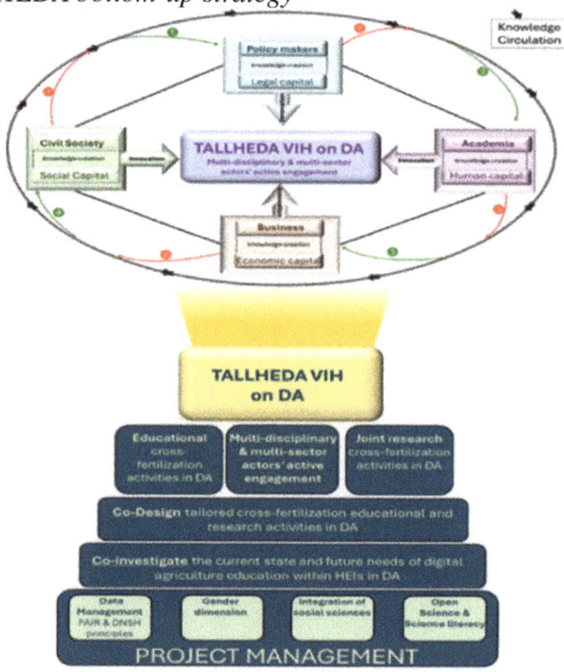

TALLHEDA's strategy is tailored to strengthen the capacity of the agricultural HEIs from the widening countries to deliver excellent education and perform cutting-edge research across two dimensions:

a) Horizontal, i.e., in key enabling digital technologies such as Artificial Intelligence (AI) and Data Analytics, Blockchain and Distributed Ledger Technology (DLT), and Digital Twins.
b) Vertical, i.e., in key agri-food topics of prominent importance, such as food transparency and traceability, sustainable farming practices, and food security and safety.

These technologies have the potential to transform the global food system by introducing important efficiency gains along value chains. All data produced under TALLHEDA's umbrella of cross-fertilization educational and joint research activities comply with the principles of DNSH and FAIR.

Methodology

Among TALLHEDA aims is to establish transnational education and research bridges. Transnational education plays a vital role in bridging DA gaps by building networks among academia and professionals for exchanging knowledge, resources, and best practices across borders, which can drive more inclusive progress in DA education and training.

The representation of the current state and future needs the integration of DA education within HEIs is fundamental to proceed to their proper educational transformation to reach research excellence in DA. The design and implementation of this research comprises the common efforts and collaboration of a group of different stakeholders involved in DA, e.g., professors, agricultural businesses, and DA specialists. The methodology implemented for identifying the current status and challenges of DA education in HEIs comprises of the following steps:

- **DA Education Literature Review:** Firstly, a literature review is conducted to identify existing critical aspects related to DA education within HEIs.
- **Identification of Research Questions:** A series of research questions is formulated based on the identified critical aspects.
- **Research Question Refinement:** The identified research questions are assessed by a group of HEI faculty members.
- **Questionnaire Development:** The final set of research questions is used to create a questionnaire regarding critical aspects related to DA education within HEIs. The sections of the questionnaire are the following:
    - General Information of the Respondents
    - Current Perceptions and Challenges
        - A. Perceptions about Digital Agriculture (DA)
        - B. Integration of Digital Agriculture (DA) Topics
        - C. Course/Curriculum Update
        - D. Program Integration
        - E. Research Capacity in Digital Agriculture (DA)
        - F. Resources for Digital Agriculture (DA)
        - G. Quality
        - H. Collaboration
        - I. Student Engagement
        - J. Integration Challenges and Reform Needs
    - Approaches for Addressing the DA Integration Challenges
- **Questionnaire Distribution and Collection:** HEI faculty members involved in DA from the University of Ghent (Belgium), the Agricultural University of

Athens-AUA (Greece), and the University of Novi Sad (Serbia) are identified based on their scientific profile, and the questionnaire is circulated to them.

- **Statistical Analysis:** After the collection of questionnaires, questionnaire data are statistically analyzed to derive meaningful insights regarding the research questions posed.

RESULTS AND DISCUSSION

The questionnaire was circulated among HEIs during the period April-May 2024. The participation in the questionnaire was anonymous and the questions were not mandatory; as a result, not all questionnaires were completed. A total of fifty (50) HEI faculty members participated in the research. Specifically, there were eleven (11) from Belgium; twenty (20) from Greece; and nineteen (19) from Serbia. The data were processed in two samples, so as to enable comparisons between non-widening and widening countries:

- Non-widening sample (N=11): documents faculty members' opinions from the non-widening country (UGent University); and
- Widening sample (N=39): documents faculty members' opinions from the widening countries (AUA and Novi Sad University).

General Information of the Respondents

The majority of the respondents (75%) are male, professors (48.98%), between 45-64 years old (53.06%) and they have been teaching for 16-20 years (27.08%) (Figure 2). Regarding their department, most of the respondents are from Agricultural Economics and Rural Development (8), Fruit Growing, Viticulture, Horticulture and Landscape Architecture (4), and Agricultural Engineering (3). They currently teach specific subjects (courses), such as precision agriculture, information systems, special fruit growing, and precision livestock farming.

Figure 2. General information of the respondents: Overall

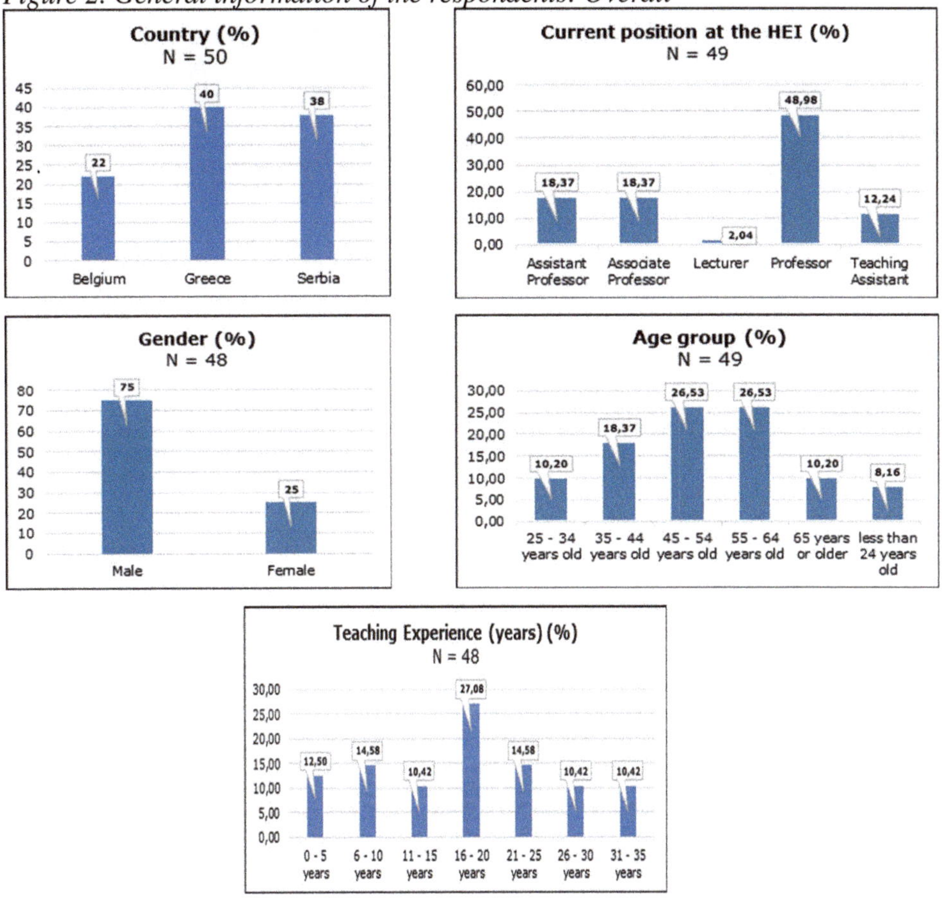

In the non-widening country (Belgium), most participants are male (81.82%), professors between 55-64 years old (45.45%) and have been teaching for 11-15 years (30%) (Figure 3).

Figure 3. General information of the respondents: Non-widening country

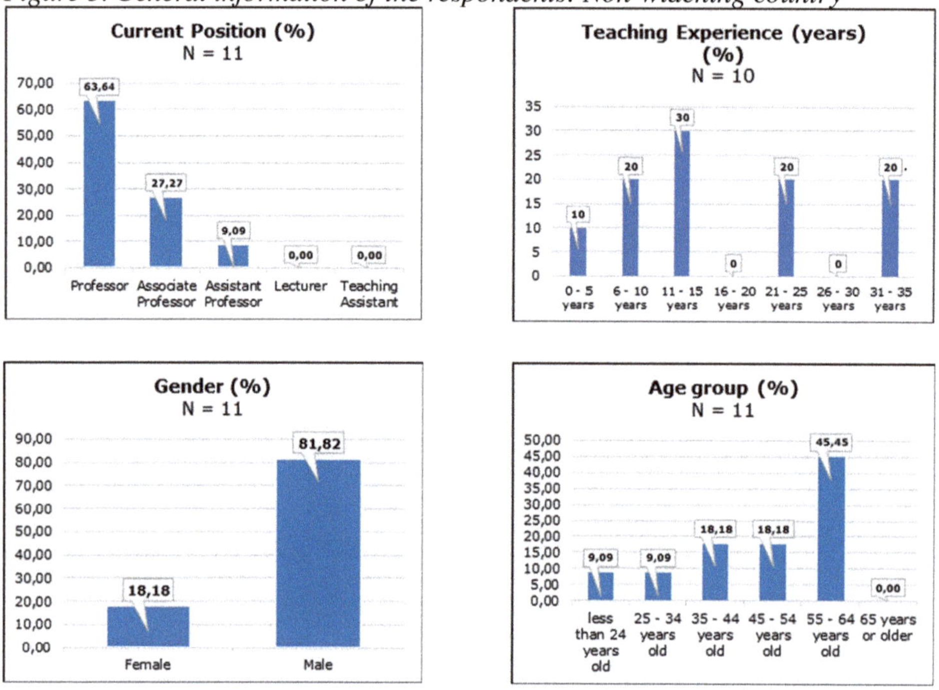

In the widening countries (Greece and Serbia), most participants are male (72.97%), professors between 45-54 years old (28.95%) and they have been teaching 16-20 years (34.21%) (Figure 4).

Figure 4. General information of the respondents: Widening countries

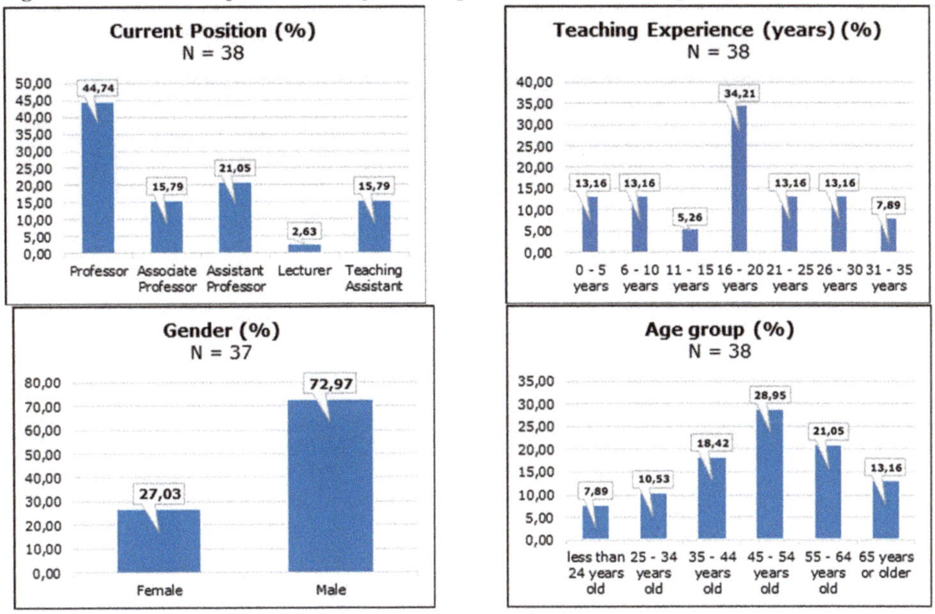

Current Perceptions and Challenges

Perceptions about Digital Agriculture

The overall impact of reinforcing the courses with DA is rated between "High" and "Very High", with the majority (40.82%) leaning towards "Very High Impact" (Figure 5). In the non-widening country, most of them (45.45%) believe that the impact is rated as "High" (Figure 6). In the widening countries, the majority (47.37%) believe that the impact on reinforcing the courses with DA is "Very High" (Figure 7).

Figure 5. Perceptions about digital agriculture: Overall

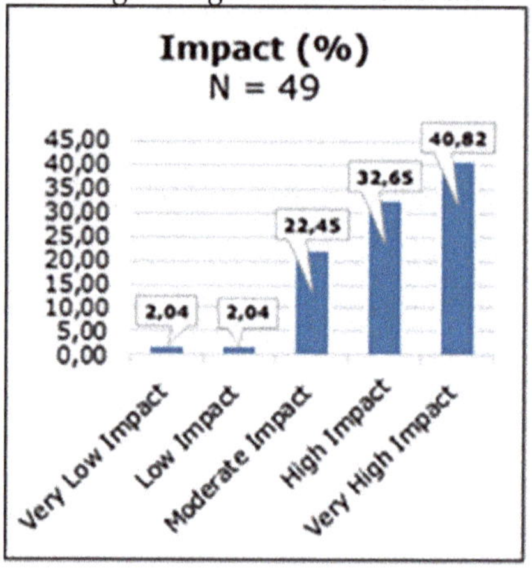

Figure 6. Perceptions about digital agriculture: Non-widening country

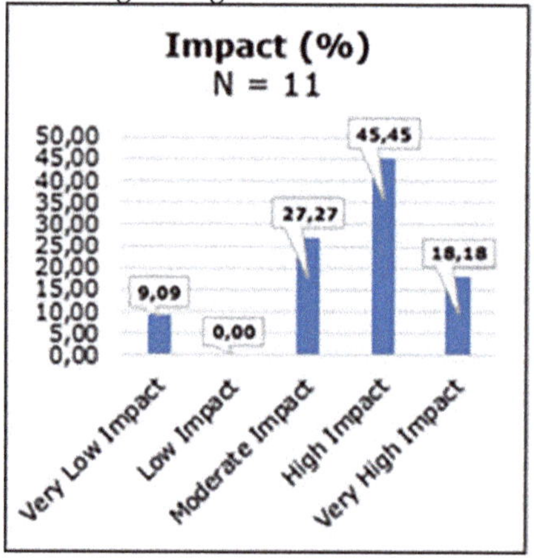

Figure 7. Perceptions about digital agriculture: Widening countries

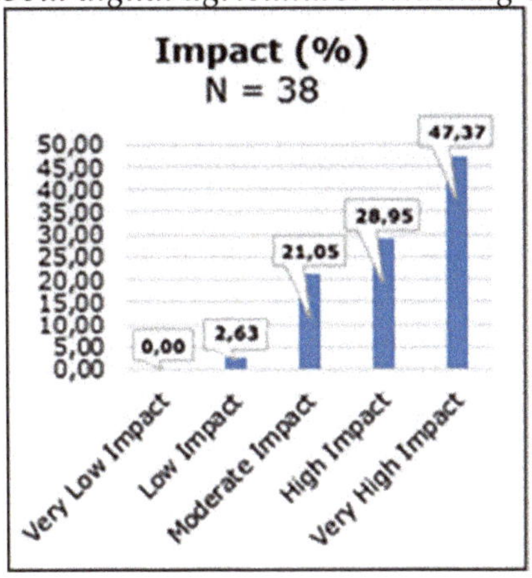

Integration of Digital Agriculture Topics

In the overall sample, most of them have incorporated sustainable agricultural practices and technologies (18) into their teaching. Additionally, they would like to incorporate blockchain for supply chain transparency and management (10) into their teaching in the future.

In the non-widening country, most of them (7) have incorporated precision agriculture and smart farming technologies into their teaching. However, in the non-widening country most would like to incorporate agricultural robotics and automation (4) in their teaching in the future.

In the widening countries, the majority has incorporated sustainable agricultural practices and technologies (11) into their teaching. Additionally, they would like to incorporate blockchain for supply chain transparency and management (10) into their teaching in the future.

Course/Curriculum Update

The majority of faculty members at HEIs (35.56%) reformed or updated the content of courses (e.g., updated slides, discussion topics, case studies) and introduced new DA topics "Within the last year". In the non-widening country, the highest percentage, 44.44% of respondents, introduced new DA topics "2 years ago". In the widening countries most (36.11%) introduced new DA topics "Within the last year".

Most (30%) feel "Moderately comfortable" of integrating DA topics into their courses. In the non-widening country, the largest group, 45.45%, feel "Moderately comfortable". More generally, in the non-widening country, there is a divide, with almost 55% feeling not at all comfortable up to moderately comfortable, and about 45% feeling comfortable or very comfortable. In the widening countries, responses vary but there also seems to be a divide, with about 56% feeling comfortable or very comfortable in integrating DA topics and 44% feeling moderately or slightly comfortable.

Program Integration

Most (41.67%) rated the integration of DA topics across the Bachelor or Master's program as "Good". In the non-widening country, the largest proportion of respondents, about 36.36%, rated the integration as "Fair". In the widening countries most (45.95%) rated the integration as "Good". The majority answered that the Bachelor or Master's program align moderately up-to-date with the latest technological developments in the field of DA. In the non-widening country, the majority of respondents, 72.73%, rated their programs as "Moderately up-to-date". In the widening countries, the majority of respondents, 31.43%, rated their programs as "Moderately up-to-date".

Research Capacity in Digital Agriculture

Most (37.50%) underlined that they are "Undecided" regarding the statement: 'My HEI has a sufficient number of research initiatives and projects related to DA'. In the non-widening country, most (45.45%) highlighted the same opinion. In the widening countries, 35.14% of them stated that they are "Undecided" regarding the above statement.

Resources for Digital Agriculture

Overall, the adequacy of educational materials and curriculum resources (40.82%), laboratory and field equipment (41.67%), Internet and computing infrastructure (32.65%), faculty expertise and support staff (39.58%), funding for DA projects (39.58%), library and research databases (41.37%), training and professional development programs (39.58%) and innovation and entrepreneurship support (38.30%) for DA are defined as "Fair". Also, technological tools and software (42.86%) are defined as "Poor" and collaboration opportunities (41.67%) are defined as "Good" (Figure 8).

Figure 8. Resources for digital agriculture: Overall

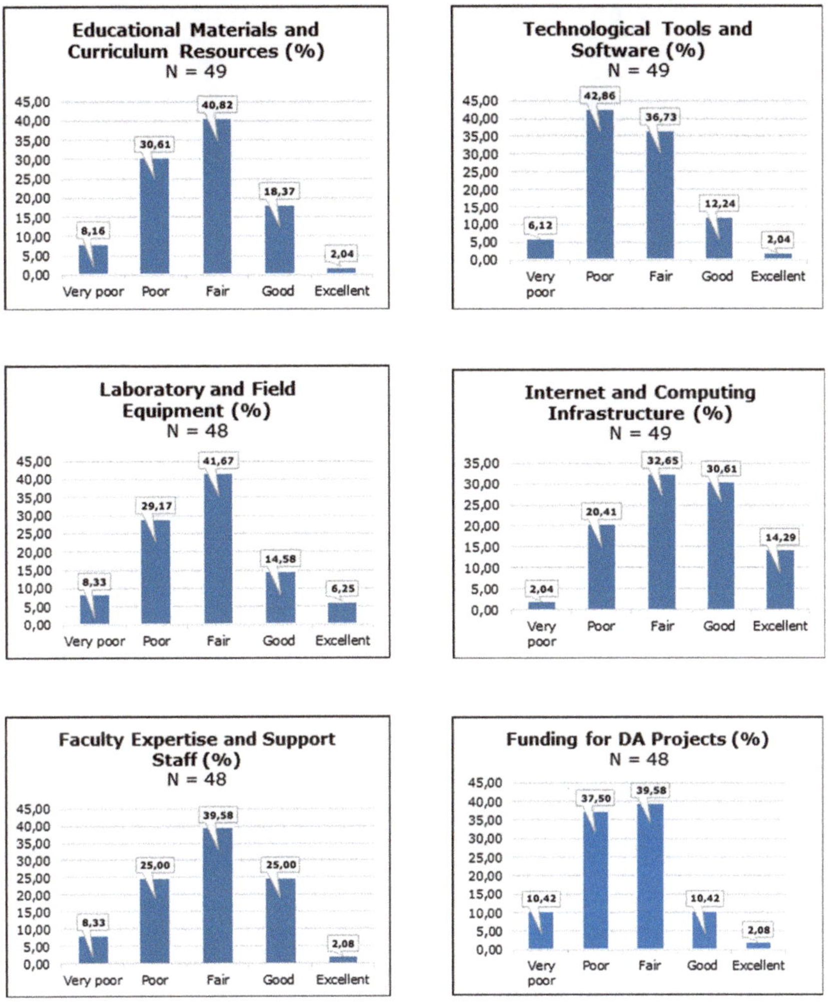

In the non-widening country, the adequacy of educational materials and curriculum resources (63.64%), technological tools and software (54.55%), funding for DA projects (70%) and training and professional development programs (72.73%) are defined as "Fair". Internet and computing infrastructure (36.36%), collaboration opportunities (72.73%) and library and research databases (45.45%) are defined as "Good". Faculty expertise and support staff (40%) is defined as "Poor". Laboratory and field equipment (90%) is defined between "Poor" and "Good" and innovation and entrepreneurship support (80%) for DA is defined between "Fair" and "Good" (Figure 9).

Figure 9. Resources for digital agriculture: Non-widening country

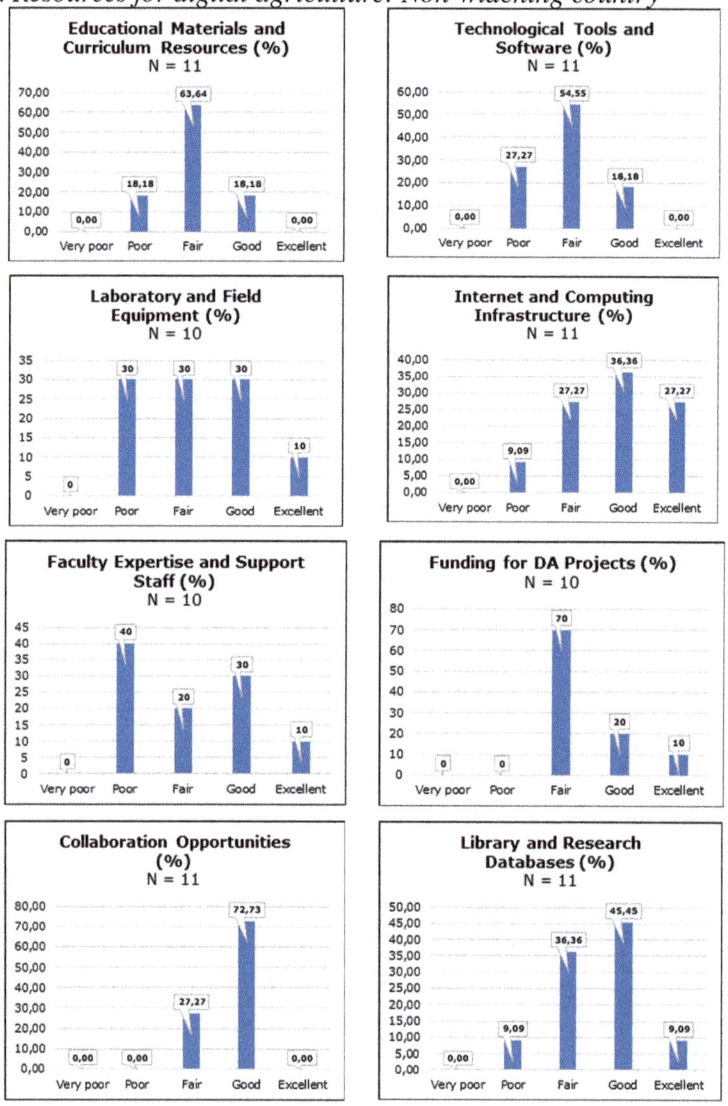

In the widening countries, the adequacy of laboratory and field equipment (44.74%), Internet and computing infrastructure (34.21%), faculty expertise and support staff (44.74%), collaboration opportunities (35.14%), library and research databases (43.24%), training and professional development programs (29.73%) and innovation and entrepreneurship support (37.84%) for DA are defined as "Fair". Technological tools and software (47.37%) and funding for DA projects (42.11%)

are defined as "Poor". Finally, the adequacy of educational materials and curriculum resources (68.42%) is defined between "Poor" and "Fair" (Figure 10).

Figure 10. Resources for digital agriculture: Widening countries

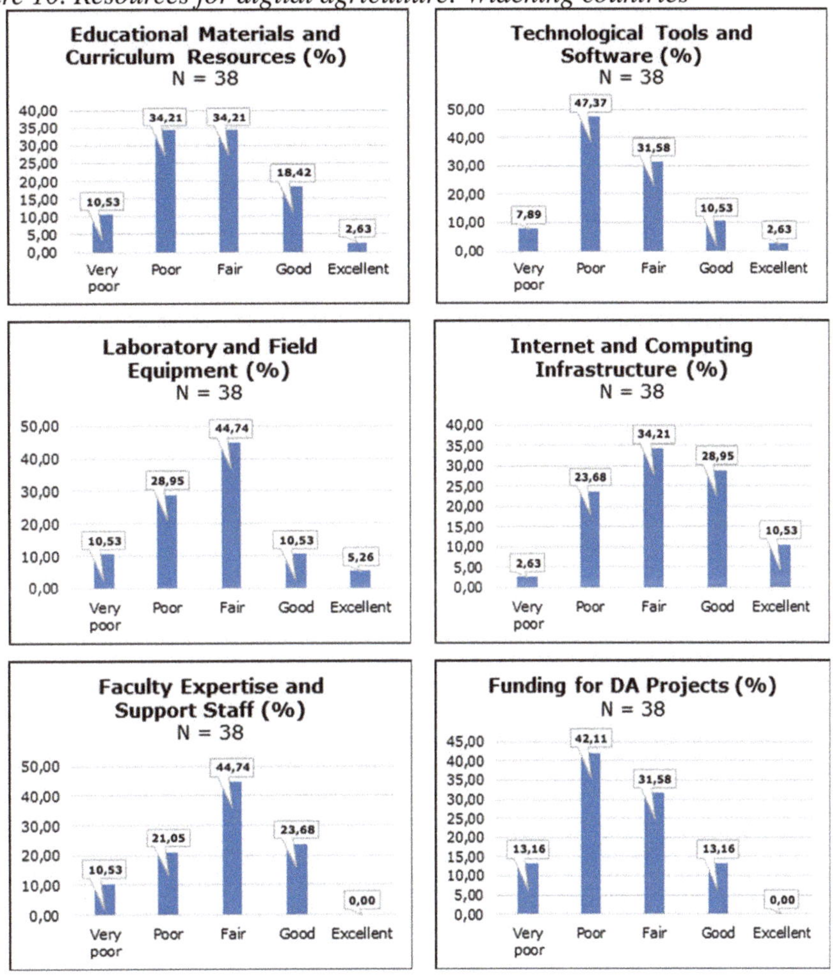

T-tests have been performed to confirm whether the difference between the non-widening country and widening countries is statistically significant for DA collaboration opportunities among HEIs, existence of DA library and research databases, as well as the provision of innovation and entrepreneurship support (Table 1 and Table 2). An independent samples t-test was conducted to compare "Collaboration Opportunities" between the two groups. Levene's test for equality of variances indicated a violation of the assumption of equal variances, $F(1, 46) = 5.108$, $p = .029$. Therefore, the t-test results for equal variances not assumed were used. The results revealed a significant difference in Collaboration Opportunities between the two groups, t(40.39) = 3.497, p = .001. The mean difference was 0.808 (SE = 0.231), with a 95% confidence interval ranging from 0.341 to 1.275.

Table 1. Group statistics for the variable "collaboration opportunities"

Group Statistics					
	Country	N	Mean	Std. Deviation	Std. Error Mean
Collaboration	Belgium	11	3.7273	.46710	.14084
	Greece and Serbia	37	2.9189	1.11501	.18331

Table 2. Independent samples test for the variable "collaboration opportunities"

Independent Samples Test										
		Levene's Test for Equality of Variances		t-test for Equality of Means						
									95% Confidence Interval of the Difference	
		F	Sig.	T	Df	Sig. (2-tailed)	Mean Difference	Std. Error Difference	Lower	Upper
Collaboration	Equal variances assumed	5.108	.029	2.330	46	.024	.80835	.34690	.11007	1.50664
	Equal variances not assumed			3.497	40.385	.001	.80835	.23116	.34130	1.27541

Additionally, an independent samples t-test was conducted to compare "Library and Research Databases" scores between the two groups. Levene's test for equality of variances indicated that the assumption of equal variances was met, F(1, 46) = 0.023, p = .880. Therefore, the t-test results for equal variances assumed were used. The results revealed a significant difference in Library scores between the two groups, t(46) = 2.138, p = .038. The mean difference was 0.627 (SE = 0.293), with a 95% confidence interval ranging from 0.037 to 1.216. (Table 3 and Table 4).

Table 3. Group statistics for the variable "library and research databases"

	Group Statistics				
	Country	**N**	**Mean**	**Std. Deviation**	**Std. Error Mean**
Library	Belgium	11	3.5455	.82020	.24730
	Greece and Serbia	37	2.9189	.86212	.14173

Table 4. Independent samples test for the variable "library and research databases"

Independent Samples Test										
		Levene's Test for Equality of Variances		t-test for Equality of Means						
		F	Sig.	T	df	Sig. (2-tailed)	Mean Difference	Std. Error Difference	95% Confidence Interval of the Difference	
									Lower	Upper
Library	Equal variances assumed	.023	.880	2.138	46	.038	.62654	.29300	.03676	1.21631
	Equal variances not assumed			2.198	17.135	.042	.62654	.28503	.02552	1.22755

Furthermore, an independent samples t-test was conducted to compare "Innovation and Entrepreneurship Support" scores between the two groups. Levene's test for equality of variances indicated that the assumption of equal variances was met, $F(1, 45) = 0.048$, $p = .828$. Therefore, the t-test results for equal variances assumed were used. The results revealed a significant difference in Innovation between the two groups, $t(45) = 2.306$, $p = .026$. The mean difference was 0.749 ($SE = 0.325$), with a 95% confidence interval ranging from 0.095 to 1.402. (Table 5 and Table 6).

Table 5. Group statistics for the variable "innovation and entrepreneurship support"

Group Statistics					
	Country	N	Mean	Std. Deviation	Std. Error Mean
Innovation	Belgium	10	3.1000	.99443	.31447
	Greece and Serbia	37	2.3514	.88870	.14610

Table 6. Independent sample test for the variable "innovation and entrepreneurship support"

Independent Samples Test										
		Levene's Test for Equality of Variances		t-test for Equality of Means						
		F	Sig.	T	Df	Sig. (2-tailed)	Mean Difference	Std. Error Difference	95% Confidence Interval of the Difference	
									Lower	Upper
Innovation	Equal variances assumed	.048	.828	2.306	45	.026	.74865	.32463	.09482	1.40248
	Equal variances not assumed			2.159	13.152	.050	.74865	.34675	.00042	1.49688

Quality

Most (47.92%), based on their experiences and observations, believed that the quality of teaching and research in DA at their HEIs compares to that of other well-recognized HEIs, i.e., as "Same quality". However, there is a difference in terms of percentages, with 63.64% of them in the non-widening country and 41.03% of them in the widening countries believing that the quality is the same. Also, in the widening countries, there is an almost equal percentage of faculty members (38.46%) who believe their HEI is moderately lower in terms of quality.

Collaboration

Most rated the level of interdisciplinary approach and collaboration in DA education between "Little" and "Average". In the non-widening country, most rated the level of interdisciplinary approach and collaboration in DA education "Little", "Average" and "Much". Most rated the level of interdisciplinary approach and collaboration in DA education between "Little" and "Average".

Student Engagement

Half of the respondents (50%) answered that students show interest in acquiring knowledge focusing on DA techniques and technologies "To some extent". Most (46.81%) answered that students have opportunities "To some extent" to engage in real-world projects or partnerships with industry partners in relation to DA. In the non-widening country, most (60%) answered that students show interest in acquiring knowledge focusing on DA techniques and technologies "To some extent". Moreover, most (54.55%) answered that students have opportunities "To some extent" to engage in real-world projects or partnerships with industry partners in relation to DA. In the widening countries, most (47.37%) answered that students show interest in acquiring knowledge focusing on DA techniques and technologies "To some extent". Also, most answered that students have opportunities "To some extent" (44.44%) or "Only a little" (41.67%) to engage in real-world projects or partnerships with industry partners in relation to DA.

Integration Challenges and Reform Needs

In both non-widening and widening countries, integrating DA into the higher education curricula is challenging. For the non-widening country, key issues include (in descending order):

- Developing relevant and up-to-date DA course content
- Adapting to rapidly changing DA technologies and methods
- Incorporating practical DA experiences and labs
- Engaging students in DA topics
- Integrating DA with existing curricular frameworks and standards
- Ensuring access to and familiarity with necessary technological tools among staff and students
- Finding appropriate educational resources and materials
- Developing case studies from our country and the international environment

In the widening countries, the most pressing challenges involve incorporating "hands-on DA experiences"; "incorporating practical DA experiences and labs"; "technological infrastructure", "industry collaboration and curriculum development"; "ensuring access to and familiarity with necessary technological tools among staff and students"; "developing relevant and up-to-date DA course content"; "adapting to rapidly changing DA technologies and methods"; "engaging students in DA topics"; "developing case studies from our country and the international environment"; "finding appropriate educational resources and materials"; and "integrating DA with existing curricular frameworks and standards ".

Approaches for Addressing Integration Challenges

Regarding the capacity-building activities (Table 7) that would most effectively address DA education and research needs at the HEIs, the overall sample considers as most important the following: "specialized DA workshops focused on emerging technologies and practices", "industry partnerships for hands-on training and real-world DA experience", and "international collaborations with other HEIs and research centers for DA". Respectively, for the non-widening country are: "specialized DA workshops focused on emerging technologies and practices", "industry partnerships for hands-on training and real-world DA experience", and "cross-disciplinary DA seminars for sharing knowledge across departments". In the widening countries, are: "specialized DA workshops focused on emerging technologies and practices", "industry partnerships for hands-on training and real-world DA experience", and "international collaborations with other HEIs and research centers for DA".

Table 7. Capacity-building activities

Topics	Overall	Non-widening Country	Widening Countries
	No.	No.	No.
Specialized DA workshops focused on emerging technologies and practices	35	8	27
Industry partnerships for hands-on training and real-world DA experience	34	8	26
International collaborations with other HEIs and research centers for DA	24	4	20
Access to DA research grants and funding opportunities for faculty and students	23	3	20
Cross-disciplinary DA seminars for sharing knowledge across departments	22	5	17
Faculty development programs in DA, including pedagogical strategies and research methods	21	5	16
Development of a DA resource center or innovation hub for faculty and students	17	0	11
Student-led DA projects and competitions to encourage practical learning and innovation	13	3	10
Online DA courses and webinars for flexible learning options	12	3	9

Regarding the types of support from the HEI (Table 8) that would be most valuable to them in feeling confident and successful, the overall sample considers as most important the following: "enhanced access to DA technologies, infrastructure, software and digital platforms", "enhanced access to DA educational resources", and "support for attracting financial resources and increased funding for DA projects". Respectively, for the non-widening country are: "opportunities for interdisciplinary and interdepartmental collaboration in DA initiatives and projects", "enhanced access to DA educational resources", and "enhanced access to DA technologies, infrastructure, software and digital platforms". In the widening countries, are: "enhanced access to DA technologies, infrastructure, software and digital platforms", "support for attracting financial resources and increased funding for DA projects", and "enhanced access to DA educational resources".

Table 8. Types of support from the HEI

Topics	Overall	Non-widening Country	Widening Countries
	No.	No.	No.
Enhanced access to DA technologies, infrastructure, software and digital platforms	32	5	27
Enhanced access to DA educational resources	26	5	21
Support for attracting financial resources and increased funding for DA projects	24	2	22
Enhanced collaboration with agri-food industry, business, and external organizations	24	4	20
Opportunities for interdisciplinary and inter-departmental collaboration in DA initiatives and projects	23	6	17
Teaching assistants and technical support	20	4	16
Mentoring for DA course/curriculum development	17	3	14
Resources and support for enhancing student engagement and hands-on experiences in DA	15	3	12
Development of appropriate institutional policies and administrative support for DA initiatives	13	1	12
Allocated time for professional development and training in DA	11	4	7
Development of an institutional culture for DA innovation and experimentation	8	0	8

Concerning the most important strategies for establishing a Culture of Excellence in DA within the HEI (Table 9), the overall sample considers as most important the following: "Research and Innovation: A robust commitment to cutting-edge research and innovative practices", "Quality in Education: High standards of teaching and learning that foster academic excellence", and "Infrastructure and Resources: Adequate physical and digital infrastructure to support educational and research activities". Respectively, for the non-widening country are: the same as previous, namely "Research and Innovation: A robust commitment to cutting-edge research and innovative practices", "Quality in Education: High standards of teaching and learning that foster academic excellence", and "Infrastructure and Resources: Adequate physical and digital infrastructure to support educational and research activities". In the widening countries, are: "Quality in Education: High standards of teaching and learning that foster academic excellence", "Research and Innovation: A robust commitment to cutting-edge research and innovative practices", and "Infrastructure and Resources: Adequate physical and digital infrastructure to support educational and research activities".

Table 9. Strategies for establishing a culture of excellence in DA

Topics	Overall	Non-widening Country	Widening Countries
	No.	No.	No.
Research and Innovation: A robust commitment to cutting-edge research and innovative practices	34	9	25
Quality in Education: High standards of teaching and learning that foster academic excellence	32	7	25
Infrastructure and Resources: Adequate physical and digital infrastructure to support educational and research activities	31	6	25
Leadership and Vision: Strong, visionary leadership that guides and inspires the entire institution	25	4	21
Student Engagement and Development: Active efforts to engage students in their own learning and development processes	23	4	19
Continuous Improvement: An ongoing commitment to evaluating and enhancing institutional practices	14	4	10
Community and Institutional Culture: A supportive and inclusive community that values collaboration and diversity	11	1	10
Global Engagement: Engagement with global challenges and opportunities through international collaboration and understanding	9	2	7

Regarding the aspects that HEI could benefit from the know-how and expertise of other HEIs (Table 10), especially more advanced HEIs, in DA, the overall sample considers as most important the following: "technology transfer and commercialization", "curriculum development and course design", and "research and innovation methodologies". Respectively, for the non-widening country are: "digital agricultural practices with emphasis on sustainability", "curriculum development and course design", and "research and innovation methodologies". In the widening countries, are: "technology transfer and commercialization", "research and innovation methodologies", and "curriculum development and course design". The establishment of a DA Innovation Hub or Community of Practice is generally perceived that it would be led to a "High Impact".

Table 10. Aspects that HEI could benefit from the know-how and expertise of other HEIs

Topics	Overall	Non-widening Country	Widening Countries
	No.	No.	No.
Technology transfer and commercialization	29	1	28
Curriculum development and course design	28	5	23
Research and innovation methodologies	28	4	24
Digital agricultural practices with emphasis on sustainability	28	6	22
Engagement and partnership models with the industry	23	2	21

Regarding the technology adoption and infrastructure enhancement, the importance of infrastructure enhancements (laboratories, computing resources, field equipment), for advancing DA at the HEI is defined as "Fairly Important". Regarding federating HEI infrastructures and facilitating open access is defined as "Important". In the non-widening country, the impact of establishing a DA Innovation Hub or Community of Practice is defined as "Moderate Impact". Regarding the technology adoption and infrastructure enhancement, the importance of infrastructure enhancements (laboratories, computing resources, field equipment), for advancing DA at the HEI is defined as "Important". Regarding federating HEI infrastructures and facilitating open access is defined as "Important".

DISCUSSION

As DA continues to grow, there is an increasing demand for professionals capable of analyzing on-farm data and effectively utilizing emerging technologies. HEIs are strategically positioned to bridge the digital divide. They will play a critical role in equipping students, as well as future agri-food educators, trainers, and farmers, with the necessary skills to thrive in an increasingly digitized agricultural landscape (Bampasidou et al., 2024). Efforts are already underway globally to develop agricultural curricula that focus on enhancing the skill sets of the next generation of professionals, including the growing need for data agronomists to support these activities (Mavridis et al., 2012).

This work focuses on depicting faculty members' view on the current state-of-the-art and future requirements of DA education in HEIs and make comparisons between widening and non-widening countries in the EU. To our knowledge, there are no identical studies in the literature.

Nevertheless, (Soma & Nuckchady, 2021) have presented a comprehensive study for understanding the emerging education and training landscape in DA technologies in British Columbia. They studied how educators communicated benefits and risks of DA, and the need for an inclusive and equitable approach to DA. One similarity with the aims of our study regards the integration of DA technologies in projects. The research revealed a wide range of DA benefits, including open-source data platforms, more precise agricultural practices, and labor-saving advantages. DA projects at the academic level focus on using software and technology for data analysis, while educators emphasize the importance of caution due to potential risks and unintended consequences. On the other hand, the majority of respondents of our study are unsure whether their HEI has enough research initiatives and projects related to data analytics. This indecision is more noticeable in non-widening countries than in widening ones, indicating a possible need for better awareness or communication about DA-related research efforts.

Moreover, our study provides valuable guidance for developing targeted strategies to improve DA education, bridging the gap between research and practice, and fostering excellence in DA across diverse educational environments. The results of our study strongly suggest that DA has the potential to significantly enhance educational curricula across both widening and non-widening countries. It indicates a broad consensus on the need of including DA in modernizing curricula, enriching course content, and equipping students and professionals with skills relevant to the evolving agricultural sector (Bampasidou et al., 2024). Also, digital technologies are considered as necessary to enhance agricultural transparency and efficiency.

DA technologies that support sustainable agricultural practices and technologies are the most frequently integrated topics in courses across widening countries which reflects a growing awareness of sustainability within agriculture. In the non-widening country, faculty members have already incorporated precision agriculture and smart farming technologies into their courses.

There is also a variation in preferred technologies across countries for future integration. While faculty members in widening countries are interested in incorporating advanced DA topics, such as blockchain for supply chain transparency, their counterparts in non-widening countries are more focused on innovations like agricultural robotics and automation. These differences illustrate a shared vison for modernization of agricultural education, while also highlighting that the academic curricula are adapting to region-specific challenges, and local food chain demands.

Across all regions, resource limitations and the rapid pace of technological advancements emerge as common hurdles. To address these, faculty members suggest increased interdisciplinary collaboration, industry partnerships, and periodic curriculum reviews to ensure content remains relevant and aligned with industry practices. Despite the identified significance and value of DA in educational process, resources limitations, particularly in widening countries, are main obstacles for full implementation of DA curricula. Challenges such as insufficient funding, lack of infrastructure, and limited access to advanced technologies remain widespread. This research emphasizes the need for targeted investments and transnational collaborations to support DA integration within HEIs, enabling future agricultural professionals to explore the digitalized world.

Faculty members in the future should focus on enhancing practical learning, fostering collaboration, and continuously updating curricula. Governmental agencies' and HEIs' future initiatives should consider and incorporate specialized education and training (e.g., webinars) in order to support faculty members. It is necessary not only to acquire DA knowledge, but also to extend one's horizons on pedagogical methods in order to be able to integrate DA in curricula. Addressing these areas will enable HEIs to equip future agricultural professionals with the necessary skills to thrive in a digitalized agricultural landscape.

CONCLUSION

This study emphasizes the transformative potential of DA in enhancing agricultural education across widening and non-widening countries. While there is strong support for integrating DA into curricula, significant regional differences in technology focus and resource availability pose challenges, particularly in widening countries. Limited infrastructure, funding, and access to advanced technologies remain key barriers. However, opportunities for collaboration, particularly through industry partnerships, can help bridge these gaps. Continuous updates to curricula are essential to keep pace with technological advancements.

This study presents specific limitations that need to be considered in future work. Firstly, the sample is very limited, particularly from the non-widening country, and should be increased to provide more solid results. Also, responses could be increased by making all questions mandatory. Following the TALLHEDA bottom-up methodological approach future work will focus on investigating the perceptions of other stakeholders, such as students, agri-technology companies, food processing companies, and farmers, for depicting DA evolution not only in academia, but also in the business world.

REFERENCES

An, I. T. U. F. A. O. (2024). Meeting the European Union's digital agriculture requirements.

Apostolopoulos, N., Petropoulos, D., Liargovas, P., & Anastasopoulou, E. (2024, September). Constraints to Green and Digital Transition in the Greek Agri-Food Sector. In *Proceedings of The 19th European Conference on Innovation and Entrepreneurship, Academic Conferences International*. DOI: 10.34190/ecie.19.1.2996

Bampasidou, M., Goldgaber, D., Gentimis, T., & Mandalika, A. (2024). Overcoming 'Digital Divides': Leveraging higher education to develop next generation digital agriculture professionals. *Computers and Electronics in Agriculture, 224*, 109181. DOI: 10.1016/j.compag.2024.109181

Daum, T., Adegbola, P. Y., Adegbola, C., Daudu, C., Issa, F., Kamau, G., Kergna, A. O., Mose, L., Ndirpaya, Y., Fatunbi, O., Zossou, R., Kirui, O., & Birner, R. (2022). Mechanization, digitalization, and rural youth-Stakeholder perceptions on three mega-topics for agricultural transformation in four African countries. *Global Food Security*, *32*, 100616. DOI: 10.1016/j.gfs.2022.100616

Deloitte. (2020). *Digital maturity study of Greek economy and Greek enterprises*. Deloitte. https://www2.deloitte.com/gr/en/pages/technology/articles/Deloitte-SEV-Digital-Observatory-Report-2nd-Edition.html

Dimitrijević, M. S. (2023). Technological progress in the function of productivity and sustainability of agriculture: The case of innovative countries and the Republic of Serbia. *Journal of Agriculture and Food Research*, *14*, 100856. DOI: 10.1016/j.jafr.2023.100856

Edwards, J. P., Kuhn-Sherlock, B., Rue, B. D., & Eastwood, C. R. (2020). Technologies and milking practices that reduce hours of work and increase flexibility through milking efficiency in pasture-based dairy farm systems. *Journal of Dairy Science*, *103*(8), 7172–7179. DOI: 10.3168/jds.2019-17941 PMID: 32505396

European Commission. (2024). *Science, research and innovation performance of the EU*. https://op.europa.eu/en/publication-detail/-/publication/c683268c-3cdc-11ef-ab8f-01aa75ed71a1/language-en

International Trade Administration. (2024, July 15). *Serbia—Country Commercial Guide*. https://www.trade.gov/country-commercial-guides/serbia-agricultural-sectors

Klerkx, L., Jakku, E., & Labarthe, P. (2019). A review of social science on digital agriculture, smart farming and agriculture 4.0: New contributions and a future research agenda. *NJAS Wageningen Journal of Life Sciences*, *90*(1), 100315. DOI: 10.1016/j.njas.2019.100315

Mavridis, A., Vlahopoulou, M., & Gertsis, A. (2012). Innovative Agribusiness In Greece. *World, 2005*, 07.

Munoz, L. A., Gómez, M. C., Jiménez, D., Hasiner, E., Schroeder, K., Treguer, D., Juergenliemk, A., Horst, A., Jarvis, A., & Kropff, W. (n.d). *Digital Agriculture Profile: Argentina*. https://openknowledge.fao.org/server/api/core/bitstreams/a7eaf9ef-37d8-4e35-b89a-b02f1491b8da/content

Muzamil, M., Rasool, S., Banday, R. U. Z., Mohiuddin, M., Rasool, S., & Reshi, I. A. (2025). Smartphone-Based Smart Agriculture: Potential Challenges and Prospects. *Handbook of Agricultural Technologies*, 1-18.

Ozdogan, B., Gacar, A., & Aktas, H. (2017). Digital agriculture practices in the context of agriculture 4.0. *Journal of Economics Finance and Accounting*, *4*(2), 186–193. DOI: 10.17261/Pressacademia.2017.448

Pogorelskaia, I., & Várallyai, L. (2020). Agriculture 4.0 and the role of education. *Journal of Agricultural Informatics/Agrárinformatika Folyóirat, 11*(1).

Saçtı, H., & Dellal, İ. (2023). Measuring the level of digitalization in agriculture: Identification indicators and scaling to determine the digitalization level of farmers. *Turkish Journal of Agriculture - Food Science and Technology*, *11*(6), 1024–1031. DOI: 10.24925/turjaf.v11i6.1024-1031.6049

Shepherd, M., Turner, J. A., Small, B., & Wheeler, D. (2020). Priorities for science to overcome hurdles thwarting the full promise of the 'digital agriculture' revolution. *Journal of the Science of Food and Agriculture*, *100*(14), 5083–5092. DOI: 10.1002/jsfa.9346 PMID: 30191570

Soma, T., & Nuckchady, B. (2021). Communicating the benefits and risks of digital agriculture technologies: Perspectives on the future of digital agricultural education and training. *Frontiers in Communication*, *6*, 762201. DOI: 10.3389/fcomm.2021.762201

Tsiforou, E. (2020). *Digital innovation for sustainable agriculture: The Greek approach.*

World Bank. (2024a). *Greece—Employment in agriculture (% of total employment)—2024 data, 2025 forecast, 1981–2022 historical*. https://tradingeconomics.com/greece/employment-in-agriculture-percent-of-total-employment-wb-data.html

World Bank. (2024b). *The World Bank in Serbia.* https://www.worldbank.org/en/country/serbia/overview

Wu, Y., & Takács-György, K. (2023, May). What is stopping Agriculture 4.0?---Examples from China. In *2023 IEEE 17th International Symposium on Applied Computational Intelligence and Informatics (SACI)* (pp. 000511-000518). IEEE.

Wu, Y., Zhong, X., & Takács-György, K. (2023). Digital agricultural education in Hungary. *Gradus, 10*(1). Advance online publication. DOI: 10.47833/2023.1.AGR.007

Chapter 8
Customer Engagement in Greek Microbreweries: Insights From Facebook Pages

Maria Ntaliani
https://orcid.org/0000-0001-7643-5009
Department of Agricultural Economics and Rural Development, Agricultural University of Athens, Greece

Effrosyni Bitakou
https://orcid.org/0009-0003-0206-8420
Department of Agricultural Economics and Rural Development, Agricultural University of Athens, Greece

Vasiliki Drakopoulou
Department of Agricultural Economics and Rural Development, Agricultural University of Athens, Greece

Filotheos Ntalianis
Department of Business Administration, University of Piraeus, Athens, Greece

ABSTRACT

Social media platforms have transformed the landscape of digital communication and interaction. Expanding from personal use to the business world, Facebook still comprises the most popular social media platform globally. Businesses have realized the potential of Facebook for building strong bonds with customers to increase customer engagement and reactions. Microbreweries have a positive perception on social media for showcasing their unique products with distinct flavors, ingredients, and brewing techniques. In this context, the objective of this research is to study customer engagement and reactions on Facebook pages. A social media analytics

DOI: 10.4018/979-8-3693-8237-0.ch008

tool, Fanpage Karma, was employed for analyzing the total of Greek microbreweries' pages for the past five years, namely from 2019 to 2023. Overall, the results indicate that Greek microbreweries maintain a noticeable presence on social media, with the majority using at least two social media platforms. However, they do not appear to have fully capitalized on the opportunities offered by Facebook.

INTRODUCTION

The proliferation of Web 2.0 and social networks has profoundly impacted how companies approach marketing and advertising (Assaad & Gómez, 2011). Worldwide, social media have innovatively enabled the connectivity, enhanced visibility, driven traffic, and sharing of information, ideas, and experiences on an unprecedented scale, representing an evolution and extension of traditional word-of-mouth networks (Sajithra & Patil, 2013). Over the past ten years, numerous social media platforms have been launched and rapidly have become popular for both personal and business purposes. Particularly, businesses utilize social media not only for professional networking, but also to raise customer engagement, increase brand awareness, promote their products, discover customers' needs and innovate. On the other hand, consumers can have access to a great amount of information for products (e.g., product details, reviews and comments), enhanced customer service, and collaboration through online communities, leading to more enriched and informed customer experience. Till today, businesses from various sectors have already explored the power of social media, such as retail, hospitality and tourism, healthcare, entertainment and media, and food and beverage. It must be noted that social media serve as a powerful tool particularly for small and medium-sized businesses, as they comprise a cost-effective marketing way to receive immediate feedback, and build strong, personalized relationships.

Customers are the heart of any business, and social networks offer a unique opportunity to build closer and more effective relationships with them (Avramidis, 2021). Customer engagement is a marketing concept that gained prominence after the turn of the millennium. The significance of customer engagement is supported by growing scholarly and industry interest (Maree & Van Heerden, 2021). It offers great value to companies by directly influencing firm performance, customer behavioral intentions, and word-of-mouth promotion (de Oliveira Santini et al., 2020). Globally, Facebook is the most widely used free social networking platform reaching three billion monthly active users in 2024 (Statista, 2024). Over the years, it has proved its ability to build strong bonds with customers and increase customer engagement (de Silva, 2020).

Beer is considered as a social product; therefore, it makes sense for microbreweries to utilize social media. According to literature, microbreweries can use social media marketing to stand out from larger breweries by highlighting their special production methods and beers (Castana, 2016). Many microbreweries host events, such as brewery tours and tastings, that social media platforms can easily promote, share details, and encourage attendance. These platforms also enable them to tailor their marketing efforts to particular demographics, including craft beer enthusiasts and local residents near the microbrewery. In addition, beer consumption follows a seasonal pattern, rising mainly in spring and lasting until early autumn, mainly due to tourism. More importantly, unforeseen events, like the COVID-19 outbreak and the war in Ukraine, have highlighted their affordability in dealing with relevant financial challenges (Dimtsoudis, 2022). Particularly for Greece, the microbrewery industry is constantly growing and some microbreweries have already seen their unique beers travelling all over the world.

In this light, the objective of this research is to examine Greek microbreweries' social media usage and customers' engagement and reactions with microbrewery Facebook pages. The sample used consisted of eighty-three Greek microbreweries. The study utilizes the social media analytics tool, Fanpage Karma, to evaluate key indicators, namely "Engagement" and "Total reactions", over the past five years, from 2019 to 2023. Specifically, this article seeks to answer the following research questions:

> RQ1: Do Greek microbreweries use a website and social media platforms for their online presence?
> RQ2: Do significant changes in consumer Facebook engagement occur from 2019 to 2023?
> RQ3: Do significant changes in consumer Facebook total reactions occur from 2019 to 2023?

The novelty of this paper lies in studying changes in consumer engagement and reactions observed on Facebook pages over five consecutive years for all existing Greek microbreweries. Therefore, the structure of this paper is as follows: Section 2 presents the background, including a description of the microbrewing business in Greece, and the research methodology to capture customer's engagement and reactions. Section 3 presents the study results, namely descriptive statistics and statistics using the SPSS software. In conclusion, the study examines insights from the findings, acknowledges its limitations, and suggests directions for future work.

BACKGROUND

This section consists of two parts. Firstly, it presents the microbrewery business in Greece, so as to provide an overview of current status and highlight the social media advantages for them. Secondly, it describes the research methodology for illustrating Facebook engagement in microbreweries.

Microbrewery in Greece

The Greek brewing industry plays a crucial role, not only in the local beverage market but also in showcasing Greek products internationally (Tsiouni et al., 2024). A small number of industries operate in the brewing sector, differentiated by size, distribution networks, and market coverage in Greece. This market is characterized by high concentration, with a few companies controlling most of the consumption. The large brewers maintain stable and organized production, offer many brands, and have extensive distribution networks.

Microbrewery is a dynamic part of the Greek brewery sector. It focuses on the production of beer in small quantities, usually much smaller than those of large breweries, and operates independently from them. The concept of a microbrewery should not be limited to production volume, but also encompasses the producers' philosophy and quality objectives for the final product. The use of high-quality raw materials, a meticulous production process, and adherence to hygiene standards are key elements that ensure the uniqueness and high quality of the beers produced by microbreweries (Stefanidakis, 2015).

The geography and the microclimate provide the adequate conditions to establish at least one microbrewery in each Greek region. Over the last decade, the number of microbreweries has risen substantially. Between 2013 and 2020, around eighty microbreweries were established and are currently operating. The emergence of new microbreweries across various regions of the country can largely be attributed to the increasing demand for unique and specialized products. This trend is reinforcedby a growing consumer preference for Greek-made goods and the rise of the "Made in Greece" movement, which reflects a commitment to supporting the nation's production base. Notably, many of these new Greek beer labels feature distinctly Greek names and packaging designs that are both striking and playful (ICAP CRIF, 2022). These labels enjoy high recognition in local retail markets within the regions they are produced, and are promoted as local products (Emmanouil, 2024).

Despite the increase in the number of microbreweries in Greece, their market share remains small. Limited production volumes and strong competition from large industries are factors that limit the expansion of their share. It is noteworthy that some microbreweries cooperate with large alcohol distribution companies, helping their products to reach restaurants and supermarkets, expanding their network beyond local shops. In addition, some microbreweries have managed to penetrate foreign markets (ICAP CRIF, 2022). Microbreweries have limited financial and human resources, as well as reduced bargaining power compared to large breweries, which restricts their presence in supermarkets. Moreover, microbreweries are facing the phenomenon of craftwashing, i.e., competing with multinational brands that are marketed as local, small-batch beers, often sold at lower prices or microbreweries that are acquired by multinationals. Most Greek microbreweries are family-run businesses and do not acquire large marketing teams like multinational companies. Therefore, usually the beer producer is the one to negotiate with intermediaries and not a salesman (Emmanouil, 2024).

Microbreweries lack budgets for advertising, so to sell craft beer successfully in the local market, they need effective communication with consumers. As a result, they rely primarily on social media and word-of-mouth marketing (Pachura, 2020). Social media is one of the main channels for gathering information about customers and competitors: who they are, what they want, what they think about the products, and how they rate them (Kuijpers, 2020). Microbreweries should take into account these facts and consider social media as a necessary marketing tool that meets their needs.

Research Methodology

The methodology used in this research work consists of three steps:

Sample Identification

To identify the sample of Greek microbreweries to be used for the current research, well-known search engines were searched from March to June 2024. Appropriate keywords, such as "microbrewery" and "microbrewery business" were used for the search. Specific data are recorded for each microbrewery found, namely its geographical location, the existence of a website, and social media pages on the most popular platforms, namely Facebook, Instagram, Twitter, YouTube, TikTok, LinkedIn, and Pinterest.

Social Media Data Collection

Fanpage Karma, a commercial online tool for social media analysis and management, was selected for data collection of the Greek microbrewery pages. Developed in 2012 and widely used in academic and professional settings, Fanpage Karma has been employed in various studies to monitor activities and content on social networks for businesses, universities, and governmental organizations (Jayasingh, 2019; Mori et al., 2021; Almeida, 2024; and Gmiterek, 2023). It enables comprehensive analysis of social media interactions of a company and its competitors, with the capability to export reports to spreadsheets for real-time trend detection, fan identification, and content analysis (Bitakou et al., 2023). Additionally, Fanpage Karma utilizes key performance indicators (KPIs) to measure and evaluate social media performance (Fanpage Karma, 2022). The tool offers a broad range of functionalities, including competitor analysis, audience insights, content scheduling and analysis, sentiment analysis, influencer identification, hashtag tracking, reporting and visualization, paid advertisement analysis, and real-time monitoring and alerts. Fanpage Karma was developed to address the needs arising in social media. Its primary aim is to assist organizations and social media managers in attracting a wide audience, creating online communities, and analyzing their competitors on social media. One of its key features is the ability to aggregate competitors into a comprehensive dashboard, allowing users to measure and compare performance across different social media platforms. Additionally, the tool provides graphs and historical data to facilitate the analysis and comparison of performance over various time periods. Overall, Fanpage Karma enables users to monitor their social media performance, respond effectively to competitive environments, and strategically reach their target audiences. Fanpage Karma was selected for this study due to its robust set of metrics, which are crucial for the research objectives, and its unique ability to perform multiple comparisons of these metrics across diverse datasets and large samples of business social media profiles. In order to be concise with the aims of this study, the following indicators were chosen:

Engagement. It shows how successful a profile is at encouraging users to interact. It shows the average number of times a fan interacts with a page's posts. This is calculated by dividing the daily number of reactions, comments, and shares by the number of fans. This metric is independent of the size of the profile because interactions are divided by the number of followers. This makes it possible to compare profiles within a specific period.

Total Reaction, Comments, and Shares. This refers to the number of interactions (i.e., reactions, comments, and shares) on page posts that were published in a specific period. For the rest of this article, this indicator is entitled as "Total Reactions" for brevity reasons.

As mentioned above, each of the indicators Fanpage Karma provides data analysis for a specific period. For comparable results, the periods under study are the following: a) 1 January to 31 December 2019; b) 1 January to 31 December 2020; c) 1 January to 31 December 2021; d) 1 January to 31 December 2022; and e) 1 January to 31 December 2023.

Statistical Analysis

Finally, the collected data were analyzed using descriptive statistics, correlation analyses, and linear regression analyses with IBM SPSS Statistics software.

RESULTS

In the following, the Facebook pages of microbreweries will be examined independently using the Fanpage Karma tool, for each of the five time periods.

Sample Identification

The sample identification resulted in eighty-three (83) microbreweries. Regarding the geographical distribution of the microbreweries, they were classified according to Greek first-level administrative entities (regions). More specifically, the distribution by region is as follows: 18.07% are located in Crete; 15.66% in Peloponnese; 13.25% in South Aegean; 12.05% in Attica; 9.64% in Thessaly; 7.23% in Central Macedonia; 4.82% in North Aegean; 4.82% in Central Greece; 4.82% in Eastern Macedonia and Thrace; 3.61% in Ionian Islands; 3.61% in Epirus; 1.20% in Western Greece; and 1.20% in Western Macedonia.

Of all microbreweries, 77.11% maintain a website, 6.02% own a domain name but their website is under-construction, and 16.87% do not possess a website. Furthermore, the presence of microbreweries on social media platforms, namely Facebook, Instagram, Twitter, YouTube, TikTok, LinkedIn, and Pinterest, was analyzed: the majority (43.40%) has pages on two platforms; 22.89% on three platforms; 10.85% on only one platform; 6.02% on four platforms; 2.40% on five platforms; 2.40% on six platforms; and 12.04% do not have any social media page (Table 1).

Table 1. Number of social media platforms used by Greek microbreweries

	Social Media Usage (N = 83)	
No. of Platforms Used	**No.**	**%**
0	10	12,04%
1	9	10.85%
2	36	43.40%
3	19	22,89%
4	5	6.02%

continued on following page

Table 1. Continued

	Social Media Usage (N = 83)	
No. of Platforms Used	No.	%
5	2	2,40%
6	2	2,40%

From the microbreweries that use social media, Facebook is the platform most preferred. Specifically, 87.95% have a Facebook page, 74.70% have Instagram, 15.66% have Twitter, 25.30% have YouTube, 8.43% have TikTok, and 4.82% have Pinterest (Table 2).

Table 2. Social media usage by Greek microbreweries

	Microbreweries Using Social Media (N = 77)	
Platforms	No.	%
Facebook	73	87.95
Instagram	62	74.70
Twitter	13	15.66
YouTube	21	25.30
TikTok	7	8.43
Pinterest	4	4.82

Social Media Data Collection

Statistical analysis focuses on Facebook which provided complete datasets. It is important to note that measurements based on historical data may have gaps (missing values) since they were not registered in the database during the periods in question, as no search was carried out for these microbrewery names (Bitakou et al., 2023). The obtained datasets were refined to exclude multiple records of the same microbreweries. Also, some microbrewery pages were registered with different names across the different time periods. Thus, all microbreweries were assigned a specific code across year to avoid data inconsistency. The study data were quantitative, and the variables "Total" and "Engagement" were operationalized as continuous variables in this analysis (i.e., descriptive, correlations, and repeated measures analysis of variance). It should be noted that in the text and tables that follow, there is a number at the end of each variable (i.e., 19, 20, 21, 22, and 23), which refers to the corresponding periods under examination, i.e., 2019, 2020, 2021, 2022, and 2023.

Statistical Analysis

Descriptive Statistics

The descriptive statistics presented in Tables 3 and 4 revealed that the data exhibited significant departures from normality, thereby invalidating the fundamental assumptions necessary for the application of parametric statistics. Therefore, we chose to conduct non-parametric statistical analyses, including Spearman's ρ (rho) correlations and Friedman's repeated measures (RM) ANOVA, to evaluate the validity of our hypotheses. The results of the correlation and RM ANOVA analyses are presented below.

Table 3. Descriptive statistics for Facebook engagement

Variables	Median	Mean	SD	$Skewness_z$	$Kurtosis_z$
Engagement19	0.000	0.000	0.003	21.828	76.577
Engagement20	0.000	0.000	0.001	13.116	24.234
Engagement21	0.000	0.000	0.002	24.526	92.348
Engagement22	0.000	0.001	0.002	8.404	12.609
Engagement23	0.000	0.001	0.002	8.530	11.265

Note: SD= Standard Deviation, $Skewness_z$= Skewness/ Std Error of Skewness, $Kurtosis_z$= Kurtosis/Std Error of Kurtosis

Table 4. Descriptive statistics for Facebook total reactions

Variables	Median	Mean	SD	$Skewness_z$	$Kurtosis_z$
Total19	805.000	4739.648	9085.916	14.709	41.794
Total20	1984.000	4883.718	7945.369	9.274	13.542
Total21	1829.000	4402.296	7941.383	14.095	34.824
Total22	1230.000	3143.958	6104.029	14.102	32.341
Total23	930.000	2862.371	4972.545	12.345	28.546

Note: SD= Standard Deviation, $Skewness_z$= Skewness/ Std Error of Skewness, $Kurtosis_z$= Kurtosis/Std Error of Kurtosis

Correlations

Table 5 presents the Spearman's rank correlation coefficients (ρ) for the Engagement variable across the years 2019 to 2023. The highest correlation is observed between the variables for the 2019 to 2020 period (ρ = .875, p < .000), while the lowest correlation is observed between the variables for the 2019 to 2023 period (ρ = .084, p = ns). In general, the majority of paired correlations for engagement demonstrate only a few significant, strong, and positive associations.

Table 5. Spearman's ρ (rho) correlations for Facebook engagement (2019–2023)

Spearman's Correlations						
Variable		**1**	**2**	**3**	**4**	**5**
1. Engagement19	ρ	—				
	p-value	—				
2. Engagement20	ρ	0.875***	—			
	p-value	< .001	—			
3. Engagement21	ρ	0.736***	0.811***	—		
	p-value	< .001	< .001	—		
4. Engagement22	ρ	0.092	0.111	0.229	—	
	p-value	0.445	0.357	0.055	—	
5. Engagement23	ρ	0.084	0.097	0.188	0.816***	—
	p-value	0.487	0.424	0.119	< .001	—

* p < .05, ** p < .01, *** p < .001, two-tailed test

Table 6 presents the total Facebook reactions correlations for the years 2019 to 2023. All paired correlations are statistically significant, indicating a robust positive association between the variables. The highest correlation is observed between 2021 and 2022 (ρ = .863, p < .000), while the lowest correlation is observed between 2019 and 2023 (ρ = .417, p < .000).

Table 6. Spearman's ρ (rho) correlations for Facebook total reactions (2019–2023)

Spearman's Correlations						
Variable		**1**	**2**	**3**	**4**	**5**
1. Total19	ρ	—				
	p-value	—				
2. Total20	ρ	0.795***	—			
	p-value	< .001	—			
3. Total21	ρ	0.579***	0.753***	—		
	p-value	< .001	< .001	—		
4. Total22	ρ	0.483***	0.626***	0.863***	—	

continued on following page

Table 6. Continued

Spearman's Correlations						
Variable		1	2	3	4	5
	p-value	< .001	< .001	< .001	—	
5. Total23	ρ	0.417***	0.554***	0.703***	0.861***	—
	p-value	< .001	< .001	< .001	< .001	—

* p < .05, ** p < .01, *** p < .001, two-tailed test

Repeated Measures ANOVA

Engagement. The Friedman test was employed to ascertain whether there is a statistically significant annual effect on users' engagement. Conover's post hoc pairwise comparisons were employed to ascertain whether there were significant differences in users' engagement levels across each year.

Table 7. Friedman test for Facebook engagement (2019–2023)

Factor	χ^2	Df	p-value	Kendall's W
YEARS	95.527	4	< .001	0.341

The Friedman test demonstrated that the Facebook engagement scores exhibited notable fluctuations from year to year, with a statistically significant result (χ^2 (4) = 95.527, p < .001; Table 7). Conover's pairwise post hoc comparisons revealed that Facebook engagement reached its maximal difference between 2021 and 2022 (T(276) = 9.056, p < .001). Additionally, considerable discrepancies were noted between the years 2019 and 2022 (T(276)= 8.073, p<.001), 2020 and 2022 (T(276)= 7.902, p<.001), and 2021 and 2023 (T(276)= 7.603, p<.001). Further changes were found to be statistically significant for the years 2019 and 2023, as well as for the years 2020 and 2023. The results of all post hoc comparisons are presented in Table 8. Figure 1 illustrates the standard error bars associated with the observed fluctuations in Facebook engagement on an annual basis.

Table 8. Conover's post hoc comparisons for Facebook engagement (2019–2023)

		T-Stat	df	W_i	W_j	p	P_{bonf}
2019	2020	0.171	276	177.500	179.500	0.864	1.000
	2021	0.982	276	177.500	166.000	0.327	1.000
	2022	8.073	276	177.500	272.000	< .001	< .001
	2023	6.621	276	177.500	255.000	< .001	< .001
2020	2021	1.153	276	179.500	166.000	0.250	1.000
	2022	7.902	276	179.500	272.000	< .001	< .001
	2023	6.450	276	179.500	255.000	< .001	< .001
2021	2022	9.056	276	166.000	272.000	< .001	< .001
	2023	7.603	276	166.000	255.000	< .001	< .001
2022	2023	1.452	276	272.000	255.000	0.148	1.000

Note: Grouped by subject, Rank-biserial correlation based on individual signed-rank tests.

Figure 1. Bar chart showing standard errors for Facebook engagement (2019–2023)

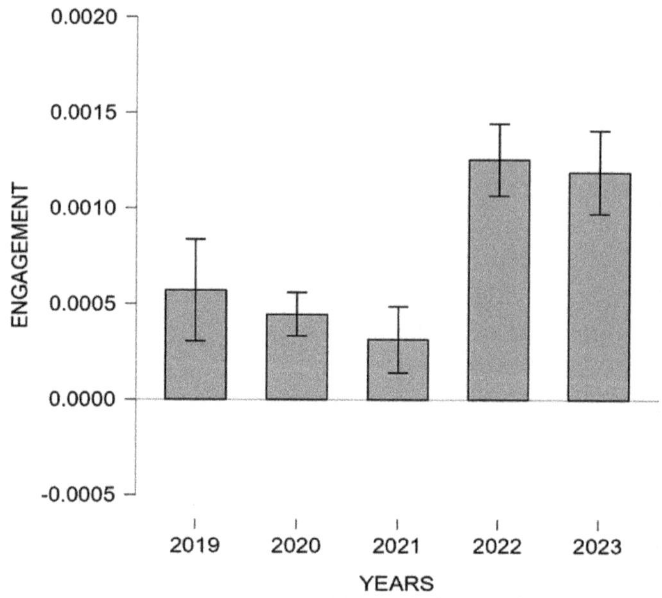

Total Reactions. In regard to Facebook total reactions, the Friedman test was employed to ascertain whether there existed any mean rank differences. As evidenced in Table 9 (χ^2 (4) = 95.527, p < .001), the results demonstrated that there were notable differences among the mean ranks.

Table 9. Friedman test for Facebook total reactions (2019–2023)

Factor	χ^2	Df	p-value	Kendall's W
YEARS	25.652	4	< .001	0.092

Conover's post hoc pairwise comparisons (see Table 10) revealed significant differences in the total number of Facebook reactions between several years. Specifically, the largest differences were observed between 2020 and 2023 ($T(276) = 4.669$, $p < 0.001$) and between 2021 and 2023 ($T(276) = 3.786$, $p < 0.01$). Additional significant differences were found between 2019 and 2023 ($T(276) = 3.131$, $p < 0.05$) and between 2020 and 2022 ($T(276) = 3.160$, $p < 0.05$). Figure 2 displays the standard error bars for the total annual reactions on Facebook.

Table 10. Conover's post hoc comparisons for Facebook total reactions (2019–2023)

		T-Stat	df	W_i	W_j	p	P_{bonf}
2019	2020	1.537	276	219.000	246.000	0.125	1.000
	2021	0.655	276	219.000	230.500	0.513	1.000
	2022	1.623	276	219.000	190.500	0.106	1.000
	2023	3.131	276	219.000	164.000	0.002	0.019
2020	2021	0.882	276	246.000	230.500	0.378	1.000
	2022	3.160	276	246.000	190.500	0.002	0.018
	2023	4.669	276	246.000	164.000	< .001	< .001
2021	2022	2.277	276	230.500	190.500	0.024	0.235
	2023	3.786	276	230.500	164.000	< .001	0.002
2022	2023	1.509	276	190.500	164.000	0.132	1.000

Note: Grouped by subject, Rank-biserial correlation based on individual signed-rank tests.

Figure 2. Bar chart showing standard errors for Facebook Total reactions (2019–2023)

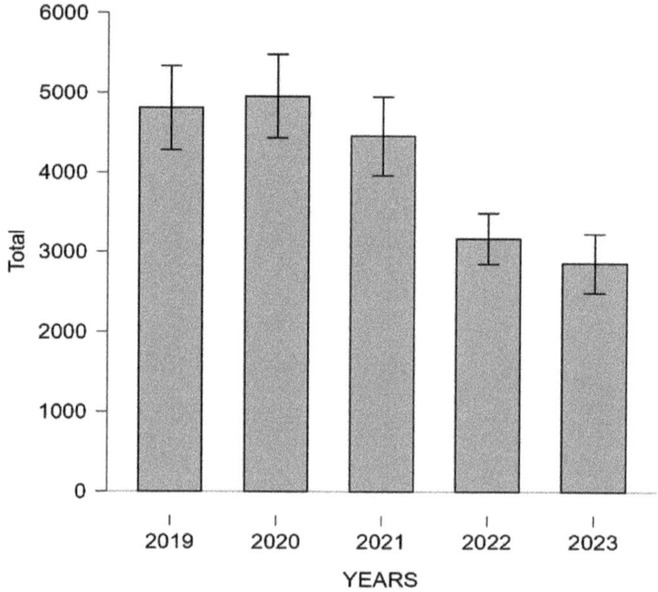

CONCLUSIONS

As the craft beer industry continues to evolve, embracing new technologies and innovative marketing techniques, such as social media will be essential for sustaining growth and enhancing their position in an increasingly competitive market, by enabling them to gain marketing insights, promote products and services, and strengthen community relations.

The unique characteristics of the Greek microbrewing landscape, including a growing craft beer culture and an emphasis on local products, provide a fertile ground for using social media. Microbreweries can cultivate a strong brand identity, engage with their audience, showcase their products in visually appealing ways, and ultimately contribute to the broader growth of the craft beer sector in the region.

This work tries to give insights into the appeal of microbreweries' Facebook pages to consumers by studying their engagement and reactions. The presence of Greek microbreweries on social media is noticeable, as the majority of them have pages on at least two social media platforms, namely Facebook and Instagram. Also, YouTube is used by one fourth of the breweries. However, Greek microbreweries appear to have yet to fully leverage the opportunities provided by Facebook. Specifically, customers' engagement appears to be in essence inactive and total reactions are at a very low level urging that communication with customers needs effective solutions to increase.

The objective of our research was twofold: firstly, to identify any differences in customer Facebook engagement and total reactions between the 2019 and 2023 years; and secondly, to ascertain whether there is a statistically significant difference between these two periods. With regard to the variable of engagement across the years 2019 to 2023, the highest correlation is observed between the 2019 and 2020 periods. In contrast, the highest observed relationship for the total reactions variable exists between the 2021 and 2022 periods. The data indicate that the Facebook engagement scores and total reactions exhibit a distinct pattern over the examined years. Specifically, while engagement scores appear to decline between 2019 and 2021, they demonstrate a notable recovery in the subsequent years, particularly in 2022 and 2023. These findings are also corroborated by post hoc analyses. With regard to customer Facebook total reactions, the observed pattern appears to be almost the opposite. Therefore, while they are relatively high for the initial three-year period (2019-2021), they subsequently follow a downward trajectory for the subsequent two-year period (2022-2023). A plausible explanation for the observed decline in total reactions in the later years may be the growing influence of Instagram, which could have diverted attention and engagement away from Facebook as a primary platform for customer interaction.

Overall, this study can provide valuable insights for Greek microbreweries. Just being present on social media disarms this powerful tool and removes all its capabilities. It is recommended that microbreweries should utilize social media more efficiently to capitalize on their positive impact it could have on their businesses. This may require a fresh glimpse and profound reorganization of their strategies. To succeed interaction with customers, they have to take into serious consideration their habits and views and adapt to emerging trends. Through effective use of platforms, microbreweries can foster a sense of community among craft beer enthusiasts but also encourage consumer loyalty, enabling microbreweries to leverage social media strategically and enhance their market competitiveness. Since the current study focuses on Facebook, further research should examine the potential of other platforms in supporting microbreweries' digital marketing and determine how breweries can select the most effective channels for reaching their audience. In doing so, it is essential for microbreweries to consider the platforms consumers themselves prefer and to deliver content within those spaces, rather than waiting for consumers to seek it out elsewhere. In particular, Instagram, that is the second popular platform used by microbreweries has strong visual orientation and popularity among younger demographics through creative, image-driven content.

Looking forward, for the better interpretation of this study results, it is necessary to identify the strategic and practical reasons that shape the current status of engagement. Thus, future research will focus on interviewing microbreweries to fully understand the challenges they face in order to adopt social media. Also, beer is a product that experiences seasonal demand, mainly from spring to early autumn. This seasonality is primarily due to tourism, which increases beer sales in the summer. Future work will study whether this seasonality influences social media interaction. In conclusion, microbreweries should revisit their social media vision for achieving a more robust online presence and tangible advantages.

REFERENCES

Almeida, C. M., Afonso, C. M., Serra, M., & António, N. (2024). The presence of Algarve wine producers on Facebook between pre-covid-19 and covid-19 periods. *Tourism and Hospitality International Journal*, *22*(1), 27–62.

Assaad, W., & Gómez, J. M. (2011). *Social network in marketing (social media marketing) opportunities and risks. International Journal of Managing Public Sector Information and Communication Technologies* (Vol. 2). IJMPICT.

Avramidis, A. (2021). *Social media marketing: What it is and why it is necessary today*. WebOne.https://shorturl.at/j0Ovn

Bitakou, E., Karetsos, S., Ntalianis, F., Ntaliani, M., & Costopoulou, C. (2023). Evaluating social media marketing in the Greek winery industry. *Sustainability (Basel)*, *16*(1), 192. DOI: 10.3390/su16010192

Castana, E. (2016). *Be social or be good. The performance of Italian microbreweries*. https://thesis.unipd.it/handle/20.500.12608/25222

de Oliveira Santini, F., Ladeira, W. J., Pinto, D. C., Herter, M. M., Sampaio, C. H., & Babin, B. J. (2020). Customer engagement in social media: A framework and meta-analysis. *Journal of the Academy of Marketing Science*, *48*(6), 1211–1228. DOI: 10.1007/s11747-020-00731-5

de Silva, T. M. (2020). Building relationships through customer engagement in Facebook brand pages. *Marketing Intelligence & Planning*, *38*(6), 713–729. DOI: 10.1108/MIP-02-2019-0085

Dimtsoudis, C. (2022). *A tool for self-produced malt*. Agronews. https://www.agronews.gr/files/PDF/Beer_bruch/04_22_beer_brunch.pdf

Emmanouil, S. (2024). Small breweries: The signal of concentration and the risk of craftwashing. PowerGame.https://www.powergame.gr/epichirisis/666993/mikres-zythopoiies-to-sima-sygkentrosis-kai-o-kindynos-craftwashing/

Fanpage Karma Academy. (2022). *Fanpage Karma Academy*.https://academy.fanpagekarma.com/en/home/

Gmiterek, G. (2023). Use of Facebook fan pages in Polish academic libraries (2009–2022). *Journal of Academic Librarianship*, *49*(2), 102678. DOI: 10.1016/j.acalib.2023.102678

ICAP CRIF. (2022). *ICAP CRIF brewing industry study*. ICAP CRIF.https://www.icapcrif.com/icap-crif-study-zythos2022/

ICAP CRIF. (2022). *Sector study for beer*. ICAP CRIF.

Jayasingh, S. (2019). Consumer brand engagement in social networking sites and its effect on brand loyalty. *Cogent Business & Management*, *6*(1), 1698793. DOI: 10.1080/23311975.2019.1698793

Kuijpers, N. (2020). *Market orientation in microbreweries* (Master's thesis, Wageningen University & Research). Wageningen University & Research. https://edepot.wur.nl/530023

Mori, E., Barabaschi, B., Cantoni, F., & Virtuani, R. (2021). Local governments' communication through Facebook. Evidences from COVID-19 pandemic in Italy. *Journal of Public Affairs*, *21*(4), e2551. DOI: 10.1002/pa.2551 PMID: 33349742

Pachura, P. (2020). The role of space in the business models of microbreweries. *Applied Geography (Sevenoaks, England)*, *125*, 102303. https://www.sciencedirect.com/science/article/pii/S014362282030326X?via%3Dihub. DOI: 10.1016/j.apgeog.2020.102303

Sajithra, K., & Patil, R. (2013). Social media–history and components. *Journal of Business and Management*, *7*(1), 69–74.

Statista (2024). *Most popular social networks worldwide as of April 2024, by number of monthly active users*. Statista. https://www.statista.com/statistics/272014/global-social-networks-ranked-by-number-of-users/

Stefanidakis, P. (2015). *Macro vs. micro breweries*. Beeroskopio. https://www.beeroskopio.com/2015/01/macro-micro.html

Tsiouni, M., Konstantinidis, C., Aggelopoulos, S., & Mamalis, S. (2024). A cluster analysis study based on financial indicators, as a competitiveness strategy of the brewery industry in Greece. In *Value chain dynamics in a biodiverse environment: Advances in biodiversity, sustainability, and agri-food supply chain development* (pp. 129–137). Springer Nature Switzerland. DOI: 10.1007/978-3-031-49845-9_7

Chapter 9
GaiaMentor AI for Farmers:
A Conversational Solution for Smarter Agriculture

Michail Petropoulos
https://orcid.org/0009-0007-5421-2024
Informatics Laboratory, Agricultural University of Athens, Greece

Sotirios Karetsos
https://orcid.org/0000-0002-3679-3896
Informatics Laboratory, Agricultural University of Athens, Greece

ABSTRACT

Chatbots increasingly support the agricultural sector by facilitating information exchange and enhancing decision-making. However, even most contemporary chatbots still do not provide contextually reliable advice to meet the day-to-day agricultural needs of farmers. This chapter presents GaiaMentor, a chatbot that provides actionable agriculture advice for two crops, aiming to assist mainly farmers and agronomists. The design and development of the chatbot are presented. The development has been based on the Botpress platform, an all-in-one platform for building chatbots powered by the latest Large Language Models (LLMs). It is sufficiently simple for users to submit any questions on crop management, pest control practices, and fertilization requirements, and it generates responses based on the user's crop type and contextual information. The GaiaMentor's knowledge base is drawn from validated scientific information to provide reliable recommendations.

DOI: 10.4018/979-8-3693-8237-0.ch009

INTRODUCTION

Recent advancements in Artificial Intelligence (AI) have significantly influenced the relationship between humans and technology and generated many intelligent innovations that can undertake actions typically accomplished by humans. Conversational Artificial Intelligence (CAI) is a field of AI that focuses on enabling machines to understand and respond to human language naturally and engagingly, similar to human-to-human conversation (Bhardwaj et al., 2024). A chatbot can be defined as a CAI application designed to simulate conversation with humans, using text, voice, or both, often through messaging apps, websites, or voice-enabled devices through Natural Language Processing (NLP) and Machine Learning (ML) processes (Nirala et al., 2022; Casheekar et al., 2024).

Early chatbot generations (e.g., ELIZA or ALICE) were constrained by rule-based logic, manual coding, keyword matching, and basic parsing techniques (Suhaili et al., 2021). Although they were effective in domain-specific applications, these systems were limited in their ability to process interactive conversations, nuanced language understanding, and adaptability. In recent decades, there has been enormous work in NLP, Deep Learning (DL), and integration with AI that has enabled chatbots to increasingly become sophisticated and capable of contextual comprehension of user input and production of coherent, human-sounding outputs (Lin et al., 2023). Today, chatbots offer tactical, near-real-time, user-centered services, with implications distinct from earlier online chat support models. These are now being applied in healthcare, education, entertainment, and agriculture (Adamopoulou & Moussiades, 2020).

Focusing on the agricultural sector, chatbots hold significant potential for addressing critical challenges, such as resource scarcity, climate change, and the need for sustainable food production (Idoje et al., 2021; FAO, 2025; European Commission, 2025). For example, they can help by delivering recommendations for optimal sowing dates based on local climate data, tailored fertilizer schedules and quantities, or pest and disease management strategies adapted to specific crop growth stages. Therefore, chatbots support the minimization of resource waste and improvement of yields. Such support is particularly valuable for small and remote farmers, who often lack formal agricultural training or access to technical expertise, by enabling them to make timely, evidence-based decisions that would otherwise require consultation with experienced agronomists (Das & Ambika, 2025).

In this context, the objective of this chapter is to present GaiaMentor, a chatbot that focuses on the needs of Greek orchard farmers. More specifically, GaiaMentor is a conversational agent that provides customized and tailored advice on key agronomic topics, including fertilization, soil preparation and properties, irrigation, plant protection, and best practices to ensure maximum productivity and fruit quality. It also considers farmers' regions, and experience level. GaiaMentor is developed on Botpress, a platform for building chatbots and virtual assistants, where users ask questions in natural language, resulting in customized and actionable advice. It intends to assist farmers where the agronomical consultancy is limited. However, GaiaMentor does not aim to substitute for agronomists.

The remainder of this chapter is structured as follows: Section 2 reviews the literature on chatbot technologies and their applications in agriculture. Section 3 details the design and development of GaiaMentor, while Section 4 focuses on its deployment. Section 5 presents the study's conclusions, followed by Section 6, which discusses its limitations. Finally, Section 7 outlines directions for future work.

BACKGROUND

Review of Chatbot Technology

One of the most evident examples of AI's impact on human–computer interaction is the emergence of chatbots. Adamopoulou and Moussiades (2020) support that chatbots are computer programs designed to imitate human conversation, enhanced by NLP, sentiment analysis, and ML. Often termed intelligent virtual assistants or conversational agents, chatbots are now integral to domains such as business, healthcare, education, and entertainment.

As noted by Shah et al. (2023), chatbots are gaining traction due to their ability to provide continuous and engaging interactions while simultaneously reducing operational costs. Moreover, some studies indicate that approximately 40% of user queries are emotional rather than purely factual, particularly with the rise of sentiment analytics and emotionally responsive AI technologies. Therefore, in contrast to static FAQ systems, chatbots offer dynamic dialogue experiences that many users find more personal and interactive (Adamopoulou & Moussiades, 2020).

The intellectual basis for chatbots stems from early inquiries in AI, notably Turing's (1950) seminal question, "Can machines think?". In his paper *Computing Machinery and Intelligence*, Turing proposed the Turing Test, which has become a foundational concept in chatbot development. This inquiry laid the groundwork for interactive AI systems, with Joseph Weizenbaum's ELIZA (1966) representing a key milestone. ELIZA employed simple pattern-matching and word-based algorithms to emulate therapist–patient conversations but lacked genuine conversational depth (Shah et al., 2023).

Chatbot functionality was built on advancements made during the late 1990s and early 2000s. ALICE, based on Artificial Intelligence Markup Language (AIML) and advanced conversational realism (Shah et al., 2023), was considered the "most human computer" of its era. Around the same period, SmarterChild pioneered chatbot implementation within instant messaging settings, improving the accessibility of conversational agents and laying the groundwork for IBM Watson, Apple Siri, Amazon Alexa, and Google Assistant, which were released in the 2010s.

Since that time, the pace of innovation has gained momentum, with advances in DL, reinforcement learning, and transformer models. The introduction of Generative Pre-trained Transformers (GPT) raised expectations for chatbot performance. ChatGPT is a well-known example of a chatbot developed by OpenAI and released in 2022. It demonstrated reasoning, multilingual understanding, content generation and manipulation, and contextual memory (Shah et al., 2023; Casheekar et al., 2024).

State-of-the-art chatbots possess a comprehensive set of capabilities enabled by a robust, multi-layered technological foundation incorporating AI, ML, NLP, and DL. While there will be some variations in the particulars of modern chatbot architecture, there are typically three primary modules: Natural Language Understanding (NLU), a Dialogue Manager, and Natural Language Generation (NLG) (Suhaili et al., 2021; Casheekar et al., 2024). In summary, the NLU module receives user inputs and determines the intent and entities from those inputs, the Dialogue Manager controls the conversational context through state management, and the NLG module generates meaningful, human-like responses for the user (Figure 1). Furthermore, with sentiment analysis and emotion detection being implemented into chatbots, responses can carry an element of being realistic (Adamopoulou & Moussiades, 2020; Shah et al., 2023). Specifically, GPT-based Large Language Models (LLMs) help chatbots to construct natural conversations (Casheekar et al., 2024). As the development of chatbots continues to evolve, new tools like Domain-Specific Languages (DSLs), based on Model-Driven Architecture (MDA), are being proposed to aid in implementation, especially in industries that require specialized communication (Ouaddi et al., 2024).

Figure 1. Interactions among core modules in chatbots

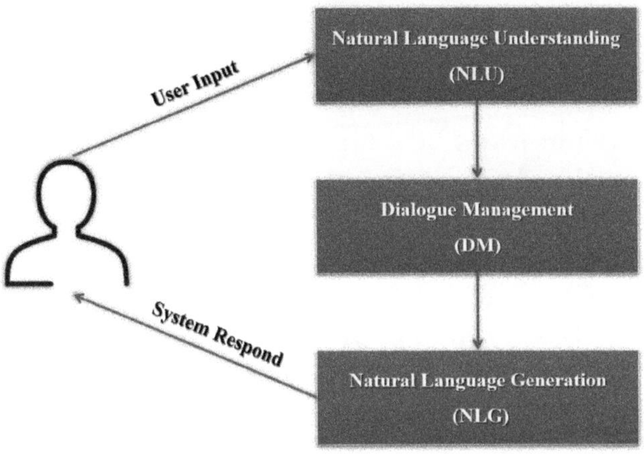

Chatbots are transforming the service delivery model. They provide operational efficiency by automating repetitive tasks, rapidly replying to questions, and offering immediate assistance for more complex inquiries, all of which can reduce costs to an organization and improve user satisfaction. One example of chatbot use is in customer service to minimize users' wait times (Fornalczyk et al., 2021). From a retail standpoint, chatbots effectively engage users through voice and chat interfaces (Idoje et al., 2021). In addition to text-based interactions, chatbots are expanding into physical spaces through voice interactions, with examples in healthcare delivery, education, and hospitality. They operate autonomously, and their deployment allows users to access them at any time (Ivanov, 2019; Vargo et al., 2021). The adoption of voice-based chatbots gained momentum during the COVID-19 pandemic, as service assistants began supporting medical receptions, telehealth consultations, and sanitation duties. There has been an important advancement in sophisticated humanoid robots. For instance, WABOT-2, Pepper, Kobian, and Sophia demonstrate significant progress in speech recognition, facial expression analysis, and social interaction capabilities (Pandey & Gelin, 2018).

Despite considerable progress, the chatbots we interact with are rarely truly empathetic or flexible. These limitations often lead to frustrating user experiences, especially when the interaction is emotionally unintelligent or involves puzzling through completely fictitious errors or mistakes (Adamopoulou & Moussiades, 2020; Fornalczyk et al., 2021). Trust, consent, and ethical design will most certainly matter to realizing the full capabilities of chatbots by improving transparency in the work they support, ensuring consistency in their functional mechanisms, and building ethical design principles that we can all follow through the probabilistic nature of activeness and interaction (Idoje et al., 2021). More design issues that make engagement more difficult include cluttered frames and distracting visual stimuli (Fornalczyk et al., 2021). It is also possible for designers of chatbots to unconsciously reinforce social stereotypes, such as gender bias in language used or in the chatbot's persona (Adamopoulou & Moussiades, 2020). Moreover, future chatbot systems with increased personalization, empathy, and context-awareness are likely to be developed around algorithms and frameworks that reflect natural language processing and transformer-model capabilities (Casheekar et al., 2024). This change will further broaden the horizons of user engagement in work and service environments (Cónego et al., 2024). To evaluate the effectiveness of chatbots, established theoretical frameworks such as the Technology Acceptance Model (TAM) and the Service Robot Acceptance Model (sRAM) are commonly employed. These models identify critical determinants, such as rapport, trust, and transparency, which influence compatibility with chatbots and virtual assistants (Idoje et al., 2021).

Chatbots in Agriculture: A Systematic Literature Review

To estimate the level of adoption of chatbots in the agricultural sector, an extensive literature review has been conducted. More specifically, the review emphasizes research involving chatbots, virtual assistants, and conversational agents for agricultural purposes, to explore their roles and potential within the sector, including weather alerts, crop recommendations, pest and disease control, and farmer education. The ultimate goal is to identify current trends and applications. The literature search was conducted on the Scopus database in July 2025. It focused on publications from the past six years based on specific research questions, namely:

> ***RQ1***: *How many studies about the use of chatbots in agriculture have been published?*
> ***RQ2***: *How many of them are providing advisory services for farmers?*

A rigorous multi-step process was employed to select and include relevant scientific literature, ensuring thematic coherence, scientific validity, and temporal relevance. The principal database query in this study was formulated as follows: "chatbot AND (agricultur* OR farm*)". The Scopus database returned 198 publications in total from the main query. Following that, a temporal filter was used to concentrate on papers released after 2019, which made up 198 of the total findings. The dataset was reduced to 196 items by restricting the language of publication to English. Thirty-three were found to be journal publications.

Only journal papers (excluding conference papers, book chapters, and reviews) were retained to filter the dataset further, resulting in 32 items. Publications were then filtered according to science categories, which were the following: computer science, engineering, business, management and accounting, engineering, agriculture and biological science, mathematics, social science, multidisciplinary, economics, econometrics and finance, and environmental science. This disciplinary filter aimed to exclude irrelevant content. The application of this filter brought the number of relevant records down to 31. Finally, this number is a focused corpus that forms the basis for the subsequent thematic analysis. Figure 2 represents the article filtering process.

Figure 2. Flow diagram of the Scopus database search process

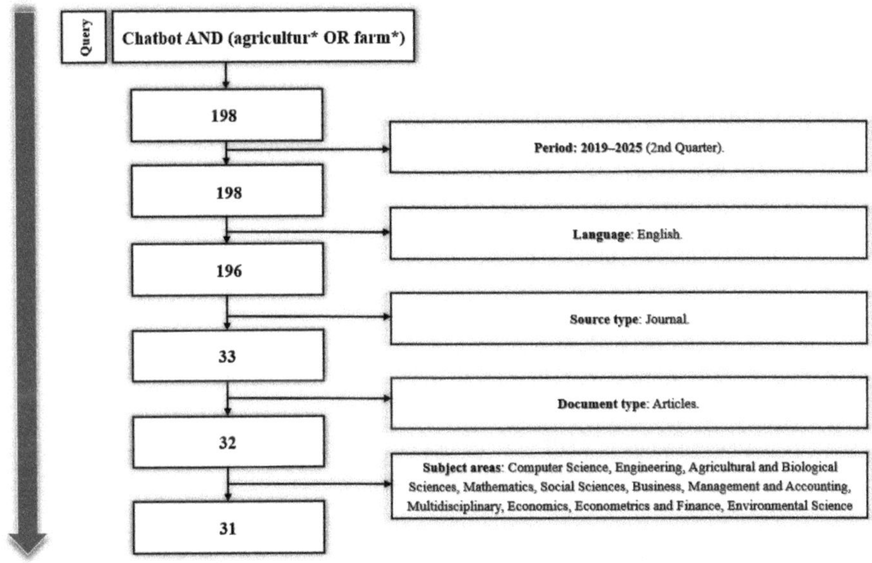

An analysis of the publication years reveals that the total number of publications occurred in the selected analysis period (i.e., since 2019). Moreover, there has been a marked increase in research interest in those years regarding this specific field of study (Figure 3).

Figure 3. Number of selected publications by year up until July 9th, 2025

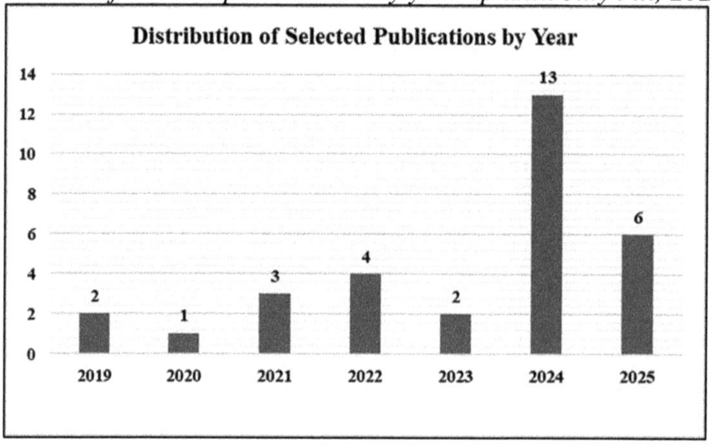

The articles that were used in this study were extracted from the Scopus database and organized with the assistance of Zotero reference management software. The final selection of sources, which met the inclusion criteria, is listed in Table 1.

Table 1. Results referring to RQ1

Authors	Titles
Patil, R.; Sinkar, Y.; Ruke, A.; Kulkarni, H.; Kadam, O.	Smart Agri-Advisor: Integrating Chatbot Technology with CNN-Based Crop Disease Classification for Enhanced Agricultural Decision-Making
Ziegfeld, L.; Scala, D.D.; Cremers, A.H.M.	The effect of preference elicitation methods on the user experience in conversational recommender systems
Al Rasyid, M.U.H.; Sukaridhoto, S.; Dzulqornain, M.I.; Rifa'i, A.	Integration of IoT and chatbot for aquaculture with natural language processing
Oumaima, S.; Mohamed, L.; Abderrahmane, B.; Iyass, M.	EnviScan+: AI-driven plant identification and ecosystem management for sustainable agriculture
Garg, P.; Rao K, S.	Building AgriRef: an AI embedded virtual reference app for farmers
Shapa, M.; Trojer, L.; Machuve, D.	Mobile-based Decision Support System for Poultry Farmers: A Case of Tanzania
Benzinho, J.; Ferreira, J.; Batista, J.; Pereira, L.; Maximiano, M.; Távora, V.; Gomes, R.; Remédios, O.	LLM Based Chatbot for Farm-to-Fork Blockchain Traceability Platform
Temniranrat, P.; Kiratiratanapruk, K.; Kitvimonrat, A.; Sinthupinyo, W.; Patarapuwadol, S.	A system for automatic rice disease detection from rice paddy images serviced via a Chatbot

continued on following page

Table 1. Continued

Authors	Titles
Sowmya, B.J.; Meeradevi, A.K.; Supreeth, S.; Pradeep Kumar, D.; Ravi Kumar, B.N.; Rohith, S.; Mishra, D.; Koushik, A.; Patil, A.U.	Leveraging machine learning for intelligent agriculture
Sivakumar, S.A.; Shankar, B.M.; Anuradha, B.; Karan, K.A.; Karthik, A.; Karthik, R.; Kumar, J.R.R.	Artificial Intelligence based Agricultural Chatbot and Virtual Assistant for Delivery of Harvested Crops
Adewopo, J.; McCampbell, M.; Mwizerwa, C.; Schut, M.	Beyond the Hype: Ten Lessons from Co-Creating and Implementing Digital Innovation in a Rwandan Smallholder Banana Farming System
Maginga, T.J.; Masabo, E.; Bakunzibake, P.; Kim, K.S.; Nsenga, J.	Using wavelet transform and hybrid CNN – LSTM models on VOC & ultrasound IoT sensor data for non-visual maize disease detection
Fanelli, C.; Giroux, J.; Moran, P.; Nayak, H.; Suresh, K.; Walter, E.	Physics event classification using Large Language Models
Fu, C.; Liu, C.; Ishi, C.T.; Yoshikawa, Y.; Iio, T.; Ishiguro, H.	Using an Android Robot to Improve Social Connectedness by Sharing Recent Experiences of Group Members in Human-Robot Conversations
Vijayalakshmi, J.; Pandimeena, K.	Agriculture talkbot using AI
Basir, M.S.; Zhang, Y.; Buckmaster, D.; Raturi, A.; Krogmeier, J.V.	Meta Ag: An automatic agricultural contextual metadata collection app
Paneru, B.; Thapa, B.; Paneru, B.	Leveraging AI in ayurvedic agriculture: A RAG chatbot for comprehensive medicinal plant insights using hybrid deep learning approaches
Wiangsamut, S.; Chomphuwiset, P.; Khummanee, S.	Chatting with Plants (Orchids) in Automated Smart Farming using IoT, Fuzzy Logic and Chatbot
Cheng, H.-H.; Dai, Y.-L.; Lin, Y.; Hsu, H.-C.; Lin, C.-P.; Huang, J.-H.; Chen, S.-F.; Kuo, Y.-F.	Identifying tomato leaf diseases under real field conditions using convolutional neural networks and a chatbot
Kothari, S.; Bagane, P.; Mishra, M.; Kulshrestha, S.; Asrani, Y.; Maheswari, V.	CropGuard: Empowering Agriculture with AI driven Plant Disease Detection Chatbot
Jain, S.; Sahni, R.; Khargonkar, T.; Gupta, H.; Verma, O.P.; Sharma, T.K.; Bhardwaj, T.; Agarwal, S.; Kim, H.	Automatic Rice Disease Detection and Assistance Framework Using Deep Learning and a Chatbot
Belghachi, M.	A Review on Explainable Artificial Intelligence Methods, Applications, and Challenges
Ibrahim, A.; Senthilkumar, K.; Saito, K.	Evaluating responses by ChatGPT to farmers' questions on irrigated lowland rice cultivation in Nigeria
Chi, N.T.K.	The Effect of AI Chatbots on Pro-environment Attitude and Willingness to Pay for Environment Protection
Jasmin, A.; Ramesh, P.; Tanveer, M.	Development of Artificial Intelligence-based chatbot for smart aquafarm practices

continued on following page

Table 1. Continued

Authors	Titles
Singh, D.; Shukla, B.; Joshi, M.	Fostering conversational artificial intelligence (CAI) for sustainable development in services: a case study of Haptik
Sulaiman, M.A.B.A.; Salman, R.; Shabbir, M.S.; Siddiqi, A.F.; Abbas, M.; Abbasi, S.G.; Ur Rehman, K.; Ashraf, M.	Research on rumors surrounding food safety based on information source differences (a review)
Nilchuen, P.; Suwanasopee, T.; Koonawootrittriron, S.	Integrating deep learning and mobile imaging for assessment of automated conformational indices and weight prediction in Brahman cattle
Richards, D.; Worden, D.; Song, X.P.; Lavorel, S.	Harnessing generative artificial intelligence to support nature-based solutions
Herrera, K.; Miranda, J.; Mauricio, D.	Milchbot: App to Support the Process of Feeding and Caring for Dairy Cows in Peru
Jutagate, A.; Pitakaso, R.; Khonjun, S.; Srichok, T.; Kaewta, C.; Luesak, P.; Gonwirat, S.; Enkvetchakul, P.; Jutagate, T.	Adaptive artificial multiple intelligence fusion system (A-AMIFS) for enhanced disease detection in Nile Tilapia

The Python programming language was utilized for text mining and keyword extraction to perform a methodical examination of the chosen literature. Thirty-one abstracts of publications totaling 6,998 words are included in the collection. The most used terms were chatbot (52 mentions), artificial intelligence (10 mentions), disease detection (10 mentions), farming (7 mentions), smartphones (7 mentions), machine learning (6 mentions), and natural language processing (4 mentions). The review underpins the growing penetration of chatbot solutions in agriculture. Therefore, regarding the RQ1, related to the overall use of chatbots in agricultural settings, 31 works could be found.

Regarding the RQ2, 3 studies are focusing on providing advisory services for farmers. The following Table 2 presents the results for the RQ2.

Table 2. Results referring to RQ2.

Authors	Titles
Jasmin, A.; Ramesh, P.; Tanveer, M.	Development of Artificial Intelligence-based chatbot for smart aquafarm practices
Sulaiman, M.A.B.A.; Salman, R.; Shabbir, M.S.; Siddiqi, A.F.; Abbas, M.; Abbasi, S.G.; Ur Rehman, K.; Ashraf, M.	Research on rumors surrounding food safety based on information source differences (a review)
Singh, D.; Shukla, B.; Joshi, M.	Fostering conversational artificial intelligence (CAI) for sustainable development in services: a case study of Haptik

The abovementioned results are further studied. More specifically, the work from Jasmin et al. (2022) presented the AquaProFAN system, in which cultivated shrimp farmers and stakeholders are supported by the AI-based chatbot "AquaGent". Implemented through social media applications (Facebook Messenger and Telegram), the system was favorably rated as an effective tool for dealing with informational queries; its ease of use contributed to the highest positive sentiment score of any model in the system functionality category.

Concurrently, Sulaiman et al. (2022) addressed the issue of food safety, which poses a threat to both consumer health and economic stability. Their study introduced the Food Safety Information Platform (FSIP), a system that combines ML with a cloud-based chatbot to deliver accurate and reliable information. The research highlights the value of disseminating trustworthy content, noting that the public often prefers to share news via peer-to-peer and social media networks. This behavior can help reduce misinformation and contribute to improved public health outcomes.

In addition, Singh et al. (2024) highlighted the broader application of CAI across a range of industries, including agriculture and aquaculture. By assisting with customer service, simplifying utility administration, and responding to inquiries, CAI promotes sustainable development in enterprises. Their study expresses the role of CAI in enhancing business operations and fostering the adoption of cutting-edge technologies across service sectors, illustrating its integration within a technopreneurship innovation framework.

Finally, these studies emphasized the life-changing role of AI and chatbots regarding communication, service provision, and productivity within agriculture. The application of such solutions driven by AI showed enormous potential towards achieving sustainability and data-informed practices within the agri-food sector. While fewer in number compared to applications in other industries (e.g., commercial industry, retail, education, tourism, health), an increasing trend has been observed in agricultural chatbots, especially to improve farmer engagement and provide context-aware advice. This phenomenon highlights the potential of chatbot solutions for the agricultural sector, as well as the growing prevalence of such technologies aimed at addressing the unique needs of farmers. Providing examples of technological implementations and studies on real-world uses of CAI systems in agriculture is considered appropriate to gain a more comprehensive understanding of the diverse chatbot applications within this field.

METHODOLOGY

The purpose of this section is to present the methodology for the design and development of GaiaMentor, a digital conversational system that aims to provide appropriate and scientifically validated agronomic advice to apple and pear growers at the right moment through continuous 24/7 availability. The motivation derives from the growing demand for real-time, accurate, and professional knowledge. The system is designed to achieve an optimal balance between scientific validity, ensuring that recommendations are evidence-based and aligned with best agronomic practices and usability, enabling broad adoption and practical impact in supporting productivity, quality, and environmental sustainability. The current focus on apple and pear tree cultivation reflects both their economic importance in Greek orchards and their suitability as pilot cases for testing the system's design. More specifically, apples and pears are significant crops in Greece, particularly in regions like Pelion and Central/Western Macedonia.

Platform Selection and Evaluation

An overview and comparison of available chatbot development platforms was performed considering a range of features, including technical flexibility, local control of the conversational flow, scalability, multilingualism, and data security. Three platforms (namely Landbot, Chatbase, and Botpress) were analyzed and assessed. Landbot supported drag-and-drop interfaces, so it was suitable for small tasks. Chatbase offered AI training capabilities and was library-based with static responses. Botpress was the most appropriate for the purposes of GaiaMentor. It is an all-in-one platform for building chatbots powered by the latest LLMs and supports rich NLU, customizable node-based architecture, modularity, multi-channel in communication, and scalability. Crucial factors also involved the support of the platform in terms of full data ownership and deeper service integrations like translation utilities and linguistic models. Table 3 provides a summary of the comparative evaluation.

Table 3. Comparative overview of chatbot development platforms

Platform	Technical Flexibility	Conversational Control	Scalability/AI Integration	Multilingual Support
Botpress	High	Full control over dialogue	Advanced NLU, dynamic AI modules	Extensive
Landbot	Limited	Minimal control	Basic workflows, no generative AI	Basic
Chatbase	Moderate	Limited (static responses)	Good for simple AI use cases	Moderate

Design and Development of GaiaMentor

Farmers' feedback during the design phase stressed the significance of simplicity, transparency, and bilingual access as facilitators of trust. They appreciated identified pathways and simplified workflows that directed them to specific advice for their crops. For questions that required expertise beyond the scope of the chatbot, farmers expressed the need to refer complex inquiries to an agronomist. Therefore, the design of GaiaMentor was guided by user-centered principles that emphasized clarity, accessibility, and actionable guidance. The primary objective was to develop a chatbot that facilitated a seamless user interaction flow. Users started by selecting a language, then navigated advisory pathways, and could escalate to expert consultation when needed. The detailed design of GaiaMentor interaction flow is presented in Figure 4. Moreover, the design ensured that AI support was always complemented by the possibility of human expertise, reinforcing trust and reliability.

Figure 4. GaiaMentor interaction flow

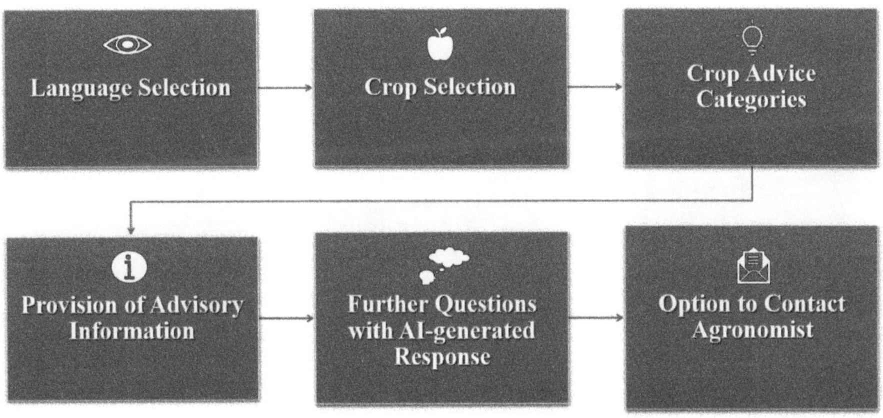

Following this conceptual design, GaiaMentor was developed and implemented using Botpress. More specifically, the development process was carried out in Botpress Studio, which allowed the use of a visual flow constructor to convert the design principles into a functional digital environment. Thematic modules were constructed as modular nodes, each aligned with agronomic topics such as fertilization, soil preparation, irrigation, plant protection, and productivity optimization. The workflow was organized into six key stages that defined the structure of the chatbot (Figures 5a–f).

Figure 5a presents the configuration of the language provision menu. At this stage, the administrator defines the available interaction languages and integrates the translation agent. The underlying node structure is designed to remain stable while still allowing additional languages to be inserted without disrupting the existing flows.

Figure 5b illustrates the crop provision menu. Each crop, initially apple and pear, is established as a branching node that directs the flow toward crop-specific advisory modules. The modular design allows for the straightforward integration of additional crops, supporting consistent scalability.

As shown in Figure 5c, the categories of crop advice are chosen. These categories correspond to major agronomic disciplines, including climate suitability, soil diagnostics, fertilization, irrigation management, pest and disease control, and best practices for productivity improvement.

Within Figure 5d, the provision of advisory information is configured. Here, nodes are directly connected to the central knowledge base, linked to curated scientific databases to ensure informational integrity. The architecture ensures that when updates are introduced, they automatically propagate across all dependent nodes, maintaining content consistency. Specifically, the knowledge base currently consists of 9 source files (6 in .txt format and 3 in .pdf format) comprising a total of 26,148 words, as well as 6 websites contributing 9,344 words. It ensures that advisory outputs are grounded on a well-documented, diverse corpus of agronomic knowledge, spanning textual files, scientific reports, and curated web sources.

Figure 5e depicts the integration of the AI-response node. At this stage, the GPT-4o mini LLM is embedded into the workflow to handle open-ended queries. Its use is defined as supplementary, with boundaries set by the administrator so that deterministic nodes govern critical agronomic recommendations, while AI provides additional flexibility.

Finally, Figure 5f shows the escalation pathway to a human agronomist. This component is constructed to capture and securely manage user details, forwarding unresolved queries to expert advisors. Administrative oversight ensures accountability and systematic follow-up.

Figure 5. GaiaMentor chatbot: System workspace

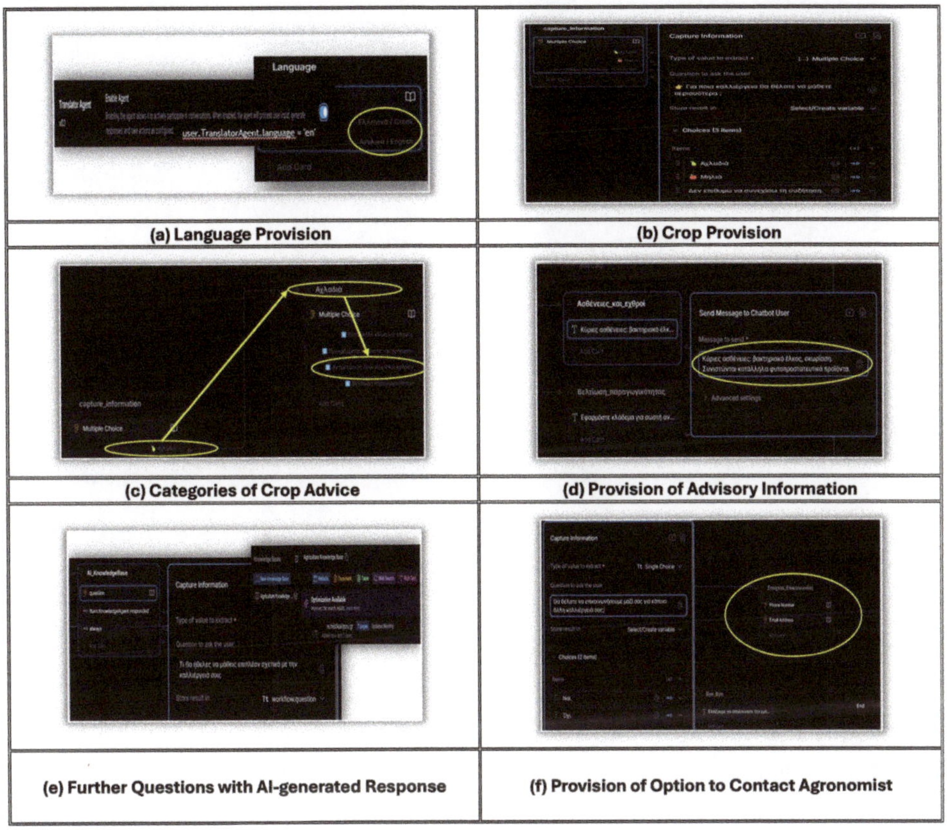

In total, GaiaMentor's structure comprised 22 interconnected nodes, spanning from the "start" to the two terminal points (Figure 6). Through this process, the chatbot evolved into a robust digital advisory tool that integrates validated knowledge, AI-driven flexibility, and human oversight.

Figure 6. GaiaMentor interaction flow diagram developed in Botpress studio

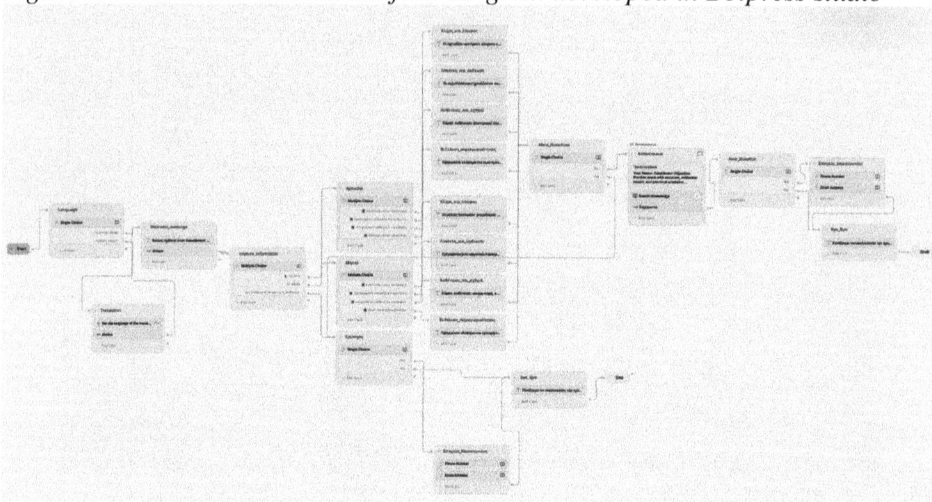

DEPLOYMENT OF GAIAMENTOR

The initial deployment provided empirical evidence of GaiaMentor's usability, adaptability, and effectiveness. The chatbot was first released in Greek, with a real-time translation agent enabling English access for a broader user base. Upon entry, users receive a personalized welcome message introducing the platform and its functions, establishing a welcoming and intuitive interface (Figure 7a). At this stage, users can select their crop of interest through a clear, intuitive menu.

Following crop selection, users choose advisory categories, including soil diagnostics, fertilization, irrigation management, pest and disease control, and productivity optimization (Figure 7b). This streamlined navigation allowed users to access the agronomic topics most relevant to their needs efficiently.

Next, GaiaMentor provides actionable advisory information, drawn from validated scientific and agronomic databases. Users could submit follow-up questions for clarification or in-depth guidance; open-ended queries were managed by the GPT-4o mini LLM, while deterministic nodes ensured reliability for critical recommendations (Figure 7c).

Finally, users were offered the option to escalate queries to a human agronomist, ensuring that complex or unresolved questions received expert attention, with secure data handling and systematic follow-up (Figure 7d). This structured onboarding flow minimizes cognitive load and promotes confident interaction, allowing smooth transitions between structured advice and AI-supported queries.

Figure 7. Indicative GaiaMentor interaction

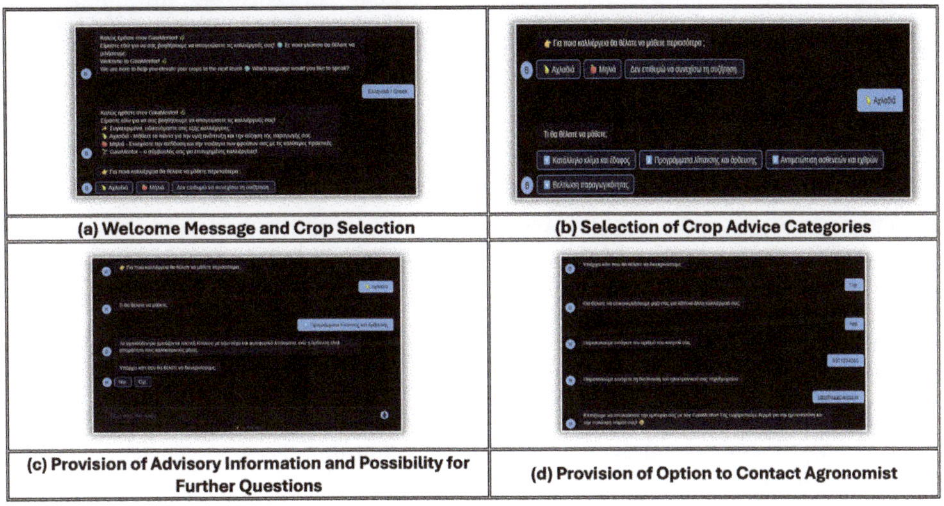

(a) Welcome Message and Crop Selection

(b) Selection of Crop Advice Categories

(c) Provision of Advisory Information and Possibility for Further Questions

(d) Provision of Option to Contact Agronomist

Early user trials of GaiaMentor with students at the Agricultural University of Athens provided the first concrete proof of concept: the system was immediately usable and its structured, evidence-based guidance measurably improved decision-making while accommodating varied user needs and preferences. Insights from these trials also surfaced workflow priorities and edge cases that informed initial refinements. Building on this foundation, a field-level validation is being designed to generate further evidence of effectiveness through on-farm pilots, usability surveys, follow-up interviews, and expert analysis of chatbot responses. This iterative feedback loop will keep farmers' perspectives at the center of ongoing updates, strengthening engagement and supporting quality management in agriculture.

CONCLUSION

GaiaMentor demonstrates how AI conversational technology can support agriculture by providing accurate, site-specific, and scientifically validated agronomic advice. By merging the specificity and precision inherent to the structured, theme-based advising modules with the flexibility of natural language interaction and access to expert consultation when needed, GaiaMentor meets a current need for easily obtained quality information among pear and apple producers.

More specifically, the system highlights the potential of integrating technologies such as chatbots, NLU, and AI with domain expertise to enhance decision-support processes in agriculture. Owing to its extensibility, modularity, and bilingualism, it is a multipurpose tool that can effectively transcend knowledge divides, particularly in regions with limited professional agronomic advice. Provisional findings imply that tools of this nature can enhance farming practices, productivity, and promote digital agriculture. Furthermore, GaiaMentor's simplicity, accessibility, and strong scientific foundation demonstrate the value of developing advisory information systems that use the latest technologies while providing practical and relevant solutions to the everyday needs and information requirements of farmers.

LIMITATIONS

Currently, GaiaMentor presents restricted functionality in terms of crops and language coverage. More specifically, it includes two fruit crops, namely apples and pears. Nevertheless, it can easily be generalized by incorporating additional crops and advisory modules. Likewise, the system currently operates only in Greek and English, which limits accessibility for agricultural communities that speak other languages beyond the current target groups. Such a language-based barrier could prevent scalability and availability globally, particularly to smallholder farmer communities in developing countries. However, GaiaMentor has been designed with a scalable structure that facilitates the seamless integration of additional crops and languages, thereby accommodating the evolving needs of its users.

Moreover, since GaiaMentor relies on a knowledge base derived from peer-reviewed agronomic literature, it may not fully capture emerging topics, site-specific variability, or tacit, experience-based field knowledge. Consequently, the output suggested by the system may neglect site-level agronomic concerns or recent developments that are not yet reflected in scholarly writing.

Because GaiaMentor is an internet-based application, reliable connectivity remains a significant barrier, especially for disadvantaged rural communities where digital infrastructure is still underdeveloped. This digital divide limits access to the system, disproportionately affecting the very populations GaiaMentor seeks to serve.

Lastly, there are critical, ethical, and socio-economic constraints that should be recognized. Many smallholder farmers experience gaps in digital literacy that may limit their use of conversational systems without introductory training. Economic considerations may also apply, as connectivity costs and access to devices present barriers to widespread diffusion. At the same time, careful attention to data ownership and compliance with regulations such as the General Data Protection Regulation (GDPR) is important for maintaining trust and protecting user data. Addressing these areas will be important if GaiaMentor is to establish and sustain adoption in diverse farming communities, especially in low- and middle-income countries.

PROSPECTS FOR FUTURE WORK

A major area for future research will be to evaluate adoption and impact under real-world conditions, including planned pilot trials that will use directed case studies with apple and pear growers, not only to measure agronomic results but also to measure user experience, trust, and willingness to re-use the system. Additional scenarios will also be developed for other crops to evaluate the generalizability of the dual-function advisory framework beyond the original target group. This type of trial will provide strong empirical evidence on what benefits conversational AI tools can provide in order to improve decision-making, overcome advisory gaps, and enhance farmer engagement in digital agriculture.

Some improvements and future research directions to enhance GaiaMentor's quality, performance, and impact include expanding system coverage to a wider range of crops and advice categories. This expansion would involve integrating additional agronomic information databases, locality-specific cultivation advice, and crop-centric advice modules. The addition of multiple languages will also be an added value. State-of-the-art translation technologies and AI-based content fine-tuning dialogues will be used to ensure proper adaptation of language according to cultural context and accuracy of dialects in their proper linguistic contexts.

Furthermore, integrating available sensors and ambient observations (e.g., weather, soil moisture, pest monitoring) is a crucial step toward delivering context-aware and personalized agronomic advice. By interfacing with agricultural monitoring networks and Internet of Things (IoT) sensors, the system can provide dynamic, context-aware decision support. Repeated and continued use and iterations of feedback will result in incremental improvements in system usability, sophistication level in the dialogue, and the end-users' experience. Care will be taken to enhance the capability of AI in answering extremely sophisticated and complex questions while maintaining scientific rigor.

Lastly, large-scale field trials and impact assessments will be conducted to evaluate the success of GaiaMentor in aiding farm decisions, improving resource application efficiency, and enhancing sustainability outcomes. These assessments will guide evidence-based improvements and form part of the literature on AI applications in agriculture. Given the advancements discussed above, GaiaMentor will serve as a scalable smart digital advisory system to be able to realize precision agriculture and support the farmers' ability to overcome current and anticipated farming situation challenges.

REFERENCES

Adamopoulou, E., & Moussiades, L. (2020). Chatbots: History, technology, and applications. *Machine Learning with Applications*, *2*, 100006. DOI: 10.1016/j.mlwa.2020.100006

Adewopo, J., McCampbell, M., Mwizerwa, C., & Schut, M. (2025). Beyond the Hype: Ten Lessons from Co-Creating and Implementing Digital Innovation in a Rwandan Smallholder Banana Farming System. *Agriculture*, *15*(2), 119. Advance online publication. DOI: 10.3390/agriculture15020119

Al Rasyid, M. U. H., Sukaridhoto, S., Dzulqornain, M. I., & Rifa'i, A. (2020). Integration of IoT and chatbot for aquaculture with natural language processing. *Telkomnika*, *18*(2), 640–648. DOI: 10.12928/telkomnika.v18i2.14788

Basir, M. S., Zhang, Y., Buckmaster, D., Raturi, A., & Krogmeier, J. V. (2025). Meta Ag: An automatic agricultural contextual metadata collection app. *Smart Agricultural Technology*, *12*, 101073. Advance online publication. DOI: 10.1016/j.atech.2025.101073

Belghachi, M. (2023). A Review on Explainable Artificial Intelligence Methods, Applications, and Challenges. *Indonesian Journal of Electrical Engineering and Informatics*, *11*(4), 1007–1024. DOI: 10.52549/ijeei.v11i4.5151

Benzinho, J., Ferreira, J., Batista, J., Pereira, L., Maximiano, M., Távora, V., Gomes, R., & Remédios, O. (2024). LLM Based Chatbot for Farm-to-Fork Blockchain Traceability Platform. *Applied Sciences (Basel, Switzerland)*, *14*(19), 8856. Advance online publication. DOI: 10.3390/app14198856

Bhardwaj, V., Kumar, M., Joshi, D., Chourasia, A., Bawaskar, B., & Sharma, S. (2024). Conversational AI—a state-of-the-Art Review. *Conversational Artificial Intelligence*, 533–555. https://doi.org/DOI: 10.1002/9781394200801.ch31

Bourdin, M., Paviot, T., Pellerin, R., & Lamouri, S. (2024). NLP in SMEs for industry 4.0: Opportunities and challenges. *Procedia Computer Science*, *239*, 396–403. DOI: 10.1016/j.procs.2024.06.186

Casheekar, A., Lahiri, A., Rath, K., Prabhakar, K. S., & Srinivasan, K. (2024). A contemporary review on chatbots, AI-powered virtual conversational agents, ChatGPT: Applications, open challenges and future research directions. *Computer Science Review*, *52*, 100632. DOI: 10.1016/j.cosrev.2024.100632

Cheng, H.-H., Dai, Y.-L., Lin, Y., Hsu, H.-C., Lin, C.-P., Huang, J.-H., Chen, S.-F., & Kuo, Y.-F. (2022). Identifying tomato leaf diseases under real field conditions using convolutional neural networks and a chatbot. *Computers and Electronics in Agriculture*, *202*, 107365. Advance online publication. DOI: 10.1016/j.compag.2022.107365

Chi, N. T. K. (2024). The Effect of AI Chatbots on Pro-environment Attitude and Willingness to Pay for Environment Protection. *SAGE Open*, *14*(1), 21582440231226001. Advance online publication. DOI: 10.1177/21582440231226001

Cónego, L., Pinto, R., & Gonçalves, G. (2024). Digital Transformation in Manufacturing: The Synergy of Chatbots and Tailored Gamification Strategies. *Procedia Computer Science*, *237*, 171–178. DOI: 10.1016/j.procs.2024.05.093

Das, S., & Ambika, M. (2025). Revolutionizing Indian Farming: Machine Learning-Powered NLP for Optimal Crop Recommendations. *Procedia Computer Science*, *258*, 2040–2049. DOI: 10.1016/j.procs.2025.04.454

European Commission. (2025). *Unlocking the potential of artificial intelligence for sustainable agriculture.* https://digital-strategy.ec.europa.eu/en/library/unlocking-potential-artificial-intelligence-sustainable-agriculture

Fanelli, C., Giroux, J., Moran, P., Nayak, H., Suresh, K., & Walter, E. (2024). Physics event classification using Large Language Models. *Journal of Instrumentation : An IOP and SISSA Journal*, *19*(7), C07011. Advance online publication. DOI: 10.1088/1748-0221/19/07/C07011

Food and Agriculture Organization (FAO). (2025). *AI can be a game-changing solution for farmers: FAO Innovation Chief.* https://www.fao.org/newsroom/detail/ai-can-be-a-game-changing-solution-for-farmers--fao-innovation-chief/en

Fornalczyk, K., Bortko, K., & Jankowski, J. (2021). Improving User Attention to Chatbots through a Controlled Intensity of Changes within the Interface. *Procedia Computer Science*, *192*, 5112–5121. DOI: 10.1016/j.procs.2021.09.289

Fu, C., Liu, C., Ishi, C. T., Yoshikawa, Y., Iio, T., & Ishiguro, H. (2021). Using an Android Robot to Improve Social Connectedness by Sharing Recent Experiences of Group Members in Human-Robot Conversations. *IEEE Robotics and Automation Letters*, *6*(4), 6670–6677. DOI: 10.1109/LRA.2021.3094779

Garg, P., & Rao, K. S. (2024). Building AgriRef: An AI embedded virtual reference app for farmers. *Library Hi Tech News*, *41*(2), 11–17. DOI: 10.1108/LHTN-02-2023-0025

Håkansson, A., & Phillips-Wren, G. (2024). Generative AI and Large Language Models—Benefits, Drawbacks, Future and Recommendations. *Procedia Computer Science, 246*, 5458–5468. DOI: 10.1016/j.procs.2024.09.689

Herrera, K., Miranda, J., & Mauricio, D. (2022). Milchbot: App to Support the Process of Feeding and Caring for Dairy Cows in Peru. *AGRIS On-Line Papers in Economics and Informatics, 14*(4), 27–37. DOI: 10.7160/aol.2022.140403

Huang, R., Kim, M., & Lennon, S. (2022). Trust as a second-order construct: Investigating the relationship between consumers and virtual agents. *Telematics and Informatics, 70*, 101811. DOI: 10.1016/j.tele.2022.101811

Ibrahim, A., Senthilkumar, K., & Saito, K. (2024). Evaluating responses by ChatGPT to farmers' questions on irrigated lowland rice cultivation in Nigeria. *Scientific Reports, 14*(1), 3407. Advance online publication. DOI: 10.1038/s41598-024-53916-1 PMID: 38341517

Idoje, G., Dagiuklas, T., & Iqbal, M. (2021). Survey for smart farming technologies: Challenges and issues. *Computers & Electrical Engineering, 92*, 107104. DOI: 10.1016/j.compeleceng.2021.107104

Jain, L., Ananthasayam, R., Gupta, U., & R, R. (2025). Comparison of Rule-based Chat Bots with Different Machine Learning Models. *Procedia Computer Science, 259*, 788–798. DOI: 10.1016/j.procs.2025.04.030

Jain, S., Sahni, R., Khargonkar, T., Gupta, H., Verma, O. P., Sharma, T. K., Bhardwaj, T., Agarwal, S., & Kim, H. (2022). Automatic Rice Disease Detection and Assistance Framework Using Deep Learning and a Chatbot. *Electronics (Basel), 11*(14), 2110. Advance online publication. DOI: 10.3390/electronics11142110

Jang, S., Lee, H., Kim, Y., Lee, D., Shin, J., & Nam, J. (2024). When, What, and how should generative artificial intelligence explain to Users? *Telematics and Informatics, 93*, 102175. DOI: 10.1016/j.tele.2024.102175

Jasmin, A., Ramesh, P., & Tanveer, M. (2024). Development of Artificial Intelligence-based chatbot for smart aquafarm practices. *Expert Systems: International Journal of Knowledge Engineering and Neural Networks, 41*(6), e13123. Advance online publication. DOI: 10.1111/exsy.13123

Jutagate, A., Pitakaso, R., Khonjun, S., Srichok, T., Kaewta, C., Luesak, P., Gonwirat, S., Enkvetchakul, P., & Jutagate, T. (2024). Adaptive artificial multiple intelligence fusion system (A-AMIFS) for enhanced disease detection in Nile Tilapia. *Aquaculture Reports, 39*, 102418. Advance online publication. DOI: 10.1016/j.aqrep.2024.102418

Kelly, S., Kaye, S.-A., & Oviedo-Trespalacios, O. (2023). What factors contribute to the acceptance of artificial intelligence? A systematic review. *Telematics and Informatics*, *77*, 101925. DOI: 10.1016/j.tele.2022.101925

Kibriya, H., Khan, W. Z., Siddiqa, A., & Khan, M. K. (2024). Privacy issues in Large Language Models: A survey. *Computers & Electrical Engineering*, *120*, 109698. DOI: 10.1016/j.compeleceng.2024.109698

Kothari, S., Bagane, P., Mishra, M., Kulshrestha, S., Asrani, Y., & Maheswari, V. (2024). CropGuard: Empowering Agriculture with AI driven Plant Disease Detection Chatbot. *International Journal of Intelligent Systems and Applications in Engineering*, *12*(12s), 530–537.

Lin, C.-C., Huang, A. Y. Q., & Yang, S. J. H. (2023). A Review of AI-Driven Conversational Chatbots Implementation Methodologies and Challenges (1999–2022). *Sustainability (Basel)*, *15*(5), 4012. DOI: 10.3390/su15054012

Longo, C. F., Longo, F., & Santoro, C. (2021). Caspar: Towards decision making helpers agents for IoT, based on natural language and first order logic reasoning. *Engineering Applications of Artificial Intelligence*, *104*, 104269. DOI: 10.1016/j.engappai.2021.104269

Maginga, T. J., Masabo, E., Bakunzibake, P., Kim, K. S., & Nsenga, J. (2024). Using wavelet transform and hybrid CNN – LSTM models on VOC & ultrasound IoT sensor data for non-visual maize disease detection. *Heliyon*, *10*(4), e26647. Advance online publication. DOI: 10.1016/j.heliyon.2024.e26647 PMID: 38420424

Mohamad Suhaili, S., Salim, N., & Jambli, M. N. (2021). Service chatbots: A systematic review. *Expert Systems with Applications*, *184*, 115461. DOI: 10.1016/j.eswa.2021.115461

Nilchuen, P., Suwanasopee, T., & Koonawootrittriron, S. (2025). Integrating deep learning and mobile imaging for assessment of automated conformational indices and weight prediction in Brahman cattle. *Smart Agricultural Technology*, *12*, 101079. Advance online publication. DOI: 10.1016/j.atech.2025.101079

Nirala, K. K., Singh, N. K., & Purani, V. S. (2022). A survey on providing customer and public administration-based services using AI: Chatbot. *Multimedia Tools and Applications*, *81*(29), 22215–22246. DOI: 10.1007/s11042-021-11458-y PMID: 35002470

Ouaddi, C., Benaddi, L., Bouziane, E. M., Jakimi, A., Chehri, A., & Saadane, R. (2024). A Sketch of DSL and Code Generator for Accelerating Chatbot Development. *Procedia Computer Science*, *246*, 3585–3594. DOI: 10.1016/j.procs.2024.09.199

Oumaima, S., Mohamed, L., Abderrahmane, B., & Iyass, M. (2025). EnviScan+: AI-driven plant identification and eco-system management for sustainable agriculture. *SoftwareX, 30*, 102182. Advance online publication. DOI: 10.1016/j.softx.2025.102182

Pahuja, S., Kukreja, S., & Singh, A. (2025). Comprehensive Review of Generative artificial Intelligence: Mechanisms, Models, and Applications. *Procedia Computer Science, 258*, 3731–3740. DOI: 10.1016/j.procs.2025.04.628

Paneru, B., Thapa, B., & Paneru, B. (2024). Leveraging AI in ayurvedic agriculture: A RAG chatbot for comprehensive medicinal plant insights using hybrid deep learning approaches. *Telematics and Informatics Reports, 16*, 100181. Advance online publication. DOI: 10.1016/j.teler.2024.100181

Patil, R., Sinkar, Y., Ruke, A., Kulkarni, H., & Kadam, O. (2024). Smart Agri-Advisor: Integrating Chatbot Technology with CNN-Based Crop Disease Classification for Enhanced Agricultural Decision-Making. *International Journal of Engineering Trends and Technology, 72*(7), 375–380. DOI: 10.14445/22315381/IJETT-V72I7P141

Pimenidis, E., & Georgiadis, C. K. (2010). Web services for rural areas—Security challenges in development and use. *Computers and Electronics in Agriculture, 70*(2), 348–354. DOI: 10.1016/j.compag.2009.07.016

Reddy, P. D., Sampath Reddy, K. S., Jayanth, P., Kakarla, B. P., & Balakrishnan, R. M. (2025). Agri Assist: An AI Integrated Farmer Assistant. *Procedia Computer Science, 258*, 3510–3522. DOI: 10.1016/j.procs.2025.04.607

Rehman, M. Z. U., Raghuvanshi, D., & Kumar, N. (2023). KisanQRS: A deep learning-based automated query-response system for agricultural decision-making. *Computers and Electronics in Agriculture, 213*, 108180. DOI: 10.1016/j.compag.2023.108180

Richards, D., Worden, D., Song, X. P., & Lavorel, S. (2024). Harnessing generative artificial intelligence to support nature-based solutions. *People and Nature, 6*(2), 882–893. DOI: 10.1002/pan3.10622

Safitra, M. F., Lubis, M., Kusumasari, T. F., & Putri, D. P. (2024). Advancements in Artificial Intelligence and Data Science: Models, Applications, and Challenges. *Procedia Computer Science, 234*, 381–388. DOI: 10.1016/j.procs.2024.03.018

Shah, I., Jhawar, S., Khater, A., Jacob, A., & Potdar, G. (2023). Chatbot development through the ages: A survey. *International Journal of Scientific Research in Computer Science. Engineering and Information Technology, 9*(3), 262–271. DOI: 10.32628/CSEIT2390329

Shapa, M., Trojer, L., & Machuve, D. (2021). Mobile-based Decision Support System for Poultry Farmers: A Case of Tanzania. *International Journal of Advanced Computer Science and Applications, 12*(2), 584–590. DOI: 10.14569/IJACSA.2021.0120273

Singh, D., Shukla, B., & Joshi, M. (2024). Fostering conversational artificial intelligence (CAI) for sustainable development in services: A case study of Haptik. *World Review of Entrepreneurship, Management and Sustainable Development, 20*(4), 397–420. DOI: 10.1504/WREMSD.2024.139371

Sivakumar, S. A., Shankar, B. M., Anuradha, B., Karan, K. A., Karthik, A., Karthik, R., & Kumar, J. R. R. (2023). Artificial Intelligence based Agricultural Chatbot and Virtual Assistant for Delivery of Harvested Crops. *International Journal of Intelligent Systems and Applications in Engineering, 11*(8), 576–583.

Sowmya, B. J., Meeradevi, A. K., Supreeth, S., Pradeep Kumar, D., Ravi Kumar, B. N., Rohith, S., Mishra, D., Koushik, A., & Patil, A. U. (2025). Leveraging machine learning for intelligent agriculture. *Discover Internet of Things, 5*(1), 33. Advance online publication. DOI: 10.1007/s43926-025-00132-6

Sulaiman, M. A. B. A., Salman, R., Shabbir, M. S., Siddiqi, A. F., Abbas, M., Abbasi, S. G., Ur Rehman, K., & Ashraf, M. (2022). Research on rumors surrounding food safety based on information source differences (a review). *Food Science and Technology (Brazil), 42*. https://doi.org/DOI: 10.1590/FST.79921

Temniranrat, P., Kiratiratanapruk, K., Kitvimonrat, A., Sinthupinyo, W., & Patarapuwadol, S. (2021). A system for automatic rice disease detection from rice paddy images serviced via a Chatbot. *Computers and Electronics in Agriculture, 185*, 106156. Advance online publication. DOI: 10.1016/j.compag.2021.106156

Vargo, D., Zhu, L., Benwell, B., & Yan, Z. (2021). Digital technology use during COVID-19 pandemic: A rapid review. *Human Behavior and Emerging Technologies, 3*(1), 13–24. DOI: 10.1002/hbe2.242

Vijayalakshmi, J., & Pandimeena, K. (2019). Agriculture talkbot using AI. *International Journal of Recent Technology and Engineering, 8*(2 Special Issue 5), 186–190. https://doi.org/DOI: 10.35940/ijrte.B1037.0782S519

Wiangsamut, S., Chomphuwiset, P., & Khummanee, S. (2019). Chatting with Plants (Orchids) in Automated Smart Farming using IoT, Fuzzy Logic and Chatbot. *Advances in Science, Technology and Engineering Systems Journal, 4*(5), 163–173. DOI: 10.25046/aj040522

Zheng, S., Yahya, Z., Wang, L., Zhang, R., & Hoshyar, A. N. (2023). Multiheaded deep learning chatbot for increasing production and marketing. *Information Processing & Management*, *60*(5), 103446. DOI: 10.1016/j.ipm.2023.103446

Ziegfeld, L., Scala, D. D., & Cremers, A. H. M. (2025). The effect of preference elicitation methods on the user experience in conversational recommender systems. *Computer Speech & Language*, *89*, 101696. Advance online publication. DOI: 10.1016/j.csl.2024.101696

Compilation of References

Abatzoglou, J. T., & Williams, A. P. (2016). Impact of anthropogenic climate change on wildfire across western US forests. *Proceedings of the National Academy of Sciences of the United States of America*, *113*(42), 11770–11775. DOI: 10.1073/pnas.1607171113

Abbasi, M. R., & Sepaskhah, A. R. (2023). Nitrogen leaching and groundwater N contamination risk in saffron/wheat intercropping under different irrigation and soil fertilizers regimes. *Scientific Reports*, *13*(1), 6587. https://doi.org/https://doi.org/10.1038/s41598-023-33817-5

Abrahim, G., & Parker, R. (2008). Assessment of heavy metal enrichment factors and the degree of contamination in marine sediments from Tamaki Estuary, Auckland, New Zealand. *Environmental Monitoring and Assessment*, *136*(1), 227–238. https://doi.org/10.1007/s10661-007-9678-2

Adamopoulou, E., & Moussiades, L. (2020). Chatbots: History, technology, and applications. *Machine Learning with Applications*, *2*, 100006. DOI: 10.1016/j.mlwa.2020.100006

Adebayo, R. P. (2024). Optimizing Food101 classification with transfer learning: A fine-tuning approach using EfficientNetB0. *International Journal of Intelligent Information Systems*, *13*(4), 59–77. DOI: 10.11648/j.ijiis.20241304.11

Adewopo, J., McCampbell, M., Mwizerwa, C., & Schut, M. (2025). Beyond the Hype: Ten Lessons from Co-Creating and Implementing Digital Innovation in a Rwandan Smallholder Banana Farming System. *Agriculture*, *15*(2), 119. Advance online publication. DOI: 10.3390/agriculture15020119

Agee, J. K., & Skinner, C. N. (2005). Basic principles of forest fuel reduction treatments. *Forest Ecology and Management*, *211*(1-2), 83–96. DOI: 10.1016/j.foreco.2005.01.034

Aggarwal, C. C. (2016). Neighborhood-based collaborative filtering. In *Recommender Systems: The Textbook* (pp. 29–70). Springer International Publishing.

Aggarwal, C. C., Philip, S. Y., Han, J., & Wang, J. (2003, January). A framework for clustering evolving data streams. In *Proceedings 2003 VLDB conference* (pp. 81-92). Morgan Kaufmann.

Akbarinia, R., & Cloez, B. (2019). Efficient matrix profile computation using different distance functions. *arXiv preprint arXiv:1901.05708.*

Al Jallad, K., Aljnidi, M., & Desouki, M. S. (2020). Anomaly detection optimization using big data and deep learning to reduce false-positive. *Journal of Big Data*, *7*(1), 68.

Al Rasyid, M. U. H., Sukaridhoto, S., Dzulqornain, M. I., & Rifa'i, A. (2020). Integration of IoT and chatbot for aquaculture with natural language processing. *Telkomnika*, *18*(2), 640–648. DOI: 10.12928/telkomnika.v18i2.14788

Alaa, A. M., & Van Der Schaar, M. (2017). Bayesian inference of individualized treatment effects using multi-task gaussian processes. *Advances in Neural Information Processing Systems*, •••, 30.

Alahe, M., Wei, L., Chang, Y., Gummi, S., Kemeshi, J., Yang, X., Won, K., & Sher, M. (2024). Cyber security in smart agriculture: Threat types, current status, and future trends. *Computers and Electronics in Agriculture*, *226*, 109401. DOI: 10.1016/j.compag.2024.109401

Albini, F. A. (1976). *Estimating Wildfire Behavior and Effects.* (INT-30).

Alengebawy, A., Abdelkhalek, S. T., Qureshi, S. R., & Wang, M.-Q. (2021). Heavy metals and pesticides toxicity in agricultural soil and plants: Ecological risks and human health implications. *Toxics*, *9*(3), 42. https://doi.org/10.3390/toxics9030042

Alkathiri, M., Abdul, J., & Potdar, M. B. (2017, August). Kluster: Application of k-means clustering to multidimensional GEO-spatial data. In *2017 International conference on information, communication, instrumentation and control (ICICIC)* (pp. 1-7). IEEE.

Alloway, B. J. (2012). *Heavy metals in soils: trace metals and metalloids in soils and their bioavailability* (Vol. 22). Springer Science & Business Media.

Almasri, M. N. (2007). Nitrate contamination of groundwater: A conceptual management framework. *Environmental Impact Assessment Review*, *27*(3), 220–242. https://doi.org/10.1016/j.eiar.2006.11.002

Almeida, C. M., Afonso, C. M., Serra, M., & António, N. (2024). The presence of Algarve wine producers on Facebook between pre-covid-19 and covid-19 periods. *Tourism and Hospitality International Journal*, *22*(1), 27–62.

Alves, T. M., Soeiro, R. O., & Cartaxo, A. V. (2019, September). Probability distribution of intercore crosstalk in weakly coupled MCFs with multiple interferers. In *2019 IEEE Photonics Conference (IPC)* (pp. 1-4). IEEE.

Alvino, A., & Marino, S. (2017). Remote Sensing for Irrigation of Horticultural Crops. *Horticulturae, 3*(2), 3020040. DOI: 10.3390/horticulturae3020040

Alwateer, M., Loke, S., & Zuchowicz, A. (2019). Drone services: Issues in drones for location-based services from human-drone interaction to information processing. *Journal of Location Based Services, 13*(2), 94–127. DOI: 10.1080/17489725.2018.1564845

An, I. T. U. F. A. O. (2024). Meeting the European Union's digital agriculture requirements.

An, W., Wu, D., Ci, S., Luo, H., Adamchuk, V., & Xu, Z. (2017). Chapter 25 - Agriculture Cyber-Physical Systems. In H. Song, D. Rawat, S. Jeschke, & C. (. Brecher, *Cyber-Physical Systems: Foundations, Principles and Applications* (pp. 399-417). Academic Press.

Angel, S., Parent, J., Civco, D. L., Blei, A., & Potere, D. (2011). The dimensions of global urban expansion: Estimates and projections for all countries, 2000 ?"2050. *Progress in Planning, 75*(2), 53 ?"107. DOI: 10.1016/j.progress.2011.04.001

Anil, R., Gupta, V., Koren, T., & Singer, Y. (2019). Memory efficient adaptive optimization. *Advances in Neural Information Processing Systems*, •••, 32.

Antić-Mladenović, S., Kresović, M., Čakmak, D., Perović, V., Saljnikov, E., Ličina, V., & Rinklebe, J. (2019). Impact of a severe flood on large-scale contamination of arable soils by potentially toxic elements (Serbia). *Environmental Geochemistry and Health, 41*(1), 249–266. https://doi.org/10.1007/s10653-018-0138-4

Antoniadis, V., & Golia, E. E. (2021). Spatial and temporal assessment of cadmium and chromium contamination in soils in the Karditsa region (Central Greece). *Environmental Science and Pollution Research International, 28*(4), 3820–3827.

Apostolopoulos, N., Petropoulos, D., Liargovas, P., & Anastasopoulou, E. (2024, September). Constraints to Green and Digital Transition in the Greek Agri-Food Sector. In *Proceedings of The 19th European Conference on Innovation and Entrepreneurship, Academic Conferences International.* DOI: 10.34190/ecie.19.1.2996

Arbinolo, M., Patimo, G., Rey, E., Stokkeland, O., Verde, J. C., Casartelli, V., Marengo, A., Melinato, S., Mysiak, J., Salpina, D., Afentoulidis, S., Brăilescu, C., & Sørensen, J. (2024). *UCPM Wildfire Peer Review Report: Greece 2024.* https://doi.org/DOI: 10.25424/CMCC-79TS-VV91

Ashibani, Y., & Mahmoud, Q. (2017). Cyber physical systems security: Analysis, challenges and solutions. *Computers & Security*, *68*, 81–97. DOI: 10.1016/j.cose.2017.04.005

Asprogerakas, E. (2003). *Characteristics and development potential of Greek medium-sized cities* [Unpublished doctoral dissertation]. University of Thessaly. [in Greek].

Assaad, W., & Gómez, J. M. (2011). *Social network in marketing (social media marketing) opportunities and risks. International Journal of Managing Public Sector Information and Communication Technologies* (Vol. 2). IJMPICT.

Athanasiadis, N. (1986). *Forest Botany (Trees and Shrubs of the Forests of Greece) part II*. Giahoudi.

Aumüller, M., Bernhardsson, E., & Faithfull, A. (2020). ANN-Benchmarks: A benchmarking tool for approximate nearest neighbor algorithms. *Information Systems*, *87*, 101374.

Avramidis, A. (2021). *Social media marketing: What it is and why it is necessary today*. WebOne.https://shorturl.at/j0Ovn

BA1/4rgi, M., Hersperger, A. M., & Schneeberger, N. (2005). Driving forces of landscape change ?"Current and new directions. *Landscape Ecology*, *19*(8), 857 ?"868. DOI: 10.1007/s10980-005-0245-3

Babakos, G. (2005). *Determination of nitrite and nitrate ion levels in soils of the Almira region, Magnesia* [Bachelor's Thesis, University of Thessaly, School of Agricultural Sciences]. Volos, Greece.

Bachelet, D., Lenihan, J. M., Daly, C., & Neilson, R. P. (2000). Interactions between fire, grazing and climate change at Wind Cave National Park, SD. *Ecological Modelling*, *134*(2-3), 229–244. DOI: 10.1016/s0304-3800(00)00343-4

Bachmann, A., & Allgöwer, B. (1998). Framework for Wildfire Risk Analysis. In Viegas, D. X. (Ed.), *Forest Fire Research*. University of Coimbra.

Baeza, M. J., De Luis, M., Raventos, J., & Escarre, A. (2002). Factors influencing fire behaviour in shrublands of different stand ages and the implications for using prescribed burning to reduce wildfire risk. *Journal of Environmental Management*, *65*(2), 199–208. DOI: 10.1006/jema.2002.0545

Baeza, M. J., Raventós, J., Escarré, A., & Vallejo, V. R. (2003). The effect of shrub clearing on the control of the fire-prone species Ulex parviflorus. *Forest Ecology and Management*, *186*(1-3), 47–59. DOI: 10.1016/s0378-1127(03)00237-8

Baldwin, V. C., Peterson, K. D., Clark, A., Ferguson, R. B., Strub, M. R., & Bower, D. R. (2000). The effects of spacing and thinning on stand and tree characteristics of 38-year-old Loblolly Pine. *Forest Ecology and Management, 137*(1-3), 91–102. DOI: 10.1016/s0378-1127(99)00340-0

Bampasidou, M., Goldgaber, D., Gentimis, T., & Mandalika, A. (2024). Overcoming 'Digital Divides': Leveraging higher education to develop next generation digital agriculture professionals. *Computers and Electronics in Agriculture, 224*, 109181. DOI: 10.1016/j.compag.2024.109181

Barbieri, M. (2016). The Importance of Enrichment Factor (EF) and Geoaccumulation Index (Igeo) to Evaluate the Soil Contamination. *Journal of Geology & Geophysics, 5*(1), 1–4. https://doi.org/10.4172/2381-8719.1000237

Bargotakis, P. (2005). *Measurement of heavy metals copper (Cu), zinc (Zn), iron (Fe) and manganese (Mn) in areas of karditsa*. University of Thessaly.

Barreto, L., & Amaral, A. (2018). Smart farming: Cyber security challenges. *2018 International Conference on Intelligent Systems (IS)* (pp. 870--876). IEEE. DOI: 10.1109/IS.2018.8710531

Barreto, L., Amaral, A., & Pereira, T. (2017). Industry 4.0 implications in logistics: An overview. *Procedia Manufacturing, 13*, 1245–1252. DOI: 10.1016/j.promfg.2017.09.045

Barron, J. T. (2019). A general and adaptive robust loss function. In *Proceedings of the IEEE/CVF conference on computer vision and pattern recognition* (pp. 4331-4339).

Barua, A., Muthirayan, D., Khargonekar, P. P., & Al Faruque, M. A. (2020, April). Hierarchical temporal memory based machine learning for real-time, unsupervised anomaly detection in smart grid: WiP abstract. In *2020 ACM/IEEE 11th International Conference on Cyber-Physical Systems (ICCPS)* (pp. 188-189). IEEE.

Basat, R. B., Einziger, G., Keslassy, I., Orda, A., Vargaftik, S., & Waisbard, E. (2018, December). Memento: Making sliding windows efficient for heavy hitters. In *Proceedings of the 14th International Conference on Emerging Networking EXperiments and Technologies* (pp. 254-266).

Basir, M. S., Zhang, Y., Buckmaster, D., Raturi, A., & Krogmeier, J. V. (2025). Meta Ag: An automatic agricultural contextual metadata collection app. *Smart Agricultural Technology, 12*, 101073. Advance online publication. DOI: 10.1016/j.atech.2025.101073

Batyuk, A., Voityshyn, V., & Verhun, V. (2018, August). Software architecture design of the real-time processes monitoring platform. In *2018 IEEE Second International Conference on Data Stream Mining & Processing (DSMP)* (pp. 98-101). IEEE.

Becker, J. (2020, February 27). *ABC News*. Retrieved from Cyber attack forces cancellation of wool sales across Australia: https://www.abc.net.au/news/rural/2020-02-27/ransomware-cyber-attack-cripples-australian-wool-sales/12007912

Belghachi, M. (2023). A Review on Explainable Artificial Intelligence Methods, Applications, and Challenges. *Indonesian Journal of Electrical Engineering and Informatics, 11*(4), 1007–1024. DOI: 10.52549/ijeei.v11i4.5151

Bellare, M., & Goldreich, O. (2011). On probabilistic versus deterministic provers in the definition of proofs of knowledge. In *Studies in Complexity and Cryptography. Miscellanea on the Interplay between Randomness and Computation: In Collaboration with Lidor Avigad, Mihir Bellare, Zvika Brakerski, Shafi Goldwasser, Shai Halevi, Tali Kaufman, Leonid Levin, Noam Nisan, Dana Ron, Madhu Sudan, Luca Trevisan, Salil Vadhan, Avi Wigderson, David Zuckerman* (pp. 114–123). Springer Berlin Heidelberg.

Bellos, D., & Sawidis, T. (2005). Chemical pollution monitoring of the river pinios (Thessalia—Greece). *Journal of Environmental Management, 76*(4), 282–292. https://doi.org/10.1016/j.jenvman.2005.01.027

Benzinho, J., Ferreira, J., Batista, J., Pereira, L., Maximiano, M., Távora, V., Gomes, R., & Remédios, O. (2024). LLM Based Chatbot for Farm-to-Fork Blockchain Traceability Platform. *Applied Sciences (Basel, Switzerland), 14*(19), 8856. Advance online publication. DOI: 10.3390/app14198856

Bhandari, R., Jadhaw, A., & Thiagarajan, V. (2024). Hybrid models for improved banana disease detection. *International Journal of Agricultural Sciences, 12*(2), 45–52.

Bhardwaj, V., Kumar, M., Joshi, D., Chourasia, A., Bawaskar, B., & Sharma, S. (2024). Conversational AI—a state-of-the-Art Review. *Conversational Artificial Intelligence*, 533–555. https://doi.org/DOI: 10.1002/9781394200801.ch31

Bhatta, B., Saraswati, S., & Bandyopadhyay, D. (2010). Urban sprawl measurement from remote sensing data. *Applied Geography (Sevenoaks, England), 30*(4), 731 ?"740. DOI: 10.1016/j.apgeog.2010.02.002

Bhattacharya, A., & Dutta, S. (2024). Deep learning for precision agriculture: Banana disease detection case study. *Agricultural Systems, 198*, 103–115.

Bifet, A., de Francisci Morales, G., Read, J., Holmes, G., & Pfahringer, B. (2015, August). Efficient online evaluation of big data stream classifiers. In *Proceedings of the 21th ACM SIGKDD international conference on knowledge discovery and data mining* (pp. 59-68).

Birot, Y. (2009). *Living with wildfires: What science can tell us - A contribution to the science-policy dialogue*.

Bitakou, E., Karetsos, S., Ntalianis, F., Ntaliani, M., & Costopoulou, C. (2023). Evaluating social media marketing in the Greek winery industry. *Sustainability (Basel), 16*(1), 192. DOI: 10.3390/su16010192

Blanco-Rodríguez, M. Á., Ameztegui, A., Gelabert, P., Rodrigues, M., & Coll, L. (2023). Short-term recovery of post-fire vegetation is primarily limited by drought in Mediterranean forest ecosystems. *Fire Ecology, 19*(1), 68–68. DOI: 10.1186/s42408-023-00228-w

Blondel, J., Aronson, J., Bodiou, J. Y., & Boeuf, G. (2010). *The Mediterranean Region: Biological Diversity in Space and Time* (2 ed.). Oxford University Press.

Bogage, J., & Reiley, L. (2021, September 21). *The Washington Post.* Retrieved from Russian hackers target Iowa grain co-op in $5.9 million ransomware attack: https://www.washingtonpost.com/business/2021/09/21/new-cooperative-hack-ransomware/

Borassi, M., Epasto, A., Lattanzi, S., Vassilvitskii, S., & Zadimoghaddam, M. (2020). Sliding window algorithms for k-clustering problems. *Advances in Neural Information Processing Systems, 33*, 8716–8727.

Bourdin, M., Paviot, T., Pellerin, R., & Lamouri, S. (2024). NLP in SMEs for industry 4.0: Opportunities and challenges. *Procedia Computer Science, 239*, 396–403. DOI: 10.1016/j.procs.2024.06.186

Bouyoucos, G. J. (1927). The hydrometer as a new method for the mechanical analysis of soils. *Soil Science, 23*(5), 343–354.

Bouyoucos, G. J. (1951). A recalibration of the hydrometer method for making mechanical analysis of soils. *Agronomy Journal, 43*(9), 434–438. https://doi.org/10.2134/agronj1951.00021962004300090005x

Bouyoucos, G. J. (1962). Hydrometer method improved for making particle size analyses of soils. *Agronomy Journal, 54*(5), 464–465. https://doi.org/10.2134/agronj1962.00021962005400050028x

Bowman, D. M. J. S., Williamson, G. J., Abatzoglou, J. T., Kolden, C. A., Cochrane, M. A., & Smith, A. M. S. (2017). Human exposure and sensitivity to globally extreme wildfire events. *Nature Ecology & Evolution, 1*(3), 0058. https://doi.org/DOI: 10.1038/s41559-016-0058

Boyatzis, R. E. (1998). Transforming qualitative information: Thematic analysis and code development. *Sage (Atlanta, Ga.).*

Bradshaw, L. S., Deeming, J. E., & Burgan, R. E. (1984). The 1978 national fire-danger rating system: technical documentation. *General Technical Report INT-169. Ogden, UT: US Department of Agriculture, Forest Service, Intermountain Forest and Range Experiment Station. 44 p., 169.*

Braun, V., & Clarke, V. (2006). Using thematic analysis in psychology. *Qualitative Research in Psychology*, *3*(2), 77 ?"101. DOI: 10.1191/1478088706qp063oa

Bremner, J. M. (1996). *Nitrogen-total* (Sparks, A. L. P. D. L., Helmke, P. A., Loeppert, R. H., Soltanpour, P. N., Tabatabai, M. A., Johnston, C. T., & Sumner, M. E., Eds.). Vol. 5)., https://doi.org/https://doi.org/10.2136/sssabookser5.3.c37

Brueckner, J. K. (2000). Urban sprawl: Diagnosis and remedies. *International Regional Science Review*, *23*(2), 160 ?"171. DOI: 10.1177/016001700761012710

Bui, H. T., Aboutorab, H., Mahboubi, A., Gao, Y., Sultan, N. H., Chauhan, A., Parvez, M. Z., Bewong, M., Islam, R., Islam, Z., Camtepe, S. A., Gauravaram, P., Singh, D., Ali Babar, M., & Yan, S. (2024). Agriculture 4.0 and beyond: Evaluating cyber threat intelligence sources and techniques in smart farming ecosystems. *Computers & Security*, *140*, 103754. DOI: 10.1016/j.cose.2024.103754

Burchell, R. W., Shad, N. A., Listokin, D., Phillips, H., Downs, A., Seskin, S., . . . Gall, M. (1998). *The costs of sprawl ?"Revisited* (TCRP Report 39). Transportation Research Board. https://digitalcommons.usf.edu/cgi/viewcontent.cgi?article=1189&context=cutr_reports

Butler, B. W. (2014). Wildland firefighter safety zones: A review of past science and summary of future needs. *International Journal of Wildland Fire*, *23*(3), 295–308. DOI: 10.1071/WF13021

Butler, B. W., & Cohen, J. D. (1998a). Firefighter safety zones: A theoretical model based on radiative heating. *International Journal of Wildland Fire*, *8*(2), 73–77. DOI: 10.1071/WF9980073

Butler, B. W., & Cohen, J. D. (1998b). Firefighter Safety Zones: How Big Is Big Enough? *Fire Management Notes*, *58*(1), 13–16.

Byram, G. M. (1959). Combustion of Forest Fuels. In Use, C. a. (Ed.), *Forest Fire*. McGraw-Hill.

Calthorpe, P., & Fulton, W. (2001). *The regional city: Planning for the end of sprawl*. Island Press.

Cameron, K., Di, H. J., & Moir, J. (2013). Nitrogen losses from the soil/plant system: A review. *Annals of Applied Biology*, *162*(2), 145–173. https://doi.org/10.1111/aab.12014

Campoverde-Molina, M., & Luján-Mora, S. (2024). Cybersecurity in smart agriculture: A systematic literature review. *Computers & Security*, •••, 104284.

Carter, M. C., & Darwin Foster, C. (2004). Prescribed burning and productivity in southern pine forests: A review. *Forest Ecology and Management*, *191*(1-3), 93–109. DOI: 10.1016/j.foreco.2003.11.006

Carvalho Ribeiro, S. M., Rajao, R., Nunes, F., Assis, D., Neto, J. A., Marcolino, C., Lima, L., Rickard, T., Salomao, C., & Filho, B. S. (2020). A spatially explicit index for mapping Forest Restoration Vocation (FRV) at the landscape scale: Application in the Rio Doce basin, Brazil. *The Science of the Total Environment, 744*, 140647. DOI: 10.1016/j.scitotenv.2020.140647

Casebeer, J., Bryan, N. J., & Smaragdis, P. (2022). Meta-AF: Meta-learning for adaptive filters. *IEEE/ACM Transactions on Audio, Speech, and Language Processing, 31*, 355–370.

Casheekar, A., Lahiri, A., Rath, K., Prabhakar, K. S., & Srinivasan, K. (2024). A contemporary review on chatbots, AI-powered virtual conversational agents, ChatGPT: Applications, open challenges and future research directions. *Computer Science Review, 52*, 100632. DOI: 10.1016/j.cosrev.2024.100632

Castana, E. (2016). *Be social or be good. The performance of Italian microbreweries*. https://thesis.unipd.it/handle/20.500.12608/25222

Catry, F. X., Rego, F. C., Bação, F. L., & Moreira, F. (2009). Modeling and mapping wildfire ignition risk in Portugal. *International Journal of Wildland Fire, 18*(8), 921–931. DOI: 10.1071/wf07123

Chambers, M. E., Fornwalt, P. J., Malone, S. L., & Battaglia, M. A. (2016). Patterns of conifer regeneration following high severity wildfire in ponderosa pine – dominated forests of the Colorado Front Range. *Forest Ecology and Management, 378*, 57–67. DOI: 10.1016/j.foreco.2016.07.001

Chen, F., Zhang, T., & Liu, R. (2021). An active learning method based on variational autoencoder and dbscan clustering. *Computational Intelligence and Neuroscience, 2021*(1), 9952596.

Cheng, H.-H., Dai, Y.-L., Lin, Y., Hsu, H.-C., Lin, C.-P., Huang, J.-H., Chen, S.-F., & Kuo, Y.-F. (2022). Identifying tomato leaf diseases under real field conditions using convolutional neural networks and a chatbot. *Computers and Electronics in Agriculture, 202*, 107365. Advance online publication. DOI: 10.1016/j.compag.2022.107365

Cheng, J., & Masser, I. (2003). Urban growth pattern modeling: A case study of Wuhan city, PR China. *Landscape and Urban Planning, 62*(4), 199 ?"217. DOI: 10.1016/S0169-2046(02)00150-0

Chen, T. H., Shang, W., Hassan, A. E., Nasser, M., & Flora, P. (2016, May). Detecting problems in the database access code of large scale systems: An industrial experience report. In *Proceedings of the 38th International Conference on Software Engineering Companion* (pp. 71-80).

Chin, N. (2002). *Unearthing the roots of urban sprawl: A critical analysis of form, function and methodology* (CASA Working Paper No. 47). University College London. https://discovery.ucl.ac.uk/id/eprint/249/1/Paper47.pdf

Chi, N. T. K. (2024). The Effect of AI Chatbots on Pro-environment Attitude and Willingness to Pay for Environment Protection. *SAGE Open*, *14*(1), 21582440231226001. Advance online publication. DOI: 10.1177/21582440231226001

Chorianopoulos, I., Pagonis, T., Koukoulas, S., & Drymoniti, S. (2010). Planning, competitiveness and sprawl in the Mediterranean city: The case of Athens. *Cities (London, England)*, *27*(4), 249 ?"259. DOI: 10.1016/j.cities.2009.12.011

Choudhury, S., & Ghosh, A. (2023). Early detection of Panama disease in bananas using deep learning techniques. *Journal of Plant Pathology*, *105*(2), 123–135.

Christakis, C. (2013). *Multifunctional Landscape; The Case of The Plain of Thessaly* [Master Thesis, University of Thessaly].

Chuvieco, E., Yebra, M., Martino, S., Thonicke, K., Gómez-Giménez, M., San-Miguel, J., Oom, D., Velea, R., Mouillot, F., Molina, J. R., Miranda, A. I., Lopes, D., Salis, M., Bugaric, M., Sofiev, M., Kadantsev, E., Gitas, I. Z., Stavrakoudis, D., Eftychidis, G., & Viegas, D. (2023). Towards an Integrated Approach to Wildfire Risk Assessment: When, Where, What and How May the Landscapes Burn. *Fire (Basel, Switzerland)*, *6*(5). Advance online publication. DOI: 10.3390/fire6050215

Ciesielczuk, T., Kusza, G., Poluszyńska, J., & Kochanowska, K. (2014). Pollution of flooded arable soils with heavy metals and polycyclic aromatic hydrocarbons (PAHs). *Water, Air, and Soil Pollution*, *225*(10), 2145. https://doi.org/10.1007/s11270-014-2145-0

Cluley, G. (2023, June 21). *Bitdefender*. Retrieved from Oreo maker Mondelez staff hit by data breach at third-party law firm: https://www.bitdefender.com/en-gb/blog/hotforsecurity/oreo-maker-mondelez-staff-hit-by-data-breach-at-third-party-law-firm

Cohen, D. J., Deeming, E.J., . (1985). *The National Fire Danger Rating System: Basic Equations* (PSW-82).

Cónego, L., Pinto, R., & Gonçalves, G. (2024). Digital Transformation in Manufacturing: The Synergy of Chatbots and Tailored Gamification Strategies. *Procedia Computer Science*, *237*, 171–178. DOI: 10.1016/j.procs.2024.05.093

Cooper, C. (2015). Cybersecurity in food and agriculture. In Leclair, J. (Ed.), *Protecting our future* (Vol. 2). Hudson Whitman/ Excelsior College Press.

Copernicus, E. (2023). *Image of floods in Thessaly, Greece*. Retrieved Accessed 28 March 2023 from https://x.com/CopernicusEU/status/1701155548744339642

Copernicus. (2024). EFFIS statistics – Estimates. https://forest-fire.emergency.copernicus.eu/apps/effis.statistics/estimates

Couch, C., & Karecha, J. (2006). Controlling urban sprawl: Some experiences from Liverpool. *Cities (London, England)*, *23*(5), 353 ?"363. DOI: 10.1016/j.cities.2006.05.003

Couch, C., Leontidou, L., & Arnstberg, K. O. (2008). Introduction: Definitions, theories and methods of comparative analysis. In Couch, C., Petschel-Held, G., & Leontidou, L. (Eds.), *Urban sprawl in Europe: Landscape, land-use change and policy* (pp. 3 ?"34). Blackwell.

Council of the European Union. (1991). *Council Directive 91/676/EEC of 12 December 1991 concerning the protection of waters against pollution caused by nitrates from agricultural sources* (CELEX:31991L0676). https://eur-lex.europa.eu/legal-content/EN/TXT/?uri=CELEX%3A31991L0676

Cuzzocrea, A. (2021, January). Big data lakes: models, frameworks, and techniques. In *2021 IEEE International Conference on Big Data and Smart Computing (BigComp)* (pp. 1-4). IEEE.

Cyber Physical Systems Public Working Group. (2016). *Framework for Cyber-Physical Systems Release 1.0*. National Institute of Standards and Technology.

Cyware Labs. (2020). *REvil Ransomware Targeted Sanitary Components Supplier | Cyware Hacker News*. Retrieved from Cyware Labs: https://social.cyware.com/news/revil-ransomware-targeted-sanitary-components-supplier-e2a41c63

DA-az, S., Settele, J., BrondA-zio, E., Ngo, H., GuA"ze, M., Agard, J., & Zayas, C. (2019). *Summary for policymakers of the global assessment report on biodiversity and ecosystem services*. IPBES. https://doi.org/DOI: 10.5281/zenodo.3553579

Dafis, S. (1986). *Forest Ecology*. Giahoudi.

Das, S., & Ambika, M. (2025). Revolutionizing Indian Farming: Machine Learning-Powered NLP for Optimal Crop Recommendations. *Procedia Computer Science*, *258*, 2040–2049. DOI: 10.1016/j.procs.2025.04.454

Das, S., & Roy, P. (2023). A deep learning approach for the identification of banana diseases. *International Journal of Computer Applications*, *183*(5), 10–15.

Daum, T., Adegbola, P. Y., Adegbola, C., Daudu, C., Issa, F., Kamau, G., Kergna, A. O., Mose, L., Ndirpaya, Y., Fatunbi, O., Zossou, R., Kirui, O., & Birner, R. (2022). Mechanization, digitalization, and rural youth-Stakeholder perceptions on three mega-topics for agricultural transformation in four African countries. *Global Food Security*, *32*, 100616. DOI: 10.1016/j.gfs.2022.100616

De Amorim, R. C. (2012, November). Constrained clustering with minkowski weighted k-means. In *2012 IEEE 13th International Symposium on Computational Intelligence and Informatics (CINTI)* (pp. 13-17). IEEE.

De Grave, A., Botija, P., Hansen, H. N., & Tang, P. T. (2006, January). Manufacturing and characterization of water repellent surfaces. In *4M 2006-Second International Conference on Multi-Material Micro Manufacture* (pp. 281-284). Elsevier.

de Oliveira Santini, F., Ladeira, W. J., Pinto, D. C., Herter, M. M., Sampaio, C. H., & Babin, B. J. (2020). Customer engagement in social media: A framework and meta-analysis. *Journal of the Academy of Marketing Science*, *48*(6), 1211–1228. DOI: 10.1007/s11747-020-00731-5

de Silva, T. M. (2020). Building relationships through customer engagement in Facebook brand pages. *Marketing Intelligence & Planning*, *38*(6), 713–729. DOI: 10.1108/MIP-02-2019-0085

Deka, B. K. (2016, March). Transformations of graph database model from multi-dimensional data model. In *2016 3rd International Conference on Computing for Sustainable Global Development (INDIACom)* (pp. 2836-2840). IEEE.

Deloitte. (2020). *Digital maturity study of Greek economy and Greek enterprises*. Deloitte. https://www2.deloitte.com/gr/en/pages/technology/articles/Deloitte-SEV-Digital-Observatory-Report-2nd-Edition.html

Demertzis, K., & Iliadis, L. (2016, January). Bio-inspired hybrid intelligent method for detecting android malware. In *Knowledge, Information and Creativity Support Systems: Selected Papers from KICSS'2014-9th International Conference, held in Limassol, Cyprus, on November 6-8, 2014* (pp. 289-304). Cham: Springer International Publishing.

Demertzis, K., Iliadis, L. S., & Anezakis, V. D. (2018). An innovative soft computing system for smart energy grids cybersecurity. *Advances in Building Energy Research*, *12*(1), 3–24.

Demertzis, K., Iliadis, L., & Anezakis, V. D. (2019, May). A machine hearing framework for real-time streaming analytics using Lambda architecture. In *International Conference on Engineering Applications of Neural Networks* (pp. 246-261). Cham: Springer International Publishing.

Demertzis, K., Iliadis, L., Kikiras, P., & Tziritas, N. (2019, May). Cyber-typhon: an online multi-task anomaly detection framework. In *IFIP International Conference on Artificial Intelligence Applications and Innovations* (pp. 19-36). Cham: Springer International Publishing.

Demertzis, K., Iliadis, L., Pimenidis, E., & Kikiras, P. (2022). Variational restricted Boltzmann machines to automated anomaly detection. *Neural Computing & Applications, 34*(18), 15207–15220.

Demertzis, K., Iliadis, L., Pimenidis, E., Tziritas, N., Koziri, M., & Kikiras, P. (2021, June). Blockchained adaptive federated auto MetaLearning BigData and DevOps CyberSecurity architecture in Industry 4.0. In *International Conference on Engineering Applications of Neural Networks* (pp. 345-363). Cham: Springer International Publishing.

Demertzis, K., Kikiras, P., Tziritas, N., Sanchez, S. L., & Iliadis, L. (2018). The next generation cognitive security operations center: Network flow forensics using cybersecurity intelligence. *Big Data and Cognitive Computing, 2*(4), 35.

Demestichas, K., Sykas, D., Zografakis, D., Kaloudis, S., Kalapodis, N., Sakkas, G., Athanasiou, M., & Costopoulou, C. (2025). Measuring Forest Resilience Against Wildfires and Climate Change: Methods and Technical Approaches. In *Paradigms on Technology Development for Security Practitioners* (pp. 53-63). Springer. https://doi.org/DOI: 10.1007/978-3-031-62083-6_5

Demestichas, K., Peppes, N., & Alexakis, T. (2020). Survey on Security Threats in Agricultural IoT and Smart Farming. *Sensors (Basel), 20*(22), 6458. DOI: 10.3390/s20226458 PMID: 33198160

Devi, N. N., Sridharan, B., & Kuiry, S. N. (2019). Impact of urban sprawl on future flooding in Chennai city, India. *Journal of Hydrology (Amsterdam), 574*, 486 ?"496. DOI: 10.1016/j.jhydrol.2019.04.041

Dias, R., Mauricio, L. A. F., & Poggi, M. (2020, June). *Toward an efficient real-time anomaly detection system for cloud datacenters. In 2020 IFIP networking conference (networking)*. IEEE.

Dimitrakopoulos, A., Baxevanis, K. E., & Mitsopoulos, I. D. (2009). Time-course correlation of forest fires and socioeconomic factors in Greece, during the period 1960-2000. 19th Panhellenic Forestry Conference, 1-4 November 2009, Patra.

Dimitrakopoulos, A. P. (2001a). A statistical classification of Mediterranean species based on their flammability components. *International Journal of Wildland Fire, 10*(2), 113–113. DOI: 10.1071/wf01004

Dimitrakopoulos, A. P. (2001b). Thermogravimetric analysis of Mediterranean plant species. *Journal of Analytical and Applied Pyrolysis, 60*(2), 123–130. DOI: 10.1016/s0165-2370(00)00164-9

Dimitrakopoulos, A. P., & Bemmerzouk, A. M. (2003). Predicting live herbaceous moisture content from a seasonal drought index. *International Journal of Biometeorology, 47*(2), 73–79. DOI: 10.1007/s00484-002-0151-1

Dimitrakopoulos, A. P., & Panov, P. I. (2001). Pyric properties of some dominant Mediterranean vegetation species. *International Journal of Wildland Fire*, *10*(1), 23–23. DOI: 10.1071/wf01003

Dimitrakopoulos, A. P., & Papaioannou, K. K. (2001). Flammability Assessment of Mediterranean Forest Fuels. *Fire Technology*, *37*(2), 143–152. DOI: 10.1023/a:1011641601076

Dimitrakopoulos, A., Bemmerzouk, A., & Mitsopoulos, I. (2011). Evaluation of the Canadian fire weather index system in an eastern Mediterranean environment. *Meteorological Applications*, *18*(1), 83–93. https://doi.org/https://doi.org/10.1002/met.214

Dimitrijević, M. S. (2023). Technological progress in the function of productivity and sustainability of agriculture: The case of innovative countries and the Republic of Serbia. *Journal of Agriculture and Food Research*, *14*, 100856. DOI: 10.1016/j.jafr.2023.100856

Dimtsoudis, C. (2022). *A tool for self-produced malt.* Agronews. https://www.agronews.gr/files/PDF/Beer_bruch/04_22_beer_brunch.pdf

Downs, A. (1999). Some realities about sprawl and urban decline. *Housing Policy Debate*, *10*(4), 955 ?"974. DOI: 10.1080/10511482.1999.9521356

Drape, T., Magerkorth, N., Sen, A., Simpson, J., Seibel, M., Murch, R., & Duncan, S. (2021). Assessing the Role of Cyber biosecurity in Agriculture: A Case study. *Frontiers in Bioengineering and Biotechnology*, •••, 742. PMID: 34490231

Duncan, B. N., Martin, R. V., Staudt, A. C., Yevich, R., & Logan, J. A. (2003). Interannual and seasonal variability of biomass burning emissions constrained by satellite observations. *Journal of Geophysical Research, D, Atmospheres*, *108*(D2). Advance online publication. DOI: 10.1029/2002jd002378

Duncan, B. W., Adrian, F. W., & Stolen, E. D. (2010). Isolating the lightning ignition regime from a contemporary background fire regime in east-central Florida, USA. *Canadian Journal of Forest Research*, *40*(2), 286–297. DOI: 10.1139/x09-193

Edwards, J. P., Kuhn-Sherlock, B., Rue, B. D., & Eastwood, C. R. (2020). Technologies and milking practices that reduce hours of work and increase flexibility through milking efficiency in pasture-based dairy farm systems. *Journal of Dairy Science*, *103*(8), 7172–7179. DOI: 10.3168/jds.2019-17941 PMID: 32505396

Efimov, K., Adamyan, L., & Spokoiny, V. (2019). Adaptive nonparametric clustering. *IEEE Transactions on Information Theory*, *65*(8), 4875–4892.

Egoh, B. N., Paracchini, M. L., Zulian, G., Schägner, J. P., Bidoglio, G., & Jones, J. (2014). Exploring restoration options for habitats, species and ecosystem services in the European Union. *Journal of Applied Ecology*, *51*(4), 899–908. DOI: 10.1111/1365-2664.12251

Ellison, M. (2013). *Biomimetic Textiles*. Engineered Biomimicry.

Emmanouil, S. (2024). Small breweries: The signal of concentration and the risk of craftwashing. PowerGame.https://www.powergame.gr/epichirisis/666993/mikres-zythopoiies-to-sima-sygkentrosis-kai-o-kindynos-craftwashing/

Enright, N. J., Fontaine, J. B., Bowman, D. M. J. S., Bradstock, R. A., & Williams, R. J. (2015). Interval squeeze: Altered fire regimes and demographic responses interact to threaten woody species persistence as climate changes. *Frontiers in Ecology and the Environment*, *13*(5), 265–272. DOI: 10.1890/140231

Erbaugh, J. T., & Oldekop, J. A. (2018). Forest landscape restoration for livelihoods and well-being. *Current Opinion in Environmental Sustainability*, *32*, 76–83. DOI: 10.1016/j.cosust.2018.05.007

Eurofound. (2012). *Medium-sized cities in Europe (summary)*. https://www.eurofound.europa.eu/en/publications/all/medium-sized-cities-europe-summary

European Commission, Joint Research Centre. (2016). *Atlas of the human planet 2016: Mapping human presence on Earth with the Global Human Settlement Layer* (EUR 28116 EN). Publications Office of the European Union. https://doi.org/DOI: 10.2788/889483

European Commission. (2024). *Science, research and innovation performance of the EU*. https://op.europa.eu/en/publication-detail/-/publication/c683268c-3cdc-11ef-ab8f-01aa75ed71a1/language-en

European Commission. (2025). *Unlocking the potential of artificial intelligence for sustainable agriculture*. https://digital-strategy.ec.europa.eu/en/library/unlocking-potential-artificial-intelligence-sustainable-agriculture

European Environment Agency, & FOEN. (2016). *Urban sprawl in Europe: Joint EEA ?"FOEN report* (EEA Report No. 11/2016). European Environment Agency & Swiss Federal Office for the Environment. https://www.eea.europa.eu/en/analysis/publications/urban-sprawl-in-europe

Ewing, R., Pendall, R., & Chen, D. (2002). *Measuring sprawl and its impact*. Smart Growth America. https://jacksontetonplan.com/DocumentCenter/View/767/Ewing-R-R-Pendall-and-D-Chen-2002-PDF

Ewing-Chow, D. (2024, September 20). *Forbes*. Retrieved from Agri-Food Sector Under Increasing Threat From Cyber Attacks: https://www.forbes.com/sites/daphneewingchow/2024/09/20/agri-food-sector-under-increasing-threat-from-cyber-attacks/

Ewing, R. (1997). Is Los Angeles-style sprawl desirable? *Journal of the American Planning Association*, *63*(1), 107 ?"126. DOI: 10.1080/01944369708975728

Falk, D. A., van Mantgem, P. J., Keeley, J. E., Gregg, R. M., Guiterman, C. H., Tepley, A. J., Jn Young, D., & Marshall, L. A. (2022). Mechanisms of forest resilience. *Forest Ecology and Management*, *512*, 120129–120129. DOI: 10.1016/j.foreco.2022.120129

Fanelli, C., Giroux, J., Moran, P., Nayak, H., Suresh, K., & Walter, E. (2024). Physics event classification using Large Language Models. *Journal of Instrumentation : An IOP and SISSA Journal*, *19*(7), C07011. Advance online publication. DOI: 10.1088/1748-0221/19/07/C07011

Fang, W., Guo, H. Y., Li, B., Li, Q., & Feng, X. Q. (2018). Revisiting the critical condition for the Cassie–Wenzel transition on micropillar-structured surfaces. *Langmuir*, *34*(13), 3838–3844.

Fanpage Karma Academy. (2022). *Fanpage Karma Academy*.https://academy.fanpagekarma.com/en/home/

FAO. (1986). *Wildland fire management terminology.*

FAO. (1998). *World Reference Base for Soil Resources*. Food and Agriculture Organization of the United Nations.

Feng, W., Liu, S., Faloutsos, C., Hooi, B., Shen, H., & Cheng, X. (2017). EagleMine: Vision-Guided Mining in Large Graphs. *arXiv preprint arXiv:1710.08756.*

Fernandes, P. M. (2013). Fire-smart management of forest landscapes in the Mediterranean basin under global change. *Landscape and Urban Planning*, *110*, 175–182. DOI: 10.1016/j.landurbplan.2012.10.014

Fernandes, P. M., & Botelho, H. S. (2003). A review of prescribed burning effectiveness in fire hazard reduction. *International Journal of Wildland Fire*, *12*(2), 117–117. DOI: 10.1071/wf02042

Fernandez, F. A. S., Rheingantz, M. L., Genes, L., Kenup, C. F., Galliez, M., Cezimbra, T., Cid, B., Macedo, L., Araujo, B. B. A., Moraes, B. S., Monjeau, A., & Pires, A. S. (2017). Rewilding the Atlantic Forest: Restoring the fauna and ecological interactions of a protected area. *Perspectives in Ecology and Conservation*, *15*(4), 308–314. DOI: 10.1016/j.pecon.2017.09.004

Ferrag, M., Shu, L., Yang, X., Derhab, A., & Maglaras, L. (2020). Security and Privacy for Green IoT-Based Agriculture: Review, Blockchain Solutions, and Challenges. *IEEE Access : Practical Innovations, Open Solutions*, *8*, 32031–32053. DOI: 10.1109/ACCESS.2020.2973178

Fijałkowski, K., Kacprzak, M., Grobelak, A., & Placek, A. (2012). The influence of selected soil parameters on the mobility of heavy metals in soils. *Inżynieria i Ochrona środowiska, 15*, 81-92.

Foley, J. A., DeFries, R., Asner, G. P., Barford, C., Bonan, G., Carpenter, S. R., Chapin, F. S., Coe, M. T., Daily, G. C., Gibbs, H. K., Helkowski, J. H., Holloway, T., Howard, E. A., Kucharik, C. J., Monfreda, C., Patz, J. A., Prentice, I. C., Ramankutty, N., & Snyder, P. K. (2005). Global consequences of land use. *Science*, *309*(5734), 570 ?"574. DOI: 10.1126/science.1111772

Food and Agriculture Organization (FAO). (2025). *AI can be a game-changing solution for farmers: FAO Innovation Chief*. https://www.fao.org/newsroom/detail/ai-can-be-a-game-changing-solution-for-farmers--fao-innovation-chief/en

Fornalczyk, K., Bortko, K., & Jankowski, J. (2021). Improving User Attention to Chatbots through a Controlled Intensity of Changes within the Interface. *Procedia Computer Science*, *192*, 5112–5121. DOI: 10.1016/j.procs.2021.09.289

Forzieri, G., Dakos, V., McDowell, N. G., Ramdane, A., & Cescatti, A. (2022). Emerging signals of declining forest resilience under climate change. *Nature*, *608*(7923), 534–539. DOI: 10.1038/s41586-022-04959-9

Founda, D., & Giannakopoulos, C. (2010). The exceptionally hot summer of 2007 in Athens, Greece — A typical summer in the future climate? *Atmospheric Research*, *98*(2-4), 245–252. DOI: 10.1016/j.atmosres.2010.07.002

Frenkel, A. (2004). The potential effect of national growth-management policy on urban sprawl and the depletion of open spaces and farmland. *Land Use Policy*, *21*(4), 357 ?"369. DOI: 10.1016/j.landusepol.2003.12.001

Friha, O., Ferrag, M., Shu, L., Maglaras, L., & Wang, X. (2021). Internet of Things for the Future of Smart Agriculture: A Comprehensive Survey of Emerging Technologies. *IEEE/CAA Journal of Automatica Sinica*, 718-752.

Friha, O., Ferrag, M., Maglaras, L., & Shu, L. (2022). Digital Agriculture Security: Aspects, Threats, Mitigation Strategies, and Future Trends. *IEEE Internet of Things Magazine*, *5*(3), 82–90. DOI: 10.1109/IOTM.001.2100164

Fu, C., Xiang, C., Wang, C., & Cai, D. (2017). Fast approximate nearest neighbor search with the navigating spreading-out graph. *arXiv preprint arXiv:1707.00143*.

Fu, C., Liu, C., Ishi, C. T., Yoshikawa, Y., Iio, T., & Ishiguro, H. (2021). Using an Android Robot to Improve Social Connectedness by Sharing Recent Experiences of Group Members in Human-Robot Conversations. *IEEE Robotics and Automation Letters*, *6*(4), 6670–6677. DOI: 10.1109/LRA.2021.3094779

Fujii, K., & Kawahara, Y. (2019). Dynamic mode decomposition in vector-valued reproducing kernel Hilbert spaces for extracting dynamical structure among observables. *Neural Networks*, *117*, 94–103.

Gaire, R., Lefort, L., Compton, M., Falzon, G., Lamb, D., & Taylor, K. (2013). Demonstration: Semantic Web Enabled Smart Farm with GSN. *International Workshop on the Semantic Web*. Retrieved from Demonstration: Semantic Web Enabled Smart Farm with GSN.

Galar, M., Fernandez, A., Barrenechea, E., Bustince, H., & Herrera, F. (2011). A review on ensembles for the class imbalance problem: Bagging-, boosting-, and hybrid-based approaches. *IEEE Transactions on Systems, Man, and Cybernetics. Part C, Applications and Reviews*, *42*(4), 463–484.

Galster, G., Hanson, R., Ratcliffe, M. R., Wolman, H., Coleman, S., & Freihage, J. (2001). Wrestling sprawl to the ground: Defining and measuring an elusive concept. *Housing Policy Debate*, *12*(4), 681 ?"717. DOI: 10.1080/10511482.2001.9521426

Garcês, A., & Pires, I. (2023). The Hell of Wildfires: The Impact on Wildlife and Its Conservation and the Role of the Veterinarian. *Conservation*, *3*(1), 96–108. DOI: 10.3390/conservation3010009

Gardizy, A. (2022, March 18). *Boston Globe*. Retrieved from Hackers hit Hood. Dairy shut down milk production this week after the event: https://www.bostonglobe.com/2022/03/18/business/school-milk-could-be-short-supply-after-hood-plants-hit-by-cyber-event/

Garg, P., & Rao, K. S. (2024). Building AgriRef: An AI embedded virtual reference app for farmers. *Library Hi Tech News*, *41*(2), 11–17. DOI: 10.1108/LHTN-02-2023-0025

Gates, A. J., & Ahn, Y. Y. (2017). The impact of random models on clustering similarity. *Journal of Machine Learning Research*, *18*(87), 1–28.

Ge, H., Zhao, X., Liu, C., Zeng, Y., Liu, Q., & Xiong, D. (2022). TGEA 2.0: A Large-Scale Diagnostically Annotated Dataset with Benchmark Tasks for Text Generation of Pretrained Language Models. *Advances in Neural Information Processing Systems*, *35*, 31612–31626.

Genton, M. G., Butry, D. T., Gumpertz, M. L., & Prestemon, J. P. (2006). Spatio-temporal analysis of wildfire ignitions in the St Johns River Water Management District, Florida. *International Journal of Wildland Fire, 15*(1), 87–97. DOI: 10.1071/wf04034

Giakoumatos, S. D., Skoulikidis, N. T., Karavoltsos, S., Sakellari, A., & Dimitriou, E. (2024). Spatiotemporal Variations in Water Physicochemical Status in Pinios River Catchment, at Eastern Mediterranean Region. *Land, 13*(11), 1959. https://doi.org/ https://doi.org/10.3390/land13111959

Giannakopoulos, C., LeSager, P., Moriondo, M., Bindi, M., Karali, A., Hatzaki, M., & Kostopoulou, E. (2012). Comparison of fire danger indices in the Mediterranean for present day conditions. *iForest-Biogeosciences and Forestry, 5*(4), 197-203. https://doi.org/https://doi.org/10.3832/ifor0622-005

Giannakopoulos, C., Bindi, M., Moriondo, M., LeSager, P., & Tin, T. (2011). An integrated assessment of climate change impacts for Greece in the near future. *Regional Environmental Change, 11*(4), 829–843. DOI: 10.1007/s10113-011-0219-8

Gibbons, P., Lindenmayer, D. B., Barry, S. C., & Tanton, M. T. (2000). The effects of slash burning on the mortality and collapse of trees retained on logged sites in south-eastern Australia. *Forest Ecology and Management, 139*(1-3), 51–61. DOI: 10.1016/s0378-1127(99)00333-3

Gimeno, M., & San-Miguel-Ayanz, J. (2004). Evaluation of RADARSAT-1 data for identification of burnt areas in Southern Europe. *Remote Sensing of Environment, 92*(3), 370–375. http://www.sciencedirect.com/science/article/B6V6V-4D7BNR7-3/2/d69d549ec6ce3112965fec443b49b287

Gkitsa, A. (2010). *Water quality and water balance of the Kalentzi River* [Master Thesis, Uinversity of Thessaly].

Gmiterek, G. (2023). Use of Facebook fan pages in Polish academic libraries (2009–2022). *Journal of Academic Librarianship, 49*(2), 102678. DOI: 10.1016/j.acalib.2023.102678

Golia, E. (2003). *Soil pollution in Thessaly by heavy metals (Zn, Cu, Mn, Fe, Cd, and Pb) and their uptake by tobacco varieties: Spatial variability of heavy metals in selected areas of the study region* [Doctoral dissertation, University of Thessaly].

Golia, E. E., & Diakoloukas, V. (2022). Soil parameters affecting the levels of potentially harmful metals in Thessaly area, Greece: A robust quadratic regression approach of soil pollution prediction. *Environmental Science and Pollution Research International, 29*(20), 29544–29561. https://doi.org/10.1007/s11356-021-14673-0

Gómez, M. C., González, M., Gómez-Aparicio, L., & Serrano, M. S. (2020). Coexistent Mediterranean woody species as a driving factor of Phytophthora cinnamomi infectivity and survival. *Annals of Applied Biology*, *177*(1), 41–50. DOI: 10.1111/aab.12599

Government Gazette. (1986). [Approval of the General Urban Plan of Patras (Municipality of Patras)] (I EK 1061/I"/07.11.1986). [in Greek].

Government Gazette. (2011). [Approval of the General Urban Plan of the Municipal Unit of Patras, Municipality of Patras, Achaia Prefecture] (I EK 358/AAI /30.12.2011). [in Greek].

Government Gazette. (2012). [Amendment of Decision No. 11247/28.12.2012 regarding the 2011 Population and Housing Census results for the permanent population of the country] (I EK 3465/BA'/28.12.2012). [in Greek].

Graham, R. T., Alan E., H., Threasa B. Jain, & Tonn, J. R. (1999). *The effects of thinning and similar stand treatments on fire behavior in western forests.*

Grammenou, A. (2020). *Heavy metal levels in soil and cereal plants in Metalleia.* Domokos. [Bachelor's Thesis, University of Thessaly]

Greenberg, A. (2022, March 8). *WIRED*. Retrieved from Chinese Spies Hacked a Livestock App to Breach US State Networks: https://www.wired.com/story/china-apt41-hacking-usaherds-log4j/

Green, K., Kempka, D., & Lackey, L. (1994). Using remote sensing to detect and monitor land-cover and land-use change. *Photogrammetric Engineering and Remote Sensing*, *60*(3), 331 ?"337. https://www.asprs.org/wp-content/uploads/pers/1994journal/mar/1994_mar_331-337.pdf

Greig, J. (2021, September 22). *ZDNET*. Retrieved from Crystal Valley Cooperative becomes latest agriculture business hit with ransomware: https://www.zdnet.com/article/crystal-valley-cooperative-becomes-latest-agriculture-business-hit-with-ransomware/

Griffin, G., Jokela, W., Ross, D., Pettinelli, D., Morris, T., & Wilf, A. (1995). *Recommended soil nitrate-N tests* (Vol. 493). Northeastern Regional Publication.

Gupta, M., Abdelsalam, M. S. K., & Mittal, S. (2020). Security and Privacy in Smart Farming: Challenges and Opportunities. *IEEE Access : Practical Innovations, Open Solutions*, *8*, 34564–34584. DOI: 10.1109/ACCESS.2020.2975142

Hagmann, R. K., Hessburg, P. F., Salter, R. B., Merschel, A. G., & Reilly, M. J. (2022). Contemporary wildfires further degrade resistance and resilience of fire-excluded forests. *Forest Ecology and Management*, *506*, 119975–119975. DOI: 10.1016/j.foreco.2021.119975

Hahsler, M., Bolanos, M., & Forrest, J. (2017). Introduction to stream: An extensible framework for data stream clustering research with R. *Journal of Statistical Software*, *76*, 1–50.

Håkansson, A., & Phillips-Wren, G. (2024). Generative AI and Large Language Models—Benefits, Drawbacks, Future and Recommendations. *Procedia Computer Science*, *246*, 5458–5468. DOI: 10.1016/j.procs.2024.09.689

Hamer, D. (2012). Probability, anti-resilience, and the weight of expectation. *Law Probability and Risk*, *11*(2-3), 135–158.

Harvey, W. S. (2011). Strategies for conducting elite interviews. *Qualitative Research*, *11*(4), 431 ?"441. DOI: 10.1177/1468794111404329

Hasim, A., Herdiyeni, Y., & Douady, S. (2016). Leaf shape recognition using centroid contour distance. [). IOP Publishing.]. *IOP Conference Series. Earth and Environmental Science*, *31*(1), 012002.

Hassani, A., & Shi, H. (2022). Dilated neighborhood attention transformer. *arXiv preprint arXiv:2209.15001*.

Hastaoglou-Martinidis, V. (1995). City form and national identity: Urban designs in nineteenth-century Greece. *Journal of Modern Greek Studies*, *13*(1), 99 ?"123. DOI: 10.1353/mgs.2010.0397

Hawkes, B., Beck, J., & Sahle, W. (1996). A wildfire threat rating system for the McGregor Model Forest. Hellenic Fire Service. (Retrieved September 27, 2024). *Datasets*. https://www.fireservice.gr/el_GR/synola-dedomenon

He, K., Zhang, X., Ren, S., & Sun, J. (2016). Deep residual learning for image recognition. *Proceedings of the IEEE Conference on Computer Vision and Pattern Recognition (CVPR)*, 770-778. https://doi.org/DOI: 10.1109/CVPR.2016.90

Hellenic National Meteorological Service. (Retrieved September 26, 2024). *Climatic data per month*. https://www.emy.gr/en/climatic-data?tab=statistics-tab

Hellenic Statistical Authority. (2023). *Permanent population by region and regional unit (2021 census)*. Retrieved 1/10/2024 from https://www.statistics.gr

Hellenic Survey for Geology and Mineral Exploration. (1983). *Geological map of Greece*. Retrieved 5/1/2025 from https://eagme.gr/site/services

Helms, J. A. (1998). *The Dictionary of Forestry*. Society of American Foresters.

Herald, N. (2020). *Fonterra Targeted by Hackers*. Retrieved from NZ Herald: https://www.nzherald.co.nz/business/fonterra-targeted-by-hackers/6PEVEZE O5AI6O2W2RS7HJ6WMTM/

Herrera, K., Miranda, J., & Mauricio, D. (2022). Milchbot: App to Support the Process of Feeding and Caring for Dairy Cows in Peru. *AGRIS On-Line Papers in Economics and Informatics, 14*(4), 27–37. DOI: 10.7160/aol.2022.140403

Hersperger, A. M., & BA1/4rgi, M. (2007). Driving forces of landscape change in the urbanizing Limmat valley, Switzerland. In Koomen, E., Stillwell, J., Bakema, A., & Scholten, H. (Eds.), *Modelling land-use change: Progress and applications* (pp. 45 ?"60). Springer., DOI: 10.1007/978-1-4020-5648-2_3

Hersperger, A. M., & BA1/4rgi, M. (2009). Going beyond landscape change description: Quantifying the importance of driving forces of landscape change in a Central Europe case study. *Land Use Policy, 26*(3), 640 ?"648. DOI: 10.1016/j.landusepol.2008.08.015

Hidalgo, J. I. G., Maciel, B. I., & Barros, R. S. (2019). Experimenting with prequential variations for data stream learning evaluation. *Computational Intelligence, 35*(4), 670–692.

Hilscherova, K., Dusek, L., Kubik, V., Cupr, P., Hofman, J., Klanova, J., & Holoubek, I. (2007). Redistribution of organic pollutants in river sediments and alluvial soils related to major floods. *Journal of Soils and Sediments, 7*(3), 167–177.

Hofman, G., & Van Cleemput, O. (2004). *Soil and Plant Nitrogen*. International Fertilizer Industry Association (IFA).

Holling, C. S. (1973). Resilience and Stability of Ecological Systems. *Annual Review of Ecology and Systematics, 4*(1), 1–23. DOI: 10.1146/annurev.es.04.110173.000245

Huang, C., Zhou, Z., Peng, C., Teng, M., & Wang, P. (2019). How is biodiversity changing in response to ecological restoration in terrestrial ecosystems? A meta-analysis in China. *The Science of the Total Environment, 650*(Pt 1), 1–9. DOI: 10.1016/j.scitotenv.2018.08.320

Huang, R., Kim, M., & Lennon, S. (2022). Trust as a second-order construct: Investigating the relationship between consumers and virtual agents. *Telematics and Informatics, 70*, 101811. DOI: 10.1016/j.tele.2022.101811

Hussain, A., Bashir, H., Zafar, S., Rehman, R., Khalid, M., Awais, M., Sadiq, M., & Amjad, I. (2023). The importance of soil organic matter (som) on soil productivity and plant growth. *Biological and Agricultural Sciences Research Journal, 2023*(1), 11-11. https://doi.org/ https://doi.org/10.54112/basrj.v2023i1.11

Hwang, W. S., Yun, J. H., Kim, J., & Kim, H. C. (2019, November). Time-series aware precision and recall for anomaly detection: considering variety of detection result and addressing ambiguous labeling. In *Proceedings of the 28th ACM International Conference on Information and Knowledge Management* (pp. 2241-2244).

Hyde, R., Angelov, P., & MacKenzie, A. R. (2017). Fully online clustering of evolving data streams into arbitrarily shaped clusters. *Information Sciences*, *382*, 96–114.

Iatrou, M., Tziouvalekas, M., Tsitouras, A., Evangelou, E., Noulas, C., Vlachostergios, D., Aschonitis, V., Arampatzis, G., Metaxa, I., & Karydas, C. (2024). Analyzing the Impact of Storm 'Daniel'and Subsequent Flooding on Thessaly's Soil Chemistry through Causal Inference. *Agriculture*, *14*(4), 549. https://doi.org/https://doi.org/10.3390/agriculture14040549

Ibe, O. (2013). Random Walk. In *Markov processes for stochastic modeling* (pp. 205–259). Elsevier.

Ibrahim, A., Senthilkumar, K., & Saito, K. (2024). Evaluating responses by ChatGPT to farmers' questions on irrigated lowland rice cultivation in Nigeria. *Scientific Reports*, *14*(1), 3407. Advance online publication. DOI: 10.1038/s41598-024-53916-1 PMID: 38341517

ICAP CRIF. (2022). *ICAP CRIF brewing industry study*. ICAP CRIF.https://www.icapcrif.com/icap-crif-study-zythos2022/

ICAP CRIF. (2022). *Sector study for beer*. ICAP CRIF.

Idoje, G., Dagiuklas, T., & Iqbal, M. (2021). Survey for smart farming technologies: Challenges and issues. *Computers & Electrical Engineering*, *92*, 107104. DOI: 10.1016/j.compeleceng.2021.107104

Igme. (1983). *Geological Map of Greece, Scale 1:50,000, Sheet Athens*. Athens, Greece, IGME.

Igme. (1993). *Explanatory Notes on the Geological Map of Greece, Sheet Athens*.

International Electrotechnical Commission. (2019). *Industrial-process measurement control and automation, Framework for functional safety and security IEC TR 63069:2019*. International Electrotechnical Commission.

International Organization for Standardization (ISO). (2016). *Information technology — Security techniques — Information security management systems, ISO 27000 series*. Geneva: International Organization for Standardization (ISO).

International Trade Administration. (2024, July 15). *Serbia—Country Commercial Guide*. https://www.trade.gov/country-commercial-guides/serbia-agricultural-sectors

Iqbal, W., Berral, J. L., & Carrera, D. (2020). Adaptive sliding windows for improved estimation of data center resource utilization. *Future Generation Computer Systems*, *104*, 212–224.

Jactel, H., Moreira, X., & Castagneyrol, B. (2021). Tree Diversity and Forest Resistance to Insect Pests: Patterns, Mechanisms, and Prospects. *Annual Review of Entomology, 66*(1), 277–296. DOI: 10.1146/annurev-ento-041720-075234

Jadhaw, A., & Bhandari, R. (2024). Detection of banana leaf disease and its analysis using different techniques. *Journal of Agriculture and Allied Sciences, 13*, 005.

Jain, L., Ananthasayam, R., Gupta, U., & R, R. (2025). Comparison of Rule-based Chat Bots with Different Machine Learning Models. *Procedia Computer Science, 259*, 788–798. DOI: 10.1016/j.procs.2025.04.030

Jain, S., Sahni, R., Khargonkar, T., Gupta, H., Verma, O. P., Sharma, T. K., Bhardwaj, T., Agarwal, S., & Kim, H. (2022). Automatic Rice Disease Detection and Assistance Framework Using Deep Learning and a Chatbot. *Electronics (Basel), 11*(14), 2110. Advance online publication. DOI: 10.3390/electronics11142110

Jang, S., Lee, H., Kim, Y., Lee, D., Shin, J., & Nam, J. (2024). When, What, and how should generative artificial intelligence explain to Users? *Telematics and Informatics, 93*, 102175. DOI: 10.1016/j.tele.2024.102175

Jasmin, A., Ramesh, P., & Tanveer, M. (2024). Development of Artificial Intelligence-based chatbot for smart aquafarm practices. *Expert Systems: International Journal of Knowledge Engineering and Neural Networks, 41*(6), e13123. Advance online publication. DOI: 10.1111/exsy.13123

Jayasingh, S. (2019). Consumer brand engagement in social networking sites and its effect on brand loyalty. *Cogent Business & Management, 6*(1), 1698793. DOI: 10.1080/23311975.2019.1698793

Jiang, S., Cordeiro, R. L., & Akoglu, L. (2022, November). D. mca: Outlier detection with explicit micro-cluster assignments. In *2022 IEEE International Conference on Data Mining (ICDM)* (pp. 987-992). IEEE.

Joffe, H., & Yardley, L. (2003). Content and thematic analysis. In D. F. Marks & L. Yardley (Eds.), *Research methods for clinical and health psychology* (pp. 56 ?"68). Sage.

Johnstone, J. F., Allen, C. D., Franklin, J. F., Frelich, L. E., Harvey, B. J., Higuera, P. E., Mack, M. C., Meentemeyer, R. K., Metz, M. R., Perry, G. L. W., Schoennagel, T., & Turner, M. G. (2016). Changing disturbance regimes, ecological memory, and forest resilience. *Frontiers in Ecology and the Environment, 14*(7), 369–378. DOI: 10.1002/fee.1311

Joint Task Force. (2020). *Security and Privacy Controls for Information Systems and Organizations. NIST SP 800-53r5*. National Institute of Standards and Technology.

Jolly, W. M., Cochrane, M. A., Freeborn, P. H., Holden, Z. A., Brown, T. J., Williamson, G. J., & Bowman, D. M. J. S. (2015). Climate-induced variations in global wildfire danger from 1979 to 2013. *Nature Communications*, *6*, 7537. DOI: 10.1038/ncomms8537

Jones, D. (2023, May 18). *CYBERSECURITY DIVE.* Retrieved from Dole incurs $10.5M in direct costs from February ransomware attack: https://www.cybersecuritydive.com/news/dole-10m-costs-ransomware/650711/

Jones, L., Sproat, R., Ishikawa, H., & Gutkin, A. (2023). Helpful Neighbors: Leveraging Neighbors in Geographic Feature Pronunciation. *Transactions of the Association for Computational Linguistics, 11*, 85–101.

Jutagate, A., Pitakaso, R., Khonjun, S., Srichok, T., Kaewta, C., Luesak, P., Gonwirat, S., Enkvetchakul, P., & Jutagate, T. (2024). Adaptive artificial multiple intelligence fusion system (A-AMIFS) for enhanced disease detection in Nile Tilapia. *Aquaculture Reports*, *39*, 102418. Advance online publication. DOI: 10.1016/j.aqrep.2024.102418

Kabata-Pendias, A. (2010). *Trace Elements in Soils and Plants* (4 ed.). CRC press. https://doi.org/10.1201/b10158

Kabata-Pendias, A., & Mukherjee, A. B. (2007). *Trace elements from soil to human.* Springer.

Kairis, O., Karamanos, A., Voloudakis, D., Kapsomenakis, J., Aratzioglou, C., Zerefos, C., & Kosmas, C. (2022). Identifying degraded and sensitive to desertification agricultural soils in Thessaly, Greece, under simulated future climate scenarios. *Land (Basel)*, *11*(3), 395. https://doi.org/https://doi.org/10.3390/land11030395

Kalabokidis, K. D., & Omi, P. N. (1998). Reduction of Fire Hazard Through Thinning/Residue Disposal in the Urban Interface. *International Journal of Wildland Fire*, *8*(1), 29–29. DOI: 10.1071/wf9980029

Kalampokas, T., Vrochidou, E., Mavridou, E., Iliadis, L., Voglitsis, D., Michalopoulou, M., & Papakostas, G. A. (2025). Empowering Kiwifruit Cultivation with AI: Leaf Disease Recognition Using AgriVision-Kiwi Open Dataset. *Electronics (Basel)*, *14*(9), 1705.

Kalapodis, N., Sakkas, G., Lazarou, A., Casciano, D., Demestichas, K., Athanasiou, M., Kaloudis, S., & Sykas, D. (2025). EU-Integrated Multifunctional Forest and Fire Management, Policies, and Practices: Challenges Between "As-Is" and "To-Be" State. In *Paradigms on Technology Development for Security Practitioners* (pp. 65-77). Springer. https://doi.org/DOI: 10.1007/978-3-031-62083-6_6

Kaloudis, S. (2008). *Decision Support System for Forest Ecosystem Management Based on Geographic Information Systems and Knowledge Systems.* [Doctoral dissertation, Agricultural University of Athens]. Athens, Greece.

Kaloudis, S., Platia, K., Pantera, A., & Galanopoulou, S. (2021). Investigation on the importance of ecosystem services based on the contingent valuation method. *MIBES Transactions, 15*(1), 87-100. https://mibes.uth.gr/vol15_issue1_2021/Kaloudis%2C%20Platia%2C%20Pantera%20%26%20Galanopoulou.pdf

Kaloudis, S., Costopoulou, C. I., Lorentzos, N. A., Sideridis, A. B., & Karteris, M. (2008). Design of forest management planning DSS for wildfire risk reduction. *Ecological Informatics, 3*(1), 122–133. DOI: 10.1016/j.ecoinf.2007.07.008

Kaloudis, S., Glykou, M., Galanopoulou, S., Fotiadis, G., Yialouris, C., & Raptis, D. (2023). Land Cover Changes in Evrytania Prefecture (Greece). *Forests, 14*(7), 1462–1462. DOI: 10.3390/f14071462

Kaloudis, S., Tocatlidou, A., Lorentzos, N. A., Sideridis, A. B., & Karteris, M. (2005). Assessing Wildfire Destruction Danger: A Decision Support System Incorporating Uncertainty. *Ecological Modelling, 181*(1), 25–38. DOI: 10.1016/j.ecolmodel.2004.06.021

Kambezidis, H. D., & Kalliampakos, G. K. (2016). Fire-Risk Assessment in Northern Greece Using a Modified Fosberg Fire-Weather Index That Includes Forest Coverage. *International Journal of Atmospheric Sciences, 2016*(1), 8108691. DOI: 10.1155/2016/8108691

Karali, A., Hatzaki, M., Giannakopoulos, C., Roussos, A., Xanthopoulos, G., & Tenentes, V. (2014). Sensitivity and evaluation of current fire risk and future projections due to climate change: The case study of Greece. *Natural Hazards and Earth System Sciences, 14*(1), 143–153. DOI: 10.5194/nhess-14-143-2014

Kavallieratos, G., Katsikas, S., & Gkioulos, V. (2020). SafeSec Tropos: Joint security and safety requirements elicitation. *Computer Standards & Interfaces, 70*, 103429. DOI: 10.1016/j.csi.2020.103429

Kayode, O., Ogunyemi, E., Odukoya, A., & Aizebeokhai, A. (2022). Assessment of chromium and nickel in agricultural soil: Implications for sustainable agriculture. *IOP Conference Series. Earth and Environmental Science.*

Keesstra, S., Nunes, J., Novara, A., Finger, D., Avelar, D., Kalantari, Z., & Cerda, A. (2018). The superior effect of nature based solutions in land management for enhancing ecosystem services. *The Science of the Total Environment, 610-611*, 997–1009. DOI: 10.1016/j.scitotenv.2017.08.077

Kelly, S., Kaye, S.-A., & Oviedo-Trespalacios, O. (2023). What factors contribute to the acceptance of artificial intelligence? A systematic review. *Telematics and Informatics, 77*, 101925. DOI: 10.1016/j.tele.2022.101925

Khandelwal, U., Levy, O., Jurafsky, D., Zettlemoyer, L., & Lewis, M. (2019). Generalization through memorization: Nearest neighbor language models. *arXiv preprint arXiv:1911.00172*.

Khannouz, M., & Glatard, T. (2020). A benchmark of data stream classification for human activity recognition on connected objects. *Sensors (Basel)*, *20*(22), 6486.

Kibriya, H., Khan, W. Z., Siddiqa, A., & Khan, M. K. (2024). Privacy issues in Large Language Models: A survey. *Computers & Electrical Engineering*, *120*, 109698. DOI: 10.1016/j.compeleceng.2024.109698

Kim, D., & Oh, A. (2022). How to find your friendly neighborhood: Graph attention design with self-supervision. *arXiv preprint arXiv:2204.04879*.

Kim, A., Wampler, B., Goppert, J., Hwang, I., & Aldridge, H. (2012). *Cyber Attack Vulnerabilities Analysis for Unmanned Aerial Vehicles. Infotech@Aerospace*. American Institute of Aeronautics and Astronautics.

Kim, S., Lim, G., & Cho, J. (2018). Drone flight scheduling under uncertainty on battery duration and air temperature. *Computers & Industrial Engineering*, *117*, 291–302. DOI: 10.1016/j.cie.2018.02.005

Kirchmeier-Young, M. C., Gillett, N. P., Zwiers, F. W., Cannon, A. J., & Anslow, F. S. (2019). Attribution of the influence of human-induced climate change on an extreme fire season. *Earth's Future*, *7*(1), 2–10. DOI: 10.1029/2018EF001050

Klerkx, L., Jakku, E., & Labarthe, P. (2019). A review of social science on digital agriculture, smart farming and agriculture 4.0: New contributions and a future research agenda. *NJAS Wageningen Journal of Life Sciences*, *90*(1), 100315. DOI: 10.1016/j.njas.2019.100315

Kloke, A. (1979). Content of Arsenic, Cadmium, Chromium, Fluorine, Lead, Mercury, and Nickel in Plants Grown on Contaminated Soils.

Klus, S., Schuster, I., & Muandet, K. (2020). Eigendecompositions of transfer operators in reproducing kernel Hilbert spaces. *Journal of Nonlinear Science*, *30*(1), 283–315.

Koduru, T., & Koduru, N. P. (2022). An Overview of Vulnerabilities in Smart Farming Systems. *Journal of Student Research*.

Konate, F.-O., Dembele, B. S., Dembele, M., & N'diaye, B. F. (2024). Evaluation of Effects of Cotton Growing on the Environment and the Population in the Koutiala Circle. *Advances in Social Sciences and Management*, *2*(10), 24–38. https://doi.org/https://doi.org/10.63002/assm.210.647

Koniavitis, G. (2024). *Forest fire risk assessment and mitigation measures: Case study of Varibombi.* Attica. [Master's thesis, Agricultural University of Athens, Department of Forestry and Natural Environment Management]

Konstantinidis, P. (2003). *Learning to Live with Forest Fires*. Christodoulides.

Konstantinova, E., Minkina, T., Sushkova, S., Konstantinov, A., Rajput, V. D., & Sherstnev, A. (2019). Urban soil geochemistry of an intensively developing Siberian city: A case study of Tyumen, Russia. *Journal of Environmental Management, 239*, 366–375. https://doi.org/https://doi.org/10.1016/j.jenvman.2019.03.095

Kosmas, C., Danalatos, N., & Moustakas, N. (1993). The soils of the broader area of Attica: Properties, classification, and land use potential. *Geoderma, 56*(1-4), 401–419. DOI: 10.1016/0016-7061(93)90131-I

Kostic, V., Novelli, P., Maurer, A., Ciliberto, C., Rosasco, L., & Pontil, M. (2022). Learning dynamical systems via Koopman operator regression in reproducing kernel Hilbert spaces. *Advances in Neural Information Processing Systems, 35*, 4017–4031.

Kothari, S., Bagane, P., Mishra, M., Kulshrestha, S., Asrani, Y., & Maheswari, V. (2024). CropGuard: Empowering Agriculture with AI driven Plant Disease Detection Chatbot. *International Journal of Intelligent Systems and Applications in Engineering, 12*(12s), 530–537.

Kou, G., Peng, Y., & Wang, G. (2014). Evaluation of clustering algorithms for financial risk analysis using MCDM methods. *Information Sciences, 275*, 1–12.

Kowalska, J., Mazurek, R., Gąsiorek, M., Setlak, M., Zaleski, T., & Waroszewski, J. (2016). Soil pollution indices conditioned by medieval metallurgical activity–A case study from Krakow (Poland). *Environmental Pollution, 218*, 1023–1036. https://doi.org/https://doi.org/10.1016/j.envpol.2016.08.053

Krestinskaya, O., & James, A. P. (2018). Feature extraction without learning in an analog spatial pooler memristive-CMOS circuit design of hierarchical temporal memory. *Analog Integrated Circuits and Signal Processing, 95*(3), 457–465.

Kuijpers, N. (2020). *Market orientation in microbreweries* (Master's thesis, Wageningen University & Research). Wageningen University & Research. https://edepot.wur.nl/530023

Kulkarni, A., Wang, Y., Gopinath, M., Sobien, D., Rahman, A., & Batarseh, F. (2024). A Review of Cybersecurity Incidents in the Food and Agriculture Sector. *arXiv:2403.08036*.

Kulshreshtha, P., & Guha, T. (2018, October). An online algorithm for constrained face clustering in videos. In *2018 25th IEEE International Conference on Image Processing (ICIP)* (pp. 2670-2674). IEEE.

Kumar, A., & Gupta, R. (2023). Leveraging deep learning for early and accurate prediction of banana diseases. *International Journal of Computer Engineering Research Trends, 11*(4), 46–57.

Kumar, M., & Mathur, R. (2014, April). Unsupervised outlier detection technique for intrusion detection in cloud computing. In *International Conference for Convergence for Technology*-2014 (pp. 1-4). IEEE.

Kunar, S., & Rani, S. (2023). MobileNetV3 for image classification in agricultural applications. *Agrárinformatika Folyóirat, 14*(2), 45–60.

Kusumaningrum, R., Rahmani, S. A., Widayatno, W. B., Wismogroho, A. S., Nugroho, D. W., Maulana, S., Rochman, N. T., & Amal, M. I. (2018). Characterization of Sumbawa Manganese Ore and Recovery of Manganese Sulfate as Leaching Products. *AIP Conference Proceedings.*

Laboratory of Architectural Technology and Spatial Design. (1986). *Methodological approach to the analysis of challenges, prospects, and planning process for the Greek urban center. Special application: Urban regeneration of Patras* [Research report, in Greek]. University of Patras.

Lambin, E. F., Geist, H. J., & Lepers, E. (2003). Dynamics of land-use and land-cover change in tropical regions. *Annual Review of Environment and Resources, 28*(1), 205 ?"241. DOI: 10.1146/annurev.energy.28.050302.105459

Lee, S. H., & Civera, J. (2020). Robust uncertainty-aware multiview triangulation. *arXiv preprint arXiv:2008.01258.*

Leontidou, L. (1990). *The Mediterranean city in transition: Social change and urban development.* Cambridge University Press. DOI: 10.1017/CBO9780511522208

Lewis, T., Verstraten, L., Hogg, B., Wehr, B. J., Swift, S., Tindale, N., Menzies, N. W., Dalal, R. C., Bryant, P., Francis, B., & Smith, T. E. (2019). Reforestation of agricultural land in the tropics: The relative contribution of soil, living biomass and debris pools to carbon sequestration. *The Science of the Total Environment, 649*, 1502–1513. DOI: 10.1016/j.scitotenv.2018.08.351

Liedloff, A. C., Coughenour, M. B., Ludwig, J. A., & Dyer, R. (2001). Modelling the trade-off between fire and grazing in a tropical savanna landscape, northern Australia. *Environment International, 27*(2-3), 173–180. DOI: 10.1016/s0160-4120(01)00079-4

Li, G., & Li, F. (2019). Urban sprawl in China: Differences and socioeconomic drivers. *The Science of the Total Environment, 673*, 367 ?"377. DOI: 10.1016/j.scitotenv.2019.04.080

Lin, T. E., & Xu, H. (2019). Deep unknown intent detection with margin loss. *arXiv preprint arXiv:1906.00434*.

Linaza, M. T., Posada, J., Bund, J., Eisert, P., Quartulli, M., Döllner, J., . . . Lucat, L. .. (2021). Data-Driven Artificial Intelligence Applications for Sustainable Precision Agriculture. *Agronomy*, no. 1227.

Lin, C.-C., Huang, A. Y. Q., & Yang, S. J. H. (2023). A Review of AI-Driven Conversational Chatbots Implementation Methodologies and Challenges (1999–2022). *Sustainability (Basel)*, *15*(5), 4012. DOI: 10.3390/su15054012

Linderman, G. C., Rachh, M., Hoskins, J. G., Steinerberger, S., & Kluger, Y. (2019). Fast interpolation-based t-SNE for improved visualization of single-cell RNA-seq data. *Nature Methods*, *16*(3), 243–245.

Liodakis, S., Bakirtzis, D., & Dimitrakopoulos, A. P. (2003). Autoignition and thermogravimetric analysis of forest species treated with fire retardants. *Thermochimica Acta*, *399*(1-2), 31–42. DOI: 10.1016/s0040-6031(02)00400-8

Liu, L., Jiang, H., He, P., Chen, W., Liu, X., Gao, J., & Han, J. (2019). On the variance of the adaptive learning rate and beyond. *arXiv preprint arXiv:1908.03265*.

Liu, L., Zheng, X., Wei, X., Kai, Z., & Xu, Y. (2021). Excessive application of chemical fertilizer and organophosphorus pesticides induced total phosphorus loss from planting causing surface water eutrophication. *Scientific Reports*, *11*(1), 23015. https://doi.org/ https://doi.org/10.1038/s41598-021-02521-7

Liu, A., Zhang, G., & Lu, J. (2014, November). Concept drift detection based on anomaly analysis. In *International Conference on Neural Information Processing* (pp. 263-270). Cham: Springer International Publishing.

Liu, W.-h., Zhao, J.-z., Ouyang, Z.-y., Söderlund, L., & Liu, G.-h. (2005). Impacts of sewage irrigation on heavy metal distribution and contamination in Beijing, China. *Environment International*, *31*(6), 805–812. https://doi.org/10.1016/j.envint.2005.05.042

Liu, Y., Shen, W., Fan, K., Pei, W., & Liu, S. (2024). Spatial distribution, source analysis, and health risk assessment of heavy metals in the farmland of Tangwang Village, Huainan City, China. *Agronomy (Basel)*, *14*(2), 394. https://doi.org/https://doi.org/10.3390/agronomy14020394

Li, Y., Yang, M., Peng, D., Li, T., Huang, J., & Peng, X. (2022). Twin contrastive learning for online clustering. *International Journal of Computer Vision*, *130*(9), 2205–2221.

Long, M., Zhu, H., Wang, J., & Jordan, M. I. (2016). Unsupervised domain adaptation with residual transfer networks. *Advances in Neural Information Processing Systems*, •••, 29.

Longo, C. F., Longo, F., & Santoro, C. (2021). Caspar: Towards decision making helpers agents for IoT, based on natural language and first order logic reasoning. *Engineering Applications of Artificial Intelligence*, *104*, 104269. DOI: 10.1016/j.engappai.2021.104269

López-Arredondo, D. L., Leyva-González, M. A., González-Morales, S. I., López-Bucio, J., & Herrera-Estrella, L. (2014). Phosphate nutrition: Improving low-phosphate tolerance in crops. *Annual Review of Plant Biology*, *65*(1), 95–123. https://doi.org/10.1146/annurev-arplant-050213-035949

Loukas, A. (2010). Surface water quantity and quality assessment in Pinios River, Thessaly, Greece. *Desalination*, *250*(1), 266–273. https://doi.org/10.1016/j.desal.2009.09.043

Lu, J., Liu, A., Dong, F., Gu, F., Gama, J., & Zhang, G. (2018). Learning under concept drift: A review. *IEEE Transactions on Knowledge and Data Engineering*, *31*(12), 2346–2363.

Luo, Y., Lu, Y., Fu, B., Zhang, Q., Li, T., Hu, W., & Comber, A. (2019). Half century change of interactions among ecosystem services driven by ecological restoration: Quantification and policy implications at a watershed scale in the Chinese Loess Plateau. *The Science of the Total Environment*, *651*(Pt 2), 2546–2557. DOI: 10.1016/j.scitotenv.2018.10.116

Lu, X., Chen, D., & Wang, Y. (2020). Is urban sprawl decoupled from the quality of economic growth? Evidence from Chinese cities. *Sustainability (Basel)*, *12*(1), 218. DOI: 10.3390/su12010218

MacLean, D. A., & Clark, K. L. (2021). Mixedwood management positively affects forest health during insect infestations in eastern North America1. *Canadian Journal of Forest Research*, *51*(7), 910–920.

Mafi, M. M. H. Matin, Sifat, R. M., Moazzam, M. G., & Uddin, M. S. (2023). *Banana disease recognition dataset*. Mendeley Data, V1. https://doi.org/DOI: 10.17632/79w2n6b4kf.1

Maginga, T. J., Masabo, E., Bakunzibake, P., Kim, K. S., & Nsenga, J. (2024). Using wavelet transform and hybrid CNN – LSTM models on VOC & ultrasound IoT sensor data for non-visual maize disease detection. *Heliyon*, *10*(4), e26647. Advance online publication. DOI: 10.1016/j.heliyon.2024.e26647 PMID: 38420424

Manninen, O. (2018). *Cybersecurity in Agricultural Communication Networks: Case Dairy Farms*. JAMK University of Applied Sciences.

Mansfield, P. A., Wang, Q., Downey, C., Wan, L., & Moreno, I. L. (2018). Links: A high-dimensional online clustering method. *arXiv preprint arXiv:1801.10123.*

Mansourian, S., Vallauri, D., & Dudley, N. (2005). *Forest Restoration in Landscapes*. Springer New York., DOI: 10.1007/0-387-29112-1

Maraveas, C., Rajarajan, M., Arvanitis, K., & Vatsanidou, A. (2024). Cybersecurity threats and mitigation measures in agriculture 4.0 and 5.0. *Smart Agricultural Technology*, *9*, 100616. DOI: 10.1016/j.atech.2024.100616

Marrugo-Negrete, J., Pinedo-Hernández, J., Combatt, E. M., Bravo, A. G., & Díez, S. (2019). Flood-induced metal contamination in the topsoil of floodplain agricultural soils: A case-study in Colombia. *Land Degradation & Development*, *30*(17), 2139–2149. https://doi.org/https://doi.org/10.1002/ldr.3398

Martín, J. A. R., Arias, M. L., & Corbí, J. M. G. (2006). Heavy metals contents in agricultural topsoils in the Ebro basin (Spain). Application of the multivariate geoestatistical methods to study spatial variations. *Environmental Pollution*, *144*(3), 1001–1012. https://doi.org/https://doi.org/10.1016/j.envpol.2006.01.045

Mavridis, A., Vlahopoulou, M., & Gertsis, A. (2012). Innovative Agribusiness In Greece. *World, 2005*, 07.

McCauley, A., Jones, C., & Jacobsen, J. (2009). Soil pH and organic matter. *Nutrient management module, 8*(2), 1-12.

McGarigal, K., & Marks, B. J. (1995). *FRAGSTATS: Spatial pattern analysis program for quantifying landscape structure* (Gen. Tech. Rep. PNW-GTR-351). U.S. Department of Agriculture, Forest Service, Pacific Northwest Research Station. https://doi.org/DOI: 10.2737/PNW-GTR-351

McGarigal, K., & Cushman, S. A. (2002). Comparative evaluation of experimental approaches to the study of habitat fragmentation effects. *Ecological Applications*, *12*(2), 335 ?"345. DOI: 10.1890/1051-0761(2002)012[0335:CEOEAT]2.0.CO;2

Medhi, E. (2022). *PSFD-Musa dataset. Mendeley Data*, V1. https://doi.org/DOI: 10.17632/4wyymrcpyz.1

Melfos, V., & Voudouris, P. C. (2012). Geological, mineralogical and geochemical aspects for critical and rare metals in Greece. *Minerals (Basel)*, *2*(4), 300–317. https://doi.org/https://doi.org/10.3390/min2040300

Mhapsekar, R., Umrani, M., Faizan, M., Ali, O., & Abraham, L. (2024). Building Trust in AI-Driven Decision Making for Cyber-Physical Systems (CPS): A Comprehensive Review. *2024 IEEE 29th International Conference on Emerging Technologies and Factory Automation (ETFA)* (pp. 1-8). Padova, Italy: IEEE.

Mhawej, M., Faour, G., Abdallah, C., & Adjizian-Gerard, J. (2016). Towards an establishment of a wildfire risk system in a Mediterranean country. *Ecological Informatics*, *32*, 167–184.

Michalak, W. Z. (1993). GIS in land use change analysis: Integration of remotely sensed data into GIS. *Applied Geography (Sevenoaks, England)*, *13*(1), 28 ?"44. DOI: 10.1016/0143-6228(93)90078-F

Mikecz, R. (2012). Interviewing elites: Addressing methodological issues. *Qualitative Inquiry*, *18*(6), 482 ?"493. DOI: 10.1177/1077800412442818

Millar, C. I., & Stephenson, N. L. (2015). Temperate forest health in an era of emerging megadisturbance. *Science*, *349*(6250), 823–826. DOI: 10.1126/science.aaa9933

Minari, G. D., Saran, L. M., Constancio, M. T. L., da Silva, R. C., Rosalen, D. L., de Melo, W. J., & Alves, L. M. C. (2020). Bioremediation potential of new cadmium, chromium, and nickel-resistant bacteria isolated from tropical agricultural soil. *Ecotoxicology and Environmental Safety*, *204*, 111038. https://doi.org/https://doi.org/10.1016/j.ecoenv.2020.111038

Mishchuk, A., Mishkin, D., Radenovic, F., & Matas, J. (2017). Working hard to know your neighbor's margins: Local descriptor learning loss. *Advances in Neural Information Processing Systems*, •••, 30.

Mishra, A., Shukla, D., Vaghela, K., & Saraf, M. (2019). Copper: Its biological role and toxicity. *Journal of the Indian Botanical Society*, *98*(1), 26–36.

Mitsopoulos, I., Mallinis, G., Dimitrakopoulos, A., Xanthopoulos, G., Eftychidis, G., & Goldammer, J. G. (2020). Vulnerability of peri-urban and residential areas to landscape fires in Greece: Evidence by wildland-urban interface data. *Data in Brief*, *31*, 106025. DOI: 10.1016/j.dib.2020.106025

Mnatzaganian, J., Fokoué, E., & Kudithipudi, D. (2017). A mathematical formalization of hierarchical temporal memory's spatial pooler. *Frontiers in Robotics and AI*, *3*, 81.

Mohamad Suhaili, S., Salim, N., & Jambli, M. N. (2021). Service chatbots: A systematic review. *Expert Systems with Applications*, *184*, 115461. DOI: 10.1016/j.eswa.2021.115461

Mohamed, A., Jose, A., Luis, G., Zuniga, M., Pedraza, M., Carvajal, G., Jara, N., & Carvajal, R. (2022). LoRa based IoT platform for remote monitoring of large-scale agriculture farms in Chile. *Sensors (Basel)*, •••, 2824. PMID: 35458808

Mohan, S., & Radhakrishnan, R. (2023). Feature fusion techniques for enhanced banana disease detection. *Computers and Electronics in Agriculture*, *201*, 107–118.

Montazer, M., & Harifi, T. (2018). *Nanofinishing of textile materials*. Woodhead Publishing.

Morais, R., Le, V., Tran, T., Saha, B., Mansour, M., & Venkatesh, S. (2019). Learning regularity in skeleton trajectories for anomaly detection in videos. In *Proceedings of the IEEE/CVF conference on computer vision and pattern recognition* (pp. 11996-12004).

Moreira, A., Moraes, L. A., de Melo, T. R., Heinrichs, R., & Moretti, L. G. (2022). Management of copper for crop production. *Advances in Agronomy*, *173*, 257–298. https://doi.org/https://doi.org/10.1016/bs.agron.2022.02.005

Mori, E., Barabaschi, B., Cantoni, F., & Virtuani, R. (2021). Local governments' communication through Facebook. Evidences from COVID-19 pandemic in Italy. *Journal of Public Affairs*, *21*(4), e2551. DOI: 10.1002/pa.2551 PMID: 33349742

Mulinka, P., Casas, P., & Vanerio, J. (2019, November). Continuous and adaptive learning over big streaming data for network security. In *2019 IEEE 8th International Conference on Cloud Networking (CloudNet)* (pp. 1-4). IEEE.

Muller, G. (1979). Schwermetalle in den Sedimenten des Rheins: Veranderungen seit 1971. *Umschau*, *24*, 778–783.

Munoz, L. A., Gómez, M. C., Jiménez, D., Hasiner, E., Schroeder, K., Treguer, D., Juergenliemk, A., Horst, A., Jarvis, A., & Kropff, W. (n.d). *Digital Agriculture Profile: Argentina*. https://openknowledge.fao.org/server/api/core/bitstreams/a7eaf9ef-37d8-4e35-b89a-b02f1491b8da/content

Muzamil, M., Rasool, S., Banday, R. U. Z., Mohiuddin, M., Rasool, S., & Reshi, I. A. (2025). Smartphone-Based Smart Agriculture: Potential Challenges and Prospects. *Handbook of Agricultural Technologies*, 1-18.

Nacke, H., Gonçalves, A.Jr, Schwantes, D., Nava, I., Strey, L., & Coelho, G. (2013). Availability of heavy metals (Cd, Pb, and Cr) in agriculture from commercial fertilizers. *Archives of Environmental Contamination and Toxicology*, *64*(4), 537–544. https://doi.org/10.1007/s00244-012-9867-z

Nadarajan, S., & Sukumaran, S. (2021). Chemistry and toxicology behind chemical fertilizers. In *Controlled Release fertilizers for sustainable agriculture* (pp. 195–229). Elsevier.

Nair, A., & Raghavan, V. (2024). AI-powered banana diseases and pest detection. *Plant Methods*, *20*(1), 47. DOI: 10.1186/s13007-019-0475-z PMID: 38515129

National Institute of Standards and Technology. (2024). *The NIST Cybersecurity Framework (CSF) 2.0*. National Institute of Standards and Technology.

Navarro, E., Costa, N., & Pereira, A. (2020). A Systematic Review of IoT Solutions for Smart Farming. *Sensors (Basel), 20*(15), 4231. DOI: 10.3390/s20154231 PMID: 32751366

Naz, M., Dai, Z., Hussain, S., Tariq, M., Danish, S., Khan, I. U., Qi, S., & Du, D. (2022). The soil pH and heavy metals revealed their impact on soil microbial community. *Journal of Environmental Management, 321*, 115770. https://doi.org/https://doi.org/10.1016/j.jenvman.2022.115770

Neupane, G., & Roberts, S. J. (2009). Quantitative comparison of heavy metals and As accumulation in agricultural and forest soils near Bowling Green, Ohio. *Water, Air, and Soil Pollution, 197*(1), 289–301. https://doi.org/10.1007/s11270-008-9811-z

Nikander, J., Manninen, O., & Laajalahti, M. (2020). Requirements for cybersecurity in agricultural communication networks. *Computers and Electronics in Agriculture, 179*, 105776. DOI: 10.1016/j.compag.2020.105776

Nikinmaa, L., de Koning, J. H. C., Derks, J., Grabska-Szwagrzyk, E., Konczal, A. A., Lindner, M., Socha, J., & Muys, B. (2024). The priorities in managing forest disturbances to enhance forest resilience: A comparison of a literature analysis and perceptions of forest professionals. *Forest Policy and Economics, 158*, 103119–103119. DOI: 10.1016/j.forpol.2023.103119

Nilchuen, P., Suwanasopee, T., & Koonawootrittriron, S. (2025). Integrating deep learning and mobile imaging for assessment of automated conformational indices and weight prediction in Brahman cattle. *Smart Agricultural Technology, 12*, 101079. Advance online publication. DOI: 10.1016/j.atech.2025.101079

Nirala, K. K., Singh, N. K., & Purani, V. S. (2022). A survey on providing customer and public administration-based services using AI: Chatbot. *Multimedia Tools and Applications, 81*(29), 22215–22246. DOI: 10.1007/s11042-021-11458-y PMID: 35002470

Nolet, C. J., Gala, D., Raff, E., Eaton, J., Rees, B., Zedlewski, J., & Oates, T. (2022). GPU semiring primitives for sparse neighborhood methods. *Proceedings of Machine Learning and Systems, 4*, 95–109.

North, M., Collins, B. M., & Stephens, S. L. (2012). Using fire to increase the scale, benefits, and future maintenance of fuels treatments. *Journal of Forestry, 110*(7), 392–401. DOI: 10.5849/jof.12-021

Nussbaum, R., & Simula, M. (2013). *The Forest Certification Handbook*. Routledge., DOI: 10.4324/9781849773003

Oliver, T. H., Heard, M. S., Isaac, N. J. B., Roy, D. B., Procter, D., Eigenbrod, F., Freckleton, R., Hector, A., Orme, C. D. L., Petchey, O. L., Proenca, V., Raffaelli, D., Suttle, K. B., Mace, G. M., Martin-Lopez, B., Woodcock, B. A., & Bullock, J. M. (2015). Biodiversity and Resilience of Ecosystem Functions. *Trends in Ecology & Evolution*, *30*(11), 673–684. DOI: 10.1016/j.tree.2015.08.009

Orsi, F., Geneletti, D., & Newton, A. C. (2011). Towards a common set of criteria and indicators to identify forest restoration priorities: An expert panel-based approach. *Ecological Indicators*, *11*(2), 337–347. DOI: 10.1016/j.ecolind.2010.06.001

Oshunsanya, S. O. (2018). Introductory chapter: Relevance of soil pH to agriculture. *Soil PH for nutrient availability and crop performance*, 3-6. https://doi.org/http://dx.doi.org/10.5772/intechopen.82551

Ouaddi, C., Benaddi, L., Bouziane, E. M., Jakimi, A., Chehri, A., & Saadane, R. (2024). A Sketch of DSL and Code Generator for Accelerating Chatbot Development. *Procedia Computer Science*, *246*, 3585–3594. DOI: 10.1016/j.procs.2024.09.199

Oumaima, S., Mohamed, L., Abderrahmane, B., & Iyass, M. (2025). EnviScan+: AI-driven plant identification and eco-system management for sustainable agriculture. *SoftwareX*, *30*, 102182. Advance online publication. DOI: 10.1016/j.softx.2025.102182

Ozdogan, B., Gacar, A., & Aktas, H. (2017). Digital agriculture practices in the context of agriculture 4.0. *Journal of Economics Finance and Accounting*, *4*(2), 186–193. DOI: 10.17261/Pressacademia.2017.448

Pachura, P. (2020). The role of space in the business models of microbreweries. *Applied Geography (Sevenoaks, England)*, *125*, 102303. https://www.sciencedirect.com/science/article/pii/S014362282030326X?via%3Dihub. DOI: 10.1016/j.apgeog.2020.102303

Paganini, P. (2021). *Distributor of Asian Food JFC International Hit by Ransomware*. Retrieved from security affairs: https://securityaffairs.com/115150/malware/jfc-international-ransomware-attack.html

Pahuja, S., Kukreja, S., & Singh, A. (2025). Comprehensive Review of Generative artificial Intelligence: Mechanisms, Models, and Applications. *Procedia Computer Science*, *258*, 3731–3740. DOI: 10.1016/j.procs.2025.04.628

Paneru, B., Thapa, B., & Paneru, B. (2024). Leveraging AI in ayurvedic agriculture: A RAG chatbot for comprehensive medicinal plant insights using hybrid deep learning approaches. *Telematics and Informatics Reports*, *16*, 100181. Advance online publication. DOI: 10.1016/j.teler.2024.100181

Papadatou-Giannopoulou, Ch. (1997). [Investigation of transfer zone designation for building coefficient allocation in the Urban Complex of Patras]. Municipality of Patras, Directorate of Urban Planning and Implementation. [in Greek].

Papagiannaki, K., Giannaros, T. M., Lykoudis, S., Kotroni, V., & Lagouvardos, K. (2020). Weather-related thresholds for wildfire danger in a Mediterranean region: The case of Greece. *Agricultural and Forest Meteorology*, *291*, 108076. DOI: 10.1016/j.agrformet.2020.108076

Papaioannou, A., Plageras, P., Dovriki, E., Minas, A., Krikelis, V., Nastos, P. T., Kakavas, K., & Paliatsos, A. (2007). Groundwater quality and location of productive activities in the region of Thessaly (Greece). *Desalination*, *213*(1-3), 209–217. https://doi.org/doi:10.1016/j.desal.2006.03.610

Papaioannou, G., Kaloudis, S., & Kerkides, P. (1998). On the proper employment of Piche evaporimeters in estimating evapotranspiration. *International Journal of Climatology*, *18*(11), 1247–1260. DOI: 10.1002/(SICI)1097-0088(199809)18:11<1247:AID-JOC310>3.0.CO;2-U

Papazotos, P., Chalkiadaki, O., Chatzistamatiou, E., Georgopoulos, G., Gkiouleka, I., Katsikis, I., Zygouri, E., Kelepertzis, E., & Argyraki, A. (2016). Heavy metals in urban park soils from Athens, Greece. *Deltio tis Ellinikis Geologikis Etaireias*, *50*(4), 2251–2260. https://doi.org/https://doi.org/10.12681/bgsg.14281

Pappas, V. (2006). *Small metropolitan areas in rapid transition: The case of Patras*. University of Patras. [in Greek]

Parker, D. B. (2002). Toward a New Framework for Information Security. In Bosworth, S., & Kabay, M. E. (Eds.), *The Computer Security Handbook* (4th ed.). John Wiley & Sons.

Patel, J., & Mehta, P. (2023). Comparative analysis of deep learning models for banana disease detection. *Journal of Machine Learning in Agriculture*, *5*(1), 1–15.

Patil, R., Sinkar, Y., Ruke, A., Kulkarni, H., & Kadam, O. (2024). Smart Agri-Advisor: Integrating Chatbot Technology with CNN-Based Crop Disease Classification for Enhanced Agricultural Decision-Making. *International Journal of Engineering Trends and Technology*, *72*(7), 375–380. DOI: 10.14445/22315381/IJETT-V72I7P141

Pausas, J. G., & Keeley, J. E. (2014). Abrupt climate-independent fire regime changes. *Ecosystems (New York, N.Y.)*, *17*(6), 1109–1120. DOI: 10.1007/s10021-014-9773-5

Pausas, J. G., & Keeley, J. E. (2019). Wildfires as an ecosystem service. *Frontiers in Ecology and the Environment*, *17*(5), 289–295. DOI: 10.1002/fee.2044

Peng, Y., Lu, T., Liu, J., Gao, Y., Guo, X., & Xie, F. (2013). Cyber-physical System Risk Assessment. *2013 Ninth International Conference on Intelligent Information Hiding and Multimedia Signal Processing* (pp. 442-447). Beijing, China: IEEE.

Piccialli, V., Sudoso, A. M., & Wiegele, A. (2022). SOS-SDP: An exact solver for minimum sum-of-squares clustering. *INFORMS Journal on Computing, 34*(4), 2144–2162.

Pidhorskyi, S., Almohsen, R., & Doretto, G. (2018). Generative probabilistic novelty detection with adversarial autoencoders. *Advances in Neural Information Processing Systems*, •••, 31.

Pielke, R. A.Sr, Pitman, A., Niyogi, D., Mahmood, R., McAlpine, C., Hossain, F., Goldewijk, K. K., Nair, U., Betts, R., Fall, S., Reichstein, M., Kabat, P., & de Noblet, N. (2011). Land use/land cover changes and climate: Modeling analysis and observational evidence. *Wiley Interdisciplinary Reviews: Climate Change, 2*(6), 828 ?"850. DOI: 10.1002/wcc.144

Pietroń, M., Wielgosz, M., & Wiatr, K. (2016, September). Formal analysis of HTM spatial pooler performance under predefined operation conditions. In *International joint conference on rough sets* (pp. 396-405). Cham: Springer International Publishing.

Pimenidis, E., & Georgiadis, C. K. (2010). Web services for rural areas—Security challenges in development and use. *Computers and Electronics in Agriculture, 70*(2), 348–354. DOI: 10.1016/j.compag.2009.07.016

Pimm, S. L. (1984). The complexity and stability of ecosystems. *Nature, 307*(5949), 321–326. DOI: 10.1038/307321a0

Plucinski, M. P. (2012). *A review of wildfire occurrence research*. Bushfire Cooperative Research Centre.

Pogorelskaia, I., & Várallyai, L. (2020). Agriculture 4.0 and the role of education. *Journal of Agricultural Informatics/Agrárinformatika Folyóirat, 11*(1).

Polat, G., Ergenc, I., Kani, H. T., Alahdab, Y. O., Atug, O., & Temizel, A. (2022, July). Class distance weighted cross-entropy loss for ulcerative colitis severity estimation. In *Annual Conference on Medical Image Understanding and Analysis* (pp. 157-171). Cham: Springer International Publishing.

Politi, N., Vlachogiannis, D., Sfetsos, A., Gounaris, N., & Varela, V. (2023). Investigation of Fire Weather Danger under a Changing Climate at High Resolution in Greece. *Sustainability, 15*(3), 2498. https://www.mdpi.com/2071-1050/15/3/2498

Pomonis, P., Tsikouras, B., & Hatzipanagiotou, K. (2005). Geological evolution of the Koziakas ophiolitic complex (western Thessaly, Greece). *Ofioliti, 30*(2), 77–86.

Ponting, J., Kelly, T. J., Verhoef, A., Watts, M. J., & Sizmur, T. (2021). The impact of increased flooding occurrence on the mobility of potentially toxic elements in floodplain soil–A review. *The Science of the Total Environment*, *754*, 142040. https://doi.org/https://doi.org/10.1016/j.scitotenv.2020.142040

Power, A. (2001). Social exclusion and urban sprawl: Is the rescue of cities possible? *Regional Studies*, *35*(8), 731 ?"742. DOI: 10.1080/00343400120084713

Prokhorenkova, L., Gusev, G., Vorobev, A., Dorogush, A. V., & Gulin, A. (2018). CatBoost: Unbiased boosting with categorical features. *Advances in Neural Information Processing Systems*, ●●●, 31.

Protopapadakis, E., Voulodimos, A., Doulamis, A., Doulamis, N., Dres, D., & Bimpas, M. (2017). Stacked autoencoders for outlier detection in over-the-horizon radar signals. *Computational Intelligence and Neuroscience*, *2017*(1), 5891417.

PTang, P., Wang, W., Lou, J., & Xiong, L. (2021). Generating adversarial examples with distance constrained adversarial imitation networks. *IEEE Transactions on Dependable and Secure Computing*, *19*(6), 4145–4155.

Pulkrabová, J., Suchanová, M., Tomaniová, M., Kocourek, V., & Hajšlová, J. (2008). Organic pollutants in areas impacted by flooding in 2002: A 4-year survey. *Bulletin of Environmental Contamination and Toxicology*, *81*(3), 299–304. https://doi.org/10.1007/s00128-008-9486-6

Quézel, P., & Médail, F. (2003). *Ecology and Biogeography of Mediterranean Basin Forests*. Springer., DOI: 10.1007/978-3-662-05238-9

Raanes, P. N., Bocquet, M., & Carrassi, A. (2019). Adaptive covariance inflation in the ensemble Kalman filter by Gaussian scale mixtures. *Quarterly Journal of the Royal Meteorological Society*, *145*(718), 53–75.

Rajasekaran, T., & Anandamurugan, S. (2019). Challenges and Applications of Wireless Sensor Networks in Smart Farming—A Survey. *Advances in Big Data and Cloud Computing. Proceedings of ICBDCC18.* (pp. 353–361). Singapore: Springer.

Ramesh, S., & Hebbar, R. (2022). Data augmentation techniques for improved banana disease detection. *International Journal of Remote Sensing*, *43*(12), 4501–4515.

Rangasami, S., Murugaragavan, R., Rakesh, S., Prabu, P., Ramesh, A. R., & Elangovan, R. (2022). Chemistry of flooded soils: A review. *The Pharma Innovation Journal*, *11*(2), 2534–2539.

Rao, K. R., & Prasad, S. (2022). Machine learning for automated classification of banana diseases. *Journal of Imaging*, *8*(6), 158. PMID: 35735957

Rassios, A. E., Ghikas, C., Dilek, Y., & Kostopoulos, D. (2019). "Rootless" Ophiolites above the Exhuming Pelagonian Core Complex, Northern Greece. *Deltio tis Ellinikis Geologikis Etaireias*, *54*(1), 60–94.

Reddy, P. D., Sampath Reddy, K. S., Jayanth, P., Kakarla, B. P., & Balakrishnan, R. M. (2025). Agri Assist: An AI Integrated Farmer Assistant. *Procedia Computer Science*, *258*, 3510–3522. DOI: 10.1016/j.procs.2025.04.607

Rehman, M. Z. U., Raghuvanshi, D., & Kumar, N. (2023). KisanQRS: A deep learning-based automated query-response system for agricultural decision-making. *Computers and Electronics in Agriculture*, *213*, 108180. DOI: 10.1016/j.compag.2023.108180

Ren, J., Fort, S., Liu, J., Roy, A. G., Padhy, S., & Lakshminarayanan, B. (2021). A simple fix to mahalanobis distance for improving near-ood detection. *arXiv preprint arXiv:2106.09022*.

Ribeiro, I., Domingos, T., McCracken, D., & Proença, V. (2023). The use of domestic herbivores for ecosystem management in Mediterranean landscapes. *Global Ecology and Conservation*, *46*, e02577–e02577. DOI: 10.1016/j.gecco.2023.e02577

Richards, D., Worden, D., Song, X. P., & Lavorel, S. (2024). Harnessing generative artificial intelligence to support nature-based solutions. *People and Nature*, *6*(2), 882–893. DOI: 10.1002/pan3.10622

Rivas-Martínez, S., Sáenz, S. R., & Penas, A. (2011). Worldwide Bioclimatic Classification System. *Global Geobotany*, *1*, 1–634. DOI: 10.5616/gg110001

Romero-Calcerrada, R., Barrio-Parra, F., Millington, J. D. A., & Novillo, C. J. (2010). Spatial modelling of socioeconomic data to understand patterns of human-caused wildfire ignition risk in the SW of Madrid (central Spain). *Ecological Modelling*, *221*(1), 34–45. DOI: 10.1016/j.ecolmodel.2009.08.008

Romero-Calcerrada, R., Novillo, C. J., Millington, J. D. A., & Gomez-Jimenez, I. (2008). GIS analysis of spatial patterns of human-caused wildfire ignition risk in the SW of Madrid (Central Spain). *Landscape Ecology*, *23*(3), 341–354. DOI: 10.1007/s10980-008-9190-2

Rothermel, R. C. (1972). *A mathematical model for predicting fire spread in wildland fuels*. F. S. Intermountain Forest & Range Experiment Station, US Department of Agriculture.

Rozos, D. E., Chatzinakos, I., & Nikolaou, N. S. (1989). Recording and classification of landslide phenomena in the mountainous network of Karditsa Prefecture. *Proceedings of the 4th Conference. Bulletin of the Greek Geological Society*, *XXIII*(3), 95–109. https://www.geosociety.gr/images/arxeio-teuxwn/GSG_XLIII_1.pdf

Rudnick, R., & Gao, S. (2003). Composition of the Continental Crust. . *Treatise on geochemistry, 3*, 1-64. https://doi.org/10.1016/B0-08-043751-6/03016-4

Ruffault, J., Curt, T., Martin-StPaul, N. K., Moron, V., & Trigo, R. M. (2018). Extreme wildfire events are linked to global-change–type droughts in the northern Mediterranean. *Nature Ecology & Evolution*, *2*(2), 279–285. DOI: 10.1038/s41559-017-0421-7

Sabau, A. S. (2016). Stream Clustering using Probabilistic Data Structures. *arXiv preprint arXiv:1612.02701.*

Saçtı, H., & Dellal, İ. (2023). Measuring the level of digitalization in agriculture: Identification indicators and scaling to determine the digitalization level of farmers. *Turkish Journal of Agriculture - Food Science and Technology*, *11*(6), 1024–1031. DOI: 10.24925/turjaf.v11i6.1024-1031.6049

Safitra, M. F., Lubis, M., Kusumasari, T. F., & Putri, D. P. (2024). Advancements in Artificial Intelligence and Data Science: Models, Applications, and Challenges. *Procedia Computer Science*, *234*, 381–388. DOI: 10.1016/j.procs.2024.03.018

Sajithra, K., & Patil, R. (2013). Social media–history and components. *Journal of Business and Management*, *7*(1), 69–74.

Salminen, R., De Vos, W., & Tarvainen, T. (2006). *Geochemical atlas of Europe*. Geological survey of Finland Espoo, Finland.

Salvati, L., & Carlucci, M. (2014). Urban growth and land-use structure in two Mediterranean regions: An exploratory spatial data analysis. *SAGE Open*, *4*(4), 2158244014561199. DOI: 10.1177/2158244014561199

Samara, T., Raptis, D., & Spanos, I. (2018). Fuel treatments and potential fire behavior in peri-urban forests in Northern Greece. *Environments*, *5*(7), 79. DOI: 10.3390/environments5070079

Sanchez Sanchez, Y., Martinez-Grana, A., Santos Frances, F., & Mateos Picado, M. (2018). Mapping Wildfire Ignition Probability Using Sentinel 2 and LiDAR (Jerte Valley, Caceres, Spain). *Sensors (Basel)*, *18*(3). Advance online publication. DOI: 10.3390/s18030826

Sanh, V., Wolf, T., & Rush, A. (2020). Movement pruning: Adaptive sparsity by fine-tuning. *Advances in Neural Information Processing Systems*, *33*, 20378–20389.

Sappa, G., Barbieri, M., & Andrei, F. (2020). Assessment of trace elements natural enrichment in topsoil by some Italian case studies. *SN Applied Sciences*, *2*(8), 1409. https://doi.org/10.1007/s42452-020-03214-y

Saqlain, A. S., Wang, L. Y., & Liu, Z. (2022, April). SL-CycleGAN: blind motion deblurring in cycles using sparse learning. In *2022 7th International Conference on Cloud Computing and Big Data Analytics (ICCCBDA)* (pp. 01-05). IEEE.

Sari, F. (2022). Identifying anthropogenic and natural causes of wildfires by maximum entropy method-based ignition susceptibility distribution models. *Journal of Forestry Research, 34*(2), 355–371. DOI: 10.1007/s11676-022-01502-4

Sarkar, S., Ganapathysubramanian, G., Singh, A., Fotouhi, F., Kar, S., Nagasubramanian, K., Chowdhary, G., Das, S. K., Kantor, G., Krishnamurthy, A., Merchant, N., & Singh, A. (2024). Cyber-agricultural systems for crop breeding and sustainable production. *Trends in Plant Science, 29*(2), 130–149. DOI: 10.1016/j.tplants.2023.08.001 PMID: 37648631

Scherer, G., Zabowski, D., Java, B., & Everett, R. (2000). Timber harvesting residue treatment. Part II. Understory vegetation response. *Forest Ecology and Management, 126*(1), 35–50. DOI: 10.1016/s0378-1127(99)00080-8

Schneeberger, N., BA1/4rgi, M., Hersperger, A. M., & Ewald, K. C. (2007). Driving forces and rates of landscape change as a promising combination for landscape change research ?"An application on the northern fringe of the Swiss Alps. *Land Use Policy, 24*(2), 349 ?"361. DOI: 10.1016/j.landusepol.2006.04.003

Schneider, M., Ertel, W., & Ramos, F. (2016). Expected similarity estimation for large-scale batch and streaming anomaly detection. *Machine Learning, 105*(3), 305–333.

Seidl, R., Thom, D., Kautz, M., Martin-Benito, D., Peltoniemi, M., Vacchiano, G., Wild, J., Ascoli, D., Petr, M., Honkaniemi, J., Lexer, M. J., Trotsiuk, V., Mairota, P., Svoboda, M., Fabrika, M., Nagel, T. A., & Reyer, C. P. O. (2017). Forest disturbances under climate change. *Nature Climate Change, 7*(6), 395–402. DOI: 10.1038/nclimate3303

Selvaraj, M. G., Vergara, A., Montenegro, F., Ruiz, H. A., Safari, N., Raymaekers, D., Ocimati, W., Ntamwira, J., Tits, L., Bonaventure, O. A., & Blomme, G. (2020). Detection of banana plants and their major diseases through aerial images and machine learning methods: A case study in DR Congo and Republic of Benin. *ISPRS Journal of Photogrammetry and Remote Sensing, 164*, 138–145. DOI: 10.1016/j.isprsjprs.2020.08.025

Sentas, A., Psilovikos, A., Karamoutsou, L., & Charizopoulos, N. (2018). Monitoring, modeling, and assessment of water quality and quantity in River Pinios, using ARIMA models. *Desalination and Water Treatment, 133*, 336-347. https://doi.org/doi: DOI: 10.5004/dwt.2018.23239

Seto, K. C., GA1/4neralp, B., & Hutyra, L. R. (2012). Global forecasts of urban expansion to 2030 and direct impacts on biodiversity and carbon pools. *Proceedings of the National Academy of Sciences of the United States of America, 109*(40), 16083 ?"16088. DOI: 10.1073/pnas.1211658109

Shah, I., Jhawar, S., Khater, A., Jacob, A., & Potdar, G. (2023). Chatbot development through the ages: A survey. *International Journal of Scientific Research in Computer Science. Engineering and Information Technology, 9*(3), 262–271. DOI: 10.32628/CSEIT2390329

Shapa, M., Trojer, L., & Machuve, D. (2021). Mobile-based Decision Support System for Poultry Farmers: A Case of Tanzania. *International Journal of Advanced Computer Science and Applications, 12*(2), 584–590. DOI: 10.14569/IJACSA.2021.0120273

Sharma, N., & Joshi, R. (2023). Machine learning applications in banana disease management. *Journal of Crop Protection, 12*(4), 225–240.

Shepherd, M., Turner, J. A., Small, B., & Wheeler, D. (2020). Priorities for science to overcome hurdles thwarting the full promise of the 'digital agriculture' revolution. *Journal of the Science of Food and Agriculture, 100*(14), 5083–5092. DOI: 10.1002/jsfa.9346 PMID: 30191570

Shin, H. K., Lee, W., Yun, J. H., & Kim, H. (2020). {HAI} 1.0:{HIL-based} augmented {ICS} security dataset. In *13Th USENIX workshop on cyber security experimentation and test (CSET 20).*

Shin, H. K., Lee, W., Yun, J. H., & Min, B. G. (2021, August). Two ICS security datasets and anomaly detection contest on the HIL-based augmented ICS testbed. In *Proceedings of the 14th Cyber Security Experimentation and Test Workshop* (pp. 36-40).

Shirtcliffe, N., Comanns, P., Hamlett, C., Roach, P., & Atherton, S. (2019). The effect of roughness geometry on superhydrophobicity and related phenomena.

Shostack, A. (2014). *Threat Modeling: Designing for Security* (1st ed.). Wiley Publishing.

Shurson, G. C., Urriola, P. E., & Hung, Y.-T. (2022). Too much of a good thing: Rethinking feed formulation and feeding practices for zinc in swine diets to achieve one health and environmental sustainability. *Animals (Basel), 12*(23), 3374. https://doi.org/https://doi.org/10.3390/ani12233374

Simonyan, K., & Zisserman, A. (2014). Very deep convolutional networks for large-scale image recognition. *arXiv preprint arXiv:1409.1556.* https://arxiv.org/abs/1409.1556

Singh, A. K., Mittal, S., Malhotra, P., & Srivastava, Y. V. (2020, March). Clustering evaluation by davies-bouldin index (dbi) in cereal data using k-means. In *2020 Fourth international conference on computing methodologies and communication (ICCMC)* (pp. 306-310). IEEE.

Singh, A., & Gupta, N. (2023). Enhancing banana disease detection using transfer learning. *Journal of Agricultural and Food Chemistry, 71*(15), 5001–5010.

Singh, D., & Sharma, R. (2023). Deep learning approaches for banana disease classification: A review. *Journal of Information Technology & Software Engineering, 13*(1), 12–25.

Singh, D., Shukla, B., & Joshi, M. (2024). Fostering conversational artificial intelligence (CAI) for sustainable development in services: A case study of Haptik. *World Review of Entrepreneurship, Management and Sustainable Development, 20*(4), 397–420. DOI: 10.1504/WREMSD.2024.139371

Sivakumar, S. A., Shankar, B. M., Anuradha, B., Karan, K. A., Karthik, A., Karthik, R., & Kumar, J. R. R. (2023). Artificial Intelligence based Agricultural Chatbot and Virtual Assistant for Delivery of Harvested Crops. *International Journal of Intelligent Systems and Applications in Engineering, 11*(8), 576–583.

SkA nes, H. M., & Bunce, R. G. H. (1997). Directions of landscape change (1741 ?"1993) in Virestad, Sweden ?"Characterised by multivariate analysis. *Landscape and Urban Planning, 38*(1 ?"2), 61 ?"75. DOI: 10.1016/S0169-2046(97)00019-4

Skordas, K. (2006). Environmental geochemical research on the bottom sediments of the old and the new bed of Pinios River (Thessaly, Greece): Assessment of the potential sources of pollution. *Hellenic Society of Mining and Metallurgical Engineers MMX, 2*, 69–80.

Snode Technologies. (2023). *Cyber Security in Smart Agriculture Technology*. Snode.

Snytsar, R., & Turakhia, Y. (2019, December). Parallel approach to sliding window sums. In *International Conference on Algorithms and Architectures for Parallel Processing* (pp. 19-26). Cham: Springer International Publishing.

Sohn, J., Choi, S., Lewis, R., & Knaap, G. (2012). Characterising urban sprawl on a local scale with accessibility measures. *The Geographical Journal, 178*(3), 230 ?"241. DOI: 10.1111/j.1475-4959.2012.00468.x

Soil Survey Staff. (2024). Field book for describing and sampling soils. *Version 4.0*. USDA, Natural Resources Conservation Service. Government Printing Office.

Soma, T., & Nuckchady, B. (2021). Communicating the benefits and risks of digital agriculture technologies: Perspectives on the future of digital agricultural education and training. *Frontiers in Communication, 6*, 762201. DOI: 10.3389/fcomm.2021.762201

Soto, M. C., & Valencia, A. P. (2021). Proposals of Rehabilitation and Management Actions for the Protection of Sclerophyllous Forests Affected by Forest Fires. *Forestist, 72*(2), 112–119. DOI: 10.5152/forestist.2021.21029

Soule, D. C., & Peirce, N. R. (2006). *Urban sprawl: A comprehensive reference guide*. Greenwood Press.

Sowmya, B. J., Meeradevi, A. K., Supreeth, S., Pradeep Kumar, D., Ravi Kumar, B. N., Rohith, S., Mishra, D., Koushik, A., & Patil, A. U. (2025). Leveraging machine learning for intelligent agriculture. *Discover Internet of Things, 5*(1), 33. Advance online publication. DOI: 10.1007/s43926-025-00132-6

Statista (2024). *Most popular social networks worldwide as of April 2024, by number of monthly active users*. Statista. https://www.statista.com/statistics/272014/global-social-networks-ranked-by-number-of-users/

Stefanidakis, P. (2015). *Macro vs. micro breweries*. Beeroskopio. https://www.beeroskopio.com/2015/01/macro-micro.html

Steinley, D. (2004). Properties of the hubert-arable adjusted rand index. *Psychological Methods, 9*(3), 386.

Stepaniants, G. (2023). Learning partial differential equations in reproducing kernel Hilbert spaces. *Journal of Machine Learning Research, 24*(86), 1–72.

Stephens, S. L., Agee, J. K., Fulé, P. Z., North, M. P., Romme, W. H., Swetnam, T. W., & Turner, M. G. (2013). Managing forests and fire in changing climates. *Science, 342*(6154), 41–42. DOI: 10.1126/science.1240294

Stevens-Rumann, C. S., Kemp, K. B., Higuera, P. E., Harvey, B. J., Rother, M. T., Donato, D. C., Morgan, P., & Veblen, T. T. (2018). Evidence for declining forest resilience to wildfires under climate change. *Ecology Letters, 21*(2), 243–252. DOI: 10.1111/ele.12889

Stouffer, K., Pease, M., Tang, C., Zimmerman, T., Pillitteri, V., Lightman, S., & Thompson, M. (2023). *Guide to industrial control systems (ICS) security. NIST special publication 800.82r3*. National Institute of Standards and Technology.

Stougiannidou, D., & Zafeiriou, E. (2022). Wildfire economic impact assessment: An empirical model-based investigation for Greek agriculture. *Modeling Earth Systems and Environment, 8*(3), 3357–3371.

Suchara, I., Sucharová, J., & Holá, M. (2021). Changes in selected physico-chemical properties of floodplain soils in three different land-use types after flooding. *Plant, Soil and Environment*, *67*(2), 99–109. https://doi.org/https://doi.org/10.17221/435/2020-PSE

Suditu, B. (2012). Urban sprawl ?"the legal context and territorial practices in Romania. *Human Geographies ?" Journal of Studies and Research in Human Geography*, *6*(1), 73 ?"77.

Sulaiman, M. A. B. A., Salman, R., Shabbir, M. S., Siddiqi, A. F., Abbas, M., Abbasi, S. G., Ur Rehman, K., & Ashraf, M. (2022). Research on rumors surrounding food safety based on information source differences (a review). *Food Science and Technology (Brazil)*, *42*. https://doi.org/DOI: 10.1590/FST.79921

Syphard, A. D., Radeloff, V. C., Keuler, N. S., Taylor, R. S., Hawbaker, T. J., Stewart, S. I., & Clayton, M. K. (2008). Predicting spatial patterns of fire on a southern California landscape. *International Journal of Wildland Fire*, *17*(5), 602–613. DOI: 10.1071/wf07087

Tangwongsan, K., Hirzel, M., & Schneider, S. (2021). In-order sliding-window aggregation in worst-case constant time. *The VLDB Journal*, *30*(6), 933–957.

Tan, M., & Le, Q. V. (2019). EfficientNet: Rethinking model scaling for convolutional neural networks. *Proceedings of the 36th International Conference on Machine Learning*, 97, 6105-6114.

Temniranrat, P., Kiratiratanapruk, K., Kitvimonrat, A., Sinthupinyo, W., & Patarapuwadol, S. (2021). A system for automatic rice disease detection from rice paddy images serviced via a Chatbot. *Computers and Electronics in Agriculture*, *185*, 106156. Advance online publication. DOI: 10.1016/j.compag.2021.106156

Teou, X. (2014). *Pollution of the Pinios River by heavy metals* [Master's Thesis, National and Kapodistrian University of Athens]. Athens.

Thiagarajan, V., Kulkarni, S. V., Jadhav, S. A., Waghe, A. A., Raja, S. P., Rajagopal, S., Poddar, H., & Subramaniam, S. (2024). Analysis of banana plant health using machine learning techniques. *Scientific Reports*, *14*(1), 63930. DOI: 10.1038/s41598-024-63930-y PMID: 38951552

Thomas, G., Sparks, D. L., Page, A. L., Helmke, P. A., Loeppert, R. H., Soltanpour, P. N., Tabatabai, M. A., Johnston, C. T., & Sumner, M. E. (1996). Soil pH and Soil Acidity. In *Methods of Soil Analysis. Part 3. Chemical Methods* (Vol. SSSA Book Series no. 5, pp. 1390). Soil Science Society of America and American Society of Agronomy.

Thompson, I., Mackey, B., & McNulty, S. G. (2009). *Forest Resilience, Biodiversity, and Climate Change. A Synthesis of the Biodiversity/Resilience/Stability Relationship in Forest Ecosystems*. https://www.researchgate.net/publication/232129022

Torrano, L., & Valderrábano, J. (2005). Grazing ability of European black pine understorey vegetation by goats. *Small Ruminant Research, 58*(3), 253–263. DOI: 10.1016/j.smallrumres.2004.11.001

Trigo, R. M., Pereira, J. M., Pereira, M. G., Mota, B., Calado, T. J., Dacamara, C. C., & Santo, F. E. (2006). Atmospheric conditions associated with the exceptional fire season of 2003 in Portugal. *International Journal of Climatology, 26*(13), 1741–1758.

Troumbis, A. Y., Gaganis, C. M., & Sideropoulos, H. (2023). Probabilistic Wildfire Risk Assessment and Modernization Transitions: The Case of Greece. *Fire (Basel, Switzerland), 6*(4), 158. https://www.mdpi.com/2571-6255/6/4/158

Tsangari, K., Karetsos, G., & Proutsos, N. (2011). *Forest fires in Greece 1983-2008* (978-960-88241-3-3). https://www.researchgate.net/publication/351330520_Forest_fires_in_Greece_1983-2008_in_greek

Tsiforou, E. (2020). *Digital innovation for sustainable agriculture: The Greek approach.*

Tsinopoulos, T. (2007). *Ecological study and mapping of wetland vegetation in the Karditsa Prefecture wetlands*. National and Kapodistrian University of Athens.

Tsiouni, M., Konstantinidis, C., Aggelopoulos, S., & Mamalis, S. (2024). A cluster analysis study based on financial indicators, as a competitiveness strategy of the brewery industry in Greece. In *Value chain dynamics in a biodiverse environment: Advances in biodiversity, sustainability, and agri-food supply chain development* (pp. 129–137). Springer Nature Switzerland. DOI: 10.1007/978-3-031-49845-9_7

Turco, M., Rosa-Cánovas, J. J., Bedia, J., Jerez, S., Montávez, J. P., Llasat, M. C., & Provenzale, A. (2018). Exacerbated fires in Mediterranean Europe due to anthropogenic warming projected with non-stationary climate-fire models. *Nature Communications, 9*(1), 3821. DOI: 10.1038/s41467-018-06358-z

Turekian, K. K., & Wedepohl, K. H. (1961). Distribution of the elements in some major units of the earth's crust. *Geological Society of America Bulletin, 72*(2), 175–192.

Turner, M. G. (2010). Disturbance and landscape dynamics in a changing world. *Ecology, 91*(10), 2833–2849. DOI: 10.1890/10-0097.1

U.S. Environmental Protection Agency. (1996). *Method 3050B: Acid digestion of sediments, sludges, and soils (Revision2)*. United States Environmental Protection Agency.

United Nations, Department of Economic and Social Affairs, Population Division. (2019). *World urbanization prospects: The 2018 revision* (ST/ESA/SER.A/420). United Nations. https://doi.org/DOI: 10.18356/b9e995fe-en

United States Agency for International Development. (2023). *Cybersecurity AGRICULTURE AND FOOD SECURITY.* Retrieved from USAID: https://www.usaid.gov/sites/default/files/2023-10/Cybersecurity%20Briefer_Agriculture%20and%20Food%20Security.pdf

Valderrábano, J., & Torrano, L. (2000). The potential for using goats to control Genista scorpius shrubs in European black pine stands. *Forest Ecology and Management, 126*(3), 377–383. DOI: 10.1016/s0378-1127(99)00108-5

Van Eetvelde, V., & Antrop, M. (2004). Analyzing structural and functional changes of traditional landscapes ?"Two examples from southern France. *Landscape and Urban Planning, 67*(1 ?"4), 79 ?"95. DOI: 10.1016/S0169-2046(03)00030-6

Van Nguyen, T., Le, N. T., An, J., & Min, K. S. (2020, November). Defect-resilient memristor crossbar of hierarchical temporal memory (HTM) spatial pooling for near-IoT-sensor cognitive computing. In *2020 27th IEEE International Conference on Electronics, Circuits and Systems (ICECS)* (pp. 1-4). IEEE.

Varela, V., Vlachogiannis, D., Sfetsos, A., Politi, N., & Karozis, S. (2020). Methodology for the Study of Near-Future Changes of Fire Weather Patterns with Emphasis on Archaeological and Protected Touristic Areas in Greece. *Forests, 11*(11), 1168. https://www.mdpi.com/1999-4907/11/11/1168

Vargo, D., Zhu, L., Benwell, B., & Yan, Z. (2021). Digital technology use during COVID-19 pandemic: A rapid review. *Human Behavior and Emerging Technologies, 3*(1), 13–24. DOI: 10.1002/hbe2.242

Varol, M. (2011). Assessment of heavy metal contamination in sediments of the Tigris River (Turkey) using pollution indices and multivariate statistical techniques. *Journal of Hazardous Materials, 195*, 355–364. https://doi.org/https://doi.org/10.1016/j.jhazmat.2011.08.051

Vatn, K., Kavallieratos, G., & Katsikas, S. (2024). Lecture Notes in Computer Science: Vol. 14398. *Threat Analysis in Dairy Farming 4.0. Computer Security. ESORICS 2023 International Workshops. ESORICS 2023.* Springer.

Vaucher, S., Pires, R., Felber, P., Pasin, M., Schiavoni, V., & Fetzer, C. (2018, July). SGX-aware container orchestration for heterogeneous clusters. In *2018 IEEE 38th International Conference on Distributed Computing Systems (ICDCS)* (pp. 730-741). IEEE.

Verma, P., & Kumar, S. (2023). Enhancing banana disease detection using hybrid deep learning models. *Computers and Electronics in Agriculture, 204*, 107–118.

Vijayalakshmi, J., & Pandimeena, K. (2019). Agriculture talkbot using AI. *International Journal of Recent Technology and Engineering, 8*(2 Special Issue 5), 186–190. https://doi.org/DOI: 10.35940/ijrte.B1037.0782S519

Vlček, V., & Pohanka, M. (2018). Adsorption of copper in soil and its dependence on physical and chemical properties. *Acta Universitatis Agriculturae et Silviculturae Mendelianae Brunensis, 66*(1), 219–224. https://doi.org/https://doi.org/10.11118/actaun201866010219

Walter, A., Finger, R., Huber, R., & Buchmann, N. (2017). Smart farming is key to developing sustainable agriculture. *Proceedings of the National Academy of Sciences U.S.A.*, 6148-6150. DOI: 10.1073/pnas.1707462114

Wang, S., Kim, H., & Chen, H. (2009, April). Contour tracking using centroid distance signature and dynamic programming method. In *2009 International Joint Conference on Computational Sciences and Optimization* (Vol. 1, pp. 274-278). IEEE.

Wang, Y., & Anderson, K. R. (2010). An evaluation of spatial and temporal patterns of lightning- and human-caused forest fires in Alberta, Canada, 1980 - 2007. *International Journal of Wildland Fire, 19*(8), 1059–1072. DOI: 10.1071/wf09085

Wang, Z., Liu, X., Liang, X., Dai, L., Li, Z., Liu, R., & Zhao, Y. (2022). Flooding-drainage regulate the availability and mobility process of Fe, Mn, Cd, and As at paddy soil. *The Science of the Total Environment, 817*, 152898. https://doi.org/http://dx.doi.org/10.1016/j.scitotenv.2021.152898

Warminsky, J. (2022, May 6). *The Record.* Retrieved from Agricultural equipment maker AGCO reports ransomware attack: https://therecord.media/agco-ransomware-attack-agricultural-sector

Weber, A., Wolf, S., Becker, N., Märker-Neuhaus, L., Bellanova, P., Brüll, C., Hollert, H., Klopries, E.-M., Schüttrumpf, H., & Lehmkuhl, F. (2023). The risk may not be limited to flooding: Polluted flood sediments pose a human health threat to the unaware public. *Environmental Sciences Europe, 35*(1), 1–19. https://doi.org/10.1186/s12302-023-00765-w

Whittaker, Z. (2020). *Home Chef Confirms 8 Million User Records Stolen in Breach.* Retrieved from techcrunch: https://techcrunch.com/2020/05/20/home-chef-data-breach/

Wiangsamut, S., Chomphuwiset, P., & Khummanee, S. (2019). Chatting with Plants (Orchids) in Automated Smart Farming using IoT, Fuzzy Logic and Chatbot. *Advances in Science, Technology and Engineering Systems Journal, 4*(5), 163–173. DOI: 10.25046/aj040522

Wielgosz, M., & Pietroń, M. (2017). Using spatial pooler of hierarchical temporal memory to classify noisy videos with predefined complexity. *Neurocomputing*, *240*, 84–97.

World Bank. (2024a). *Greece—Employment in agriculture (% of total employment)—2024 data, 2025 forecast, 1981–2022 historical.* https://tradingeconomics.com/greece/employment-in-agriculture-percent-of-total-employment-wb-data.html

World Bank. (2024b). *The World Bank in Serbia.* https://www.worldbank.org/en/country/serbia/overview

Wu, Y., & Takács-György, K. (2023, May). What is stopping Agriculture 4.0?---Examples from China. In *2023 IEEE 17th International Symposium on Applied Computational Intelligence and Informatics (SACI)* (pp. 000511-000518). IEEE.

Wu, M., Ma, Y., Fan, F., Mei, X., & Huang, J. (2020). Infrared and visible image fusion via joint convolutional sparse representation. *Journal of the Optical Society of America. A, Optics, Image Science, and Vision*, *37*(7), 1105–1115.

Wu, Y., Zhong, X., & Takács-György, K. (2023). Digital agricultural education in Hungary. *Gradus*, *10*(1). Advance online publication. DOI: 10.47833/2023.1.AGR.007

Xia, F., Zhu, Y., Hu, B., Chen, X., Li, H., Shi, K., & Xu, L. (2021). Pollution characteristics, spatial patterns, and sources of toxic elements in soils from a typical industrial city of Eastern China. *Land*, *10*(11), 1126. https://doi.org/ https://doi.org/10.3390/land10111126

Xu, T., Yang, H., Yu, A., Bao, B., Guo, H., & Jiang, Y. (2020, October). Traffic optimization and congestion prevention strategy based on hierarchical temporal memory of data center optical network. In *2020 Opto-Electronics and Communications Conference (OECC)* (pp. 1-3). IEEE.

Xu, Z., Kakde, D., & Chaudhuri, A. (2019, December). Automatic hyperparameter tuning method for local outlier factor, with applications to anomaly detection. In *2019 IEEE International Conference on Big Data (Big Data)* (pp. 4201-4207). IEEE.

Yang, Y. (2017). Temporal Data Clustering. In *Temporal Data Mining Via Unsupervised Ensemble Learning* (pp. 19–34). Elsevier.

Yaroshevsky, A. A. (2006). Abundances of chemical elements in the Earth's crust. *Geochemistry International*, *44*(1), 48–55.

Yassoglou, N., Kosmas, C., Moustakas, N., & Detsis, V. (1987). *The Soils of Greece.* Soil Survey Division, Ministry of Agriculture.

Yazdinejad, A., Zolfaghari, B., Azmoodeh, A., Dehghantanha, A., Karimipour, H., Fraser, E., Green, A. G., Russell, C., & Duncan, E. (2021). A Review on Security of Smart Farming and Precision Agriculture: Security Aspects, Attacks, Threats and Countermeasures. *Applied Sciences (Basel, Switzerland), 11*(16), 7518. DOI: 10.3390/app11167518

Ye, J., Wu, M., Deng, Z., Xu, S., Zhou, R., & Clarke, K. C. (2017). Modeling the spatial patterns of human wildfire ignition in Yunnan province, China. *Applied Geography (Sevenoaks, England), 89*, 150–162. DOI: 10.1016/j.apgeog.2017.09.012

Yu, H., Liu, T., Lu, J., & Zhang, G. (2021). Automatic learning to detect concept drift. *arXiv preprint arXiv:2105.01419.*

Zabowski, D., Java, B., Scherer, G., Everett, R. L., & Ottmar, R. (2000). Timber Harvesting Residue Treatment: Part I. Responses of Conifer Seedlings, Soils and Microclimate. *Forest Ecology and Management, 126*(1), 25–34. http://www.sciencedirect.com/science/article/B6T6X-3Y9H9FT-3/2/7b40d498ed732b1e6a8ab8d9a09f2149

Zacharakis, I., & Tsihrintzis, V. A. (2023). Environmental Forest Fire Danger Rating Systems and Indices around the Globe: A Review. *Land (Basel), 12*. Advance online publication. DOI: 10.3390/land12010194

Zagas, T., Raptis, D., Zagas, D., & Karamanolis, D. (2013). Planning and assessing the effectiveness of traditional silvicultural treatments for mitigating wildfire hazard in pine woodlands of Greece. *Natural Hazards, 65*(1), 545–561. DOI: 10.1007/s11069-012-0380-5

Zhang, J., & Liu, C. (2002). Riverine composition and estuarine geochemistry of particulate metals in China—Weathering features, anthropogenic impact and chemical fluxes. *Estuarine, Coastal and Shelf Science, 54*(6), 1051–1070. https://doi.org/https://doi.org/10.1006/ecss.2001.0879

Zhang, Y., Chen, Y., Bai, X., Yu, S., Yu, K., Li, Z., & Yang, K. (2020, April). Adaptive unimodal cost volume filtering for deep stereo matching. *Proceedings of the AAAI Conference on Artificial Intelligence, 34*(07), 12926–12934.

Zhang, Z. X., Zhang, H. Y., & Zhou, D. W. (2010). Using GIS spatial analysis and logistic regression to predict the probabilities of human-caused grassland fires. *Journal of Arid Environments, 74*(3), 386–393. DOI: 10.1016/j.jaridenv.2009.09.024

Zhang, Z., Li, J., Tsui, C.-C., & Chen, Z.-S. (2020). The study of gaining more detailed variability information of soil organic carbon in surface soils and its significance to enriching the existing soil database. *Sustainability, 12*(12), 4866.

Zhao, Q., Xu, M., & Fränti, P. (2009, April). Sum-of-squares based cluster validity index and significance analysis. In *International conference on adaptive and natural computing algorithms* (pp. 313-322). Berlin, Heidelberg: Springer Berlin Heidelberg.

Zhao, J., Wei, S., Wen, X., & Qiu, X. (2020). Analysis and prediction of big stream data in real-time water quality monitoring system. *Journal of Ambient Intelligence and Smart Environments*, *12*(5), 393–406.

Zheng, S., Yahya, Z., Wang, L., Zhang, R., & Hoshyar, A. N. (2023). Multiheaded deep learning chatbot for increasing production and marketing. *Information Processing & Management*, *60*(5), 103446. DOI: 10.1016/j.ipm.2023.103446

Zheng, Z., Wang, P., Liu, W., Li, J., Ye, R., & Ren, D. (2020, April). Distance-IoU loss: Faster and better learning for bounding box regression. *Proceedings of the AAAI Conference on Artificial Intelligence*, *34*(07), 12993–13000.

Zhou, S., Zhang, J., Pan, J., Xie, H., Zuo, W., & Ren, J. (2019). Spatio-temporal filter adaptive network for video deblurring. In *Proceedings of the IEEE/CVF international conference on computer vision* (pp. 2482-2491).

Ziegfeld, L., Scala, D. D., & Cremers, A. H. M. (2025). The effect of preference elicitation methods on the user experience in conversational recommender systems. *Computer Speech & Language*, *89*, 101696. Advance online publication. DOI: 10.1016/j.csl.2024.101696

About the Contributors

Sotirios Karetsos is Assistant Professor at the Informatics Laboratory of the Agricultural University of Athens (AUA). He received his BSc and an integrated Master from the AUA, MSc from the National Technical University of Athens, and PhD from Informatics Laboratory of the AUA. His research interests mainly focus on information systems, electronic and mobile services, electronic and mobile government, electronic and mobile commerce, social media analytics. He has published more than 60 scientific publications in refereed journals, conferences and books. He has also participated in national and European research and development projects.

Constantina Costopoulou is the Director of the Informatics Laboratory at the Department of Agricultural Economics and Rural Development, Agricultural University of Athens (Greece). Her research focuses on the design and implementation of digital technologies in the agricultural sector, with particular emphasis on web services, e-commerce, e-government, e-learning, mobile applications, blockchain, data analytics, digital marketing, and digital transformation.

***

Effrosyni Bitakou holds an Integrated Master's degree in Agricultural Economics and Rural Development from the Agricultural University of Athens, completed at the university's Informatics Laboratory. She is currently a PhD candidate at the same university, focusing on "Extended Reality Marketing in the Agrifood Sector". Her research interests include augmented reality, digital marketing, data analytics, blockchain and digital transformation, with an emphasis on artificial intelligence applications in agriculture. She has published in scientific journals and conferences and has participated in funded European and national research projects in the above research areas.

Konstantinos Demertzis was a Postdoctoral Researcher in Cyber Security of Critical Infrastructures at School of Civil Engineering, Democritus University of Thrace - until his death in 2024.

Konstantinos Demestichas received his Diploma (2005) and PhD (2009) in Telecommunications from NTUA, an MBA (2012) in Techno-Economic Systems (NTUA–Univ. of Piraeus), and an MSc (2015) in Quality Assurance (HOU). Since 2021, he has been Assistant Professor at the Agricultural University of Athens, Dept. of Agricultural Economics and Rural Development, and member of the Informatics Lab. He has lectured at NTUA, Univ. of Western Macedonia, Univ. of West Attica, and HOU. Active in EU and national R&D since 2005, he has coordinated projects, served in TPCs and journals, authored 200+ publications, and acted as EC technical expert. His primary research interests include Artificial intelligence and Big Data analytics, as well as secure digital network technologies.

Vasiliki Drakopoulou is a graduate of the Agricultural University of Athens, Department of Agricultural Economics and Rural Development (Class of 2024). During my studies, I completed my internship at the Laboratory of Informatics of the university, where I also carried out the research for my undergraduate thesis. My academic interests focus on statistical analysis and digital marketing, particularly on the application of data-driven methods in assessing employee performance and organizational efficiency. I am passionate about integrating quantitative approaches into decision-making processes within the agri-business and corporate sectors. Beyond my academic and professional pursuits, I enjoy hiking, exploring nature, and discovering new places.

Stavroula Galanopoulou is an Assistant Professor at the Department of Forestry and Natural Environment Management of the Agricultural University of Athens. She holds a B.Sc. in Geology from the University of Patras, Greece, an M.Sc. from the Université de Droit, d'Économie et des Sciences d'Aix-Marseille III, France, and a Ph.D. from the National Technical University of Athens, Greece. Her research interests focus on environmental geochemistry and pollution, mainly of soils and sediments, as well as on mineralogy. She has considerable experience in national and European research projects and has published several papers in peer-reviewed journals and conference proceedings. Her main teaching work focuses on Geology–Mineralogy–Petrography, Environmental Chemistry, and Forest Soil Science.

Chrysoula Georgouli holds a degree in Forestry from the Technological Educational Institute of Lamia, Greece (1997), and an MSc in Ecology and Environmental Management from the Agricultural University of Athens, Greece

(2024). From 1999 to 2019, she worked in Forest Services under the Ministry of Environment and Energy of Greece. She currently serves as a Forest Technician at the same Ministry. She has also worked at the Ministry of Rural Development and Food of Greece, conducting field inspections and managing EU agricultural support schemes. She additionally holds certificates in Computer Science and English language.

Lazaros Iliadis is Professor of "Applied Informatics" Head of the Department of Civil Engineering. Head of the Lab of Mathematics & Informatics, Democritus University of Thrace. Coordinator and Consultant-Educational staff of the Program of Studies "Informatics" in the Hellenic Open University since 2001. General chair/ Program Chair of 27 International Conferences in AI-Machine Learning Guest Editor in 24 Volumes of more than 18 International scientific journals Editor of 50 Volumes of International Conference Proceedings Scientifically responsible of: A/ Several competitive research projects, B/ Private project for Brainchip US (California) & Australia (Canberra) company. C/ Researcher in the HORIZON MORPHEMIC, 2021-2022 5 million Euros.

Spyridon Kaloudis is a Professor in the Department of Forestry and Natural Environment Management at the Agricultural University of Athens, Greece. He holds a Ph.D. in Decision Support Systems for Forest Ecosystem Management from the Agricultural University of Athens and a postgraduate specialization in Environmental Physics from the National and Kapodistrian University of Athens. His academic foundation is rooted in a degree in Forestry and Natural Environment from the Aristotle University of Thessaloniki. Prof. Kaloudis's research interests include the application of Geographic Information Systems (GIS) and Knowledge Systems in forestry and environmental management, forest fire risk assessment, and ecosystem services. His contributions span decision support system development and sustainable forest management. An active researcher and educator, Prof. Kaloudis has authored numerous publications in esteemed international journals and conferences. He has also participated in EU-funded projects like SILVANUS and AGFORWARD, which aim to enhance forest resilience and rural development. In addition to his academic achievements, he has a longstanding commitment to teaching and mentoring undergraduate and postgraduate students. He has organized and equipped specialized laboratories and supervised numerous theses, furthering applied learning and innovation.

Sokratis K. Katsikas is the Director of the Norwegian Centre for Cybersecurity in Critical Sectors and Professor with the Department of Information Security and Communication Technology, Norwegian University of Science and Technology.

He is also Professor Emeritus of the Department of Digital Systems, University of Piraeus, Greece, and Member of the Board of the University of Patras, Greece. In 2019 he was awarded a Doctorate Honoris Causa from the Department of Production and Management Engineering, Democritus University of Thrace, Greece. In May-June 2023 he served as Minister of Digital Governance in the interim (caretaking) government of the Hellenic Republic.

Georgios Kavallieratos was awarded his Ph.D. in Information Security and Communication Technology by the Norwegian University of Science and Technology – NTNU in 2021. Moreover, he received the B.Sc. degree in computer science and the M.Sc. degree in digital systems security in 2016 and 2018, respectively, from the University of Piraeus, Piraeus, Greece.

Grigorios Koniavitis holds an M.Sc. in Ecology and Environmental Management from the Agricultural University of Athens (2024, GPA: 8.46/10), an M.Sc. in Language Technology jointly awarded by the National and Kapodistrian University of Athens and the National Technical University of Athens (2007, GPA: 8.32/10), and a B.A. in Greek Philology with a major in Linguistics from the National and Kapodistrian University of Athens (2001, GPA: 7.15/10). He currently works for the Greek Ministry of Justice. He is fluent in English (C2, New York Language Center Certificate) and has intermediate-level Spanish (B2, KPG Certificate, Greek Ministry of Education & Religious Affairs).

Katerina Konstantinou is a biologist with a PhD in Human and Animal Physiology. She has extensive research experience in human metabolic regulation and its implications in metabolic disorders and cancer, gained through her work at leading research institutions including the Foundation for Research and Technology – Hellas (FORTH) and Harvard Medical School. Her research interests include the effects of medicinal plants on brain metabolic function. She has authored over 40 peer-reviewed articles and presented more than 50 abstracts at national and international conferences. Her biomedical research background is complemented by substantial academic teaching experience in Molecular Biology and Physiology at the Hellenic Open University, where she serves as an adjunct Ass. Prof. of Biology.

Branislava Lalic is a professor of Meteorology and Biophysics, specialising in modelling biosphere-atmosphere interactions and the development of climate-smart technologies. Fulbright Visiting Scholar at the ASRC (SUNY, Albany). With a strong background in physics, biometeorology, and environmental modelling, she actively mentors students and early-career researchers, integrating education and innovation. Her teaching emphasises systems thinking, data-driven analysis, and creative problem-solving. Through interdisciplinary collaboration, she transforms scientific knowledge into practical, scalable tools that foster innovation and promote climate justice.

Sophia Leni is a member of the Laboratory Teaching Staff in the Department of Agricultural Economics and Rural Development at the Agricultural University of Athens, Greece, and a member of the Informatics Laboratory. She holds a PhD in Land Policy and Spatial Analysis, with a focus on GIS applications in urban and rural systems. Her research focuses on Geographic Information Systems (GIS), spatial analysis, urban and regional planning, land use change modeling, and digital tools for sustainable development.

Nenad Magazin is a full professor of fruit growing that earned his PhD degree in 2011 at the Faculty of Agriculture, University of Novi Sad, Serbia, in the field of biotechnical sciences. He has more than 20 years of teaching, research and extension service experience in temperate fruit species production and postharvest technologies. The majority of his research work is dedicated to pre and post harvest treatments on fruits. He is author and co-author of 7 monographs and manuals in the field of fruit growing and fruit post harvest, 33 papers in international scientific journals with impact factor, 165 scientific papers in other categories of journals and scientific meetings proceedings. He is co-author of 3 apricot cultivars: „Buda", "Zaharije", "Jelena".

Angeliki Milioti studied Agronomy-Food Science and Technology (MSc) at Aristotle University of Thessaloniki and Biotechnology at the Catholic University of Porto. She also obtained an MSc in Entrepreneurship and Consulting in Rural Development from the Agricultural University of Athens. She speaks Greek, English, French, Spanish, and Polish. She works as a Dissemination, Exploitation and Communication Manager of Horizon and Digital Europe Agri-Food Projects at Smart Agro Hub, a competence center in the field of smart farming initiated by the Agricultural University of Athens.

Anshit Mukherjee is presently pursuing Bachelor of Technology (Hons.) in Computer Science and Engineering from the Abacus Institute of Engineering and Management, which is affiliated with Maulana Abul Kalam Azad University of Technology, and is presently in final year. He was the zonal topper of the AMBIKA KALNA zone in I.C.S.E. 2019 (10 TH STANDARD) and the second zonal topper in I.S.C. 2021 (12 TH STANDARD). He is also awarded with NPTEL Discipline Star in July–December 2023 and Motivated Learner in January-June 2024.

Charlotte Niyomizero is a project manager at ILVO's Technology & Food Sciences department, working on horizon Europe research and innovation projects in digital and sustainable agriculture. Her background is in Environmental Science and she has experience of 11 years in agroforestry project management. Through ILVO she is facilitating cross-continental collaboration in EU-Africa initiatives. Her work focuses on accelerating responsible deployment of agri-food technologies, and knowledge systems to support resilient, data-driven farming.

Maria Ntaliani is an Associate Professor at the Informatics Laboratory in the Department of Agricultural Economics and Rural Development at the Agricultural University of Athens. Her research has been published in peer-reviewed journals, special volumes, and conference proceedings. She investigates the intersections of technology and society, focusing on digital government, digital marketing, e-business, e-learning, and mobile applications, for driving innovation in both academic and practical contexts.

Filotheos Ntalianis is a Professor in the Department of Business Administration at the University of Piraeus, Greece. He holds a Ph.D. from Concordia University, Montreal, Canada, and completed his graduate studies at Florida Atlantic University, Boca Raton, Florida, USA. His research focuses on motivation, personality, psychological contracts, new technologies, and employee health.

Michail Petropoulos is a Postgraduate Student at the Technical University of Athens, enrolled in the Interdepartmental MSc Program "Data Science and Machine Learning." He holds an Integrated Master's degree in Agricultural Economics and Rural Development from the Agricultural University of Athens, completed at the university's Informatics Laboratory. He has strong expertise in informatics and data analytics, with research experience in applying machine learning to crop yield forecasting. His research interests include digital and precision agriculture, AI-driven agricultural solutions, chatbots, and data-informed decision support systems.

Jan Pieters is senior full professor in the field of biosystems engineering at the Ghent University. In 1991 he graduated as an agricultural engineer and in 1995 he obtained his PhD in Applied Biological Sciences at the same university. His teaching activities so far include all aspects of agricultural engineering (a.o. agricultural and horticultural constructions, agricultural mechanisation) and environmental engineering for master students, as well as fundamental courses on heat and mass transfer for bachelor students. Actual research themes include vertical horticulture and precision agriculture. Up to know, he was supervisor of more than 100 master students and more than 20 PhD students, while he (co-)authored more than 120 peer reviewed journal articles.

Elias Pimenidis is a Senior Lecturer in Computer Science at the University of the West of England, Bristol in the United Kingdom . His research areas are, Electronic Governance, Web Services, Security, Neural Networks, Collaborative Filtering and Explainable Artificial Intelligence in supporting sustainable supply chains. He holds a PhD from Abertay University, a Master of Science (Systems Engineering) from Cardiff University and a BSc in Mechanical Engineering from the University of Strathclyde, all in the United Kingdom.

Olivera Stojilović Trivunić is the Head of Ecosystem Development at Foodscale Hub and a sustainability scientist. She has over six years of experience in science communication and STEM empowerment currently working in the domain of bioeconomy, capacity raising and ecosystem building. Her experience in applied and basic science research and project implementation allowed her to work with both aquatic and land systems with organization such as the Nature Conservation of Vojvodina Province and Prince of Songkla University in areas of socio-ecological systems. Olivera has experience in running large-scale projects of international EU-funded origin focused on increasing R&I capabilities and food system innovations.

Koen Uyttenhove holds a PhD in Microelectronics and has extensive experience across different industrial sectors. Over the past eight years, he has focused on digital transformation in agriculture and founded ODYC, a digital innovation hub in Belgium. Through ODYC, he leads European and regional R&D projects on smart farming, data sharing, and sustainability, fostering collaboration between farmers, researchers, and agri-tech companies.

George Vlahos is currently an associate professor in the Agricultural University of Athens and has collaborated in various research projects focusing on rural and agri-environmental policies. He is author of various articles on these subjects and collaborated in books. He has also been employed as project manager in the Greek Organisation for the Certification and Inspection of Agricultural Products.

Index

M

N

P

R

S

T

U

V

W

www.ingramcontent.com/pod-product-compliance
Lightning Source LLC
LaVergne TN
LVHW070904300926
849162LV00011B/1847

* 9 7 9 8 3 6 9 3 8 2 3 7 0 *